Polymer Characterization

ADVANCES IN CHEMISTRY SERIES **227**

Polymer Characterization
Physical Property, Spectroscopic, and Chromatographic Methods

Clara D. Craver, EDITOR
Chemir Laboratories

Theodore Provder, EDITOR
The Glidden Company
(Member of ICI Paints)

Developed from a symposium sponsored
by the Division of Polymeric Materials:
Science and Engineering
at the 196th National Meeting
of the American Chemical Society,
Los Angeles, CA,
September 25–30, 1988

American Chemical Society, Washington, DC 1990

Library of Congress Cataloging-in-Publication Data

Polymer Characterization: Physical Property, Spectroscopic, and Chromatographic Methods
    Theodore Provder, editor; Clara D. Craver, editor.

    p.   cm.—(Advances in Chemistry Series, ISSN 0065–2393; 227)

    Developed from a symposium sponsored by the Division of Polymeric Materials: Science
and Engineering at the 196th Meeting of the American Chemical Society, Los Angeles, CA,
September 25–30, 1988.

    Includes bibliographical references and index.

    ISBN 0–8412–1651–7

    1. Polymers—Congresses

I. Provder, Theodore. 1939–   II. Craver, Clara D.   III. American Chemical Society.
Division of Polymeric Materials: Science and Engineering.   IV. American Chemical Society.
Meeting (196th: 1988: Los Angeles, Calif.). V. Series.

QD1.A355 no. 227
[QD380]
540 s—dc20   90–47157
                                                   CIP

Advances in Chemistry Series

M. Joan Comstock, *Series Editor*

1990 ACS Books Advisory Board

FOREWORD

The ADVANCES IN CHEMISTRY SERIES was founded in 1949 by the American Chemical Society as an outlet for symposia and collections of data in special areas of topical interest that could not be accommodated in the Society's journals. It provides a medium for symposia that would otherwise be fragmented because their papers would be distributed among several journals or not published at all. Papers are reviewed critically according to ACS editorial standards and receive the careful attention and processing characteristic of ACS publications. Volumes in the ADVANCES IN CHEMISTRY SERIES maintain the integrity of the symposia on which they are based; however, verbatim reproductions of previously published papers are not accepted. Papers may include reports of research as well as reviews, because symposia may embrace both types of presentation.

ABOUT THE EDITORS

CLARA D. CRAVER is the president of Chemir Laboratories, which she founded in 1958, and vice president of Polytech Laboratories. She received her B.Sc. degree in chemistry from Ohio State University in 1945 and was awarded the honorary degree of Doctor of Science by Fisk University in 1974. Her early work at Esso Research Laboratories, 1945–1949, resulted in patents on characterization of complex hydrocarbon mixtures. She established the molecular spectroscopy laboratory at Battelle Memorial Institute in 1949, and as group leader through 1958, she conducted spectroscopic studies for Battelle's research projects and original research in the areas of drying oils, asphalts, paper, and rosins. Her research on organic coatings won her the 1955 Carbide and Carbon Award of the ACS Division of Organics Coatings and Plastics Chemistry. In 1975 she became chairman of that division and has been active in ACS governance continually since then.

In the early 1980s, Craver initiated workshops on Fourier transform infrared spectroscopy at ACS national meetings to offer educational opportunities along with symposia on new developments in polymer characterization. She began the Coblentz Society's spectral publication program and is editor of five books of Special Collections of Infrared Spectra and of an 11-volume spectral data collection. She served as consultant to ASTM for their evaluated infrared spectral publication program, supported by the Office of Standard Reference Data of the National Bureau of Standards. She was chairman of the Joint Committee on Atomic and Molecular Physical Data, on which she now serves as Executive Committee member-at-large. She is past chairman of the ASTM Committee on Molecular Spectroscopy, and was named a Fellow of ASTM in 1982 when she received that organization's highest award, the Award of merit. She is a Fellow of the American Institute of Chemists and a Certified Professional Chemist. She was honored by being named the 1989 National Honorary Member of the Women's Professional Honorary Chemistry Sorority, Iota Sigma Pi.

She is director of the Fisk Infrared Institute held annually at Vanderbilt University on IR, Raman, and FTIR spectroscopy. She also lectures at other short courses and is an ACS tour speaker. She continues to direct short courses including FTIR laboratories at national ACS meetings for the Division of Polymeric Materials: Science and Engineering.

THEODORE PROVDER is Principal Scientist at The Glidden Company's Research Center and is responsible for the research activities of the Materials Science Department. He received a B.S. degree in Chemistry from the University of Miami in 1961 and a Ph.D. degree in Physical Chemistry from the University of Wisconsin in 1965. After receiving his doctorate, he joined the Monsanto Company in St. Louis as a Senior Research Chemist and carried out research on the characterization and material properties of exploratory polymers and composites. While at Monsanto, his research interests focused on molecular weight characterization, particularly by size exclusion chromatography. Recently, Provder's research has focused on size exclusion chromatography, particle size distribution analysis, cure chemistry and physics, and the application of computers in the polymer laboratory. He is the author of more than 85 publications, is credited with three patents, and has edited seven ACS Symposium Series volumes.

Provder was past chairman of the ACS Division of Polymeric Materials: Science and Engineering and has served on the Advisory Board for the ACS Books Department and on the Editorial Advisory Board for ACS's *Industrial & Engineering Chemistry Product Research and Development* journal. He currently is a member of the editorial boards for the *Journal of Coatings Technology* and *Progress in Organic Coatings*. He is also the current treasurer for the Joint Polymer Education Committee of the Division of Polymeric Materials and Polymer Chemistry. Provder is a recipient of an SCM Corporation Scientific and Technical Award in the area of computer modeling and two Glidden Awards for Technical Excellence for advanced latex particle size analysis methods and instrumentation development. In addition, Provder received the coatings industry's highest honor by being awarded the 1987 Joseph J. Mattiello Lecture at the annual meeting of the Federation of Societies for Coatings Technology. In 1989, Provder was awarded the ACS Division of Polymeric Materials: Science and Engineering Roy W. Tess Award in coatings.

CONTENTS

MORPHOLOGY

INDEXES

PREFACE

T HE CURRENT TECHNOLOGICAL DIRECTIONS of polymer-related industries have been shaped by the operative business and societal driving forces of the past several years. The resultant technological directions affect the product development cycle and shape the required materials characterization needs. The role of polymer characterization in the product development cycle is shown in Figure 1. Product development is no longer a simple straight-line process from product design to product performance, bridged by polymer and product characterization analysis and testing. The product development cycle must take into account the many constraints produced by the operative business and societal driving forces. These constraints include

- product development costs
- raw material supply
- energy conservation
- safety, health, and environmental considerations
- public consumerism
- product quality
- emphasis on customer needs
- shorter product development and market introduction cycles
- improved product–process–customer economics
- global competition

As can be seen from Figure 1, the role of polymer characterization is to facilitate product development subject to these constraints. For example, the development of polymer products has been strongly influenced by environmental considerations and government regulations. Coatings are now being developed with significantly lower volatile organic content. Plastic packaging is being developed with built-in environmental degradability. For containers, it has become very desirable to use plastics that can be readily recycled.

On a more basic level, very few new commodity building blocks (monomers) are expected to be developed because of economic and environmental considerations. Increased strategic use of low levels of specialty building blocks is expected in order to add product value.

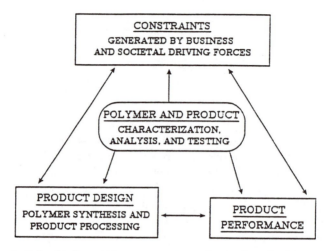

Figure 1. Role of polymer characterization in the product development cycle.

In general, structure–morphology–property considerations are becoming of paramount importance in the development of new products. Examples include high-performance engineering plastics and composites that require strategically designed polymers, polymer alloys, and blends, as well as strategically designed polymers for electronic and biopolymer applications. It is not a question of what building blocks are put together, but how to put them together to make unique polymer products. This approach implies polymer structure–morphology control down to the molecular level to enhance properties.

The polymer product development process in Figure 2 shows that polymer characterization methodology is required in each step of the process to acheive two goals: (1) to characterize the molecular architecture and physical properties produced by a particular polymerization method and mechanism, and (2) to characterize the polymer product resulting from product processing to relate surface and bulk properties and morphologies to application and end-use properties. Polymer characterization methodology is an important component of the product development cycle and process. On the basis of the ever-expanding list of constraints to product development produced by the operative business and societal driving forces, in the future polymer characterization will most likely assume an even greater role in the product development cycle.

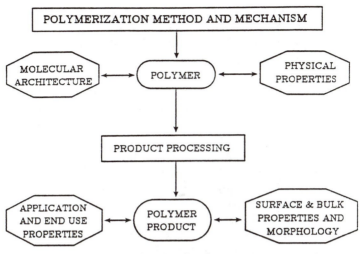

Figure 2. Product development process.

About the Book

This book covers significant advances in polymer characterization methodology and is organized into four main areas: (1) polymer fractionation and particle size distribution, (2) dynamic mechanical analysis and rheology, (3) spectroscopy, and (4) morphology. Many of the chapters report on the combined use of several characterization methods in order to elucidate the relationship between polymer structure–morphology and polymer performance.

The chromatographic method of thermal field flow fractionation (FFF) is complementary to gel permeation chromatography (GPC) for fractionating polymer molecules. It depends upon the thermal diffusion coefficient of the polymer molecule and is sensitive to both the polymer chemical composition and the molecular weight. Giddings et al. show that temperature gradient programming can improve the speed and resolution of the method. The thermal FFF method shows promise for fractionating and elucidating the molecular weight distribution of very high molecular weight polymers up to 60,000,000 daltons that cannot be adequately fractionated by GPC. Recent advances in high-temperature GPC include the use of laboratory robotics in the automation of instrumentation, discussed by Moldovan and Polemenakos, as well as the use of on-line viscosity detection, along with conventional refractive index detection to elucidate the long-chain branching distributions in polyethylene discussed by Mirabella and Wild. The separation and characterization of charge-containing synthetic polymers is a difficult problem that is amenable to solution by gel electrophoresis, a novel and promising technique that is discussed by Smisek and Hoagland.

A variety of methods are available to characterize particle size and particle size distribution of latex particles. These can be categorized as fractionation techniques or nonfractionation (counting) techniques. In this book, three counting techniques based on light-scattering methods are discussed: zero-angle depolarized light scattering, turbidity, and dynamic light scattering. Eliçabe and Garciá-Rubio revitalized the turbidimetric method for obtaining particle size distribution information by the application of sophisticated mathematical regularization techniques. Kourti et al. and Nicoli and co-workers, in an unusual academic–industrial collaboration, have shown that dynamic light scattering can be used to monitor on-line, in real time, the particle growth and particle size distribution during the emulsion polymerization of vinyl acetate in a pilot scale reactor.

The nine-chapter section on dynamic mechanical analysis and rheology involves the use of a variety of methods applied to study the kinetics and cure of polymerizing systems, as well as resulting polymer properties. The chapter by Wisanrakkit and Gillham is an excellent example of the use of the time–temperature–transformation principle applied to the glass transition temperature for monitoring thermoset cure.

The relatively new techniques of thermally stimulated current and relaxation map analysis spectroscopy, discussed by Ibar et al. and Demont et al., represents a significant advance in dynamic methods to study the molecular response of materials such as semicrystalline polymers, copolymers, polyblends, polymer complexes, composites, and coatings. The use of dielectric thermal analysis (DETA) methods to study thermoset cure ex situ by Martin and co-workers and in situ by Kranbuehl et al. illustrates the complimentary nature of this technique to standard rheological measurements and to other methods for elucidating cure such as differential scanning calorimetry (DSC) and dynamic mechanical analysis (DMA). DETA uniquely monitors the ionic mobility of a curing system and, therefore, can monitor the microviscosity (local viscosity) of a curing system using mobile ions as a probe. Therefore, DETA has excellent potential as a process monitoring method through the use of remote sensors.

The chapter by Ishida and Nigro illustrates that the combination of chemical (Fourier transform infrared) and physical (dynamic mechanical testing) methods to study cure can provide a very complete picture of the cure process. Biesenberger and Rosendale demonstrate that a specifically designed rheocalorimeter can follow step and chain polymerization for a variety of polymerizing systems to provide viscosity–conversion data.

The nine-chapter section on spectroscopy includes aspects of the many subtechniques of IR spectroscopy, Raman, fluorescence, and NMR spectroscopy. The IR spectroscopic techniques discussed include transmission Fourier transform IR (FTIR), photoacoustic (PA) FTIR, polarized attenuated total reflectance, and evolved-gas analysis. Unfortunately, the PA FTIR technique has generally been underutilized. In an overview chapter, Urban et al. demonstrate the wide range of useful applications of the PA FTIR technique, which include (1) the depth of profiling capability to study surface-treated fibers, (2) utility to studying cure in a thermoset system, and (3) the ability to perform rheo-optical measurements in a photoacoustic IR cell on polymer systems. Kuo and Provder illustrate the utility of evolved-gas analysis by FTIR to study the kinetics of curing systems through the detection of evolved volatile compounds and demonstrate the complementary nature of the kinetic information to that obtained from thin film transmission FTIR measurements. Schwab and Levy were able to uniquely monitor physical aging of an epoxy resin by using a fluorescent molecular probe to follow the decrease in free volume with time by the increases in fluorescence intensity levels.

Chapters that combine spectroscopic and physical methods to elucidate microstructure of polymers are reported. Mirabella simultaneously monitors the crystalline melting of polyolefinic blends with DSC and FTIR spectroscopy. Mandelkern combines Raman spectroscopy with DSC to characterize crystalline polymers. Tonelli et al. combine solid-state NMR spectroscopy with DSC and X-ray diffraction studies to elucidate polymer microstructure–morphology.

The last section deals with two unique studies involving morphological characterization. Sperling et al. studied the morphology of multicomponent and heterogeneous polymer systems, such as block copolymers, latex dispersions, blends, and interpenetrating networks with small-angle neutron scattering. This technique is able to provide information on the physical size of micromorphological domains. Hair and Letts were able to elucidate the morphological structure of gels and foams made from ultra-high-molecular-weight polyethylene by using a range of characterization methods including, DSC, viscometry, optical microscopy, scanning electron microscopy, X-ray diffraction, and cloud point measurements.

Acknowledgments

We are grateful to the authors for their effective oral and written communications and for the effort they have expended to provide well-balanced coverage of the characterization methods included in this book. We also acknowledge the many peer reviewers for their critiques and constructive comments.

THEODORE PROVDER
The Glidden Company
(Member of ICI Paints)
Strongsville, OH 44136

CLARA D. CRAVER
Chemir Laboratories
Glendale, MO 63122

August 14, 1990

POLYMER FRACTIONATION AND PARTICLE SIZE DISTRIBUTION

Polymer Separation by Thermal Field-Flow Fractionation

High-Speed Power Programming

J. Calvin Giddings, Vijay Kumar, P. Stephen Williams, and Marcus N. Myers

Field-Flow Fractionation Research Center, Department of Chemistry, University of Utah, Salt Lake City, UT 84112

In this chapter we describe the mechanism of thermal field-flow fractionation (thermal FFF), explain its uses and advantages for polymer analysis, and report its first implementation in the form of a high-speed power-programmed system. Thermal FFF is characterized by high resolving power and remarkable adaptability such that a single system can be readily tuned to work effectively for almost any molecular weight, polymer type, and solvent. Because shear degradation and surface interaction effects are minimal, the method is applicable to ultra-high-molecular-weight polymers. The method exhibits selectivity with respect to both molecular weight and polymer composition. By combining a thin (76-μm) thermal FFF channel with power-programmed operation, in which the temperature drop is decreased according to a specified power function during the run, we can resolve six polymer standards with molecular weights from 9×10^3 to 5.5×10^6 in approximately 8–20 min. The effects of changes in various operating parameters on these programmed separations are reported.

FIELD-FLOW FRACTIONATION (FFF) is a suite of separation techniques carried out in thin flow channels. These techniques are especially applicable to the separation and characterization of macromolecular and particulate species (1–4). Different subtechniques of FFF have specific characteristics that make them advantageous for particular categories of macromaterials.

0065–2393/90/0227–0003$06.00/0

Although we first reported polymer separation by the subtechnique of thermal FFF 20 years ago, the mechanism was not well understood and the separation not efficient. Thermal FFF has continuously improved in the intervening years and, as a prime tool for polymer characterization, it has found a particular niche in the separation of synthetic polymeric materials (5, 6).

In this chapter, the nature of thermal FFF and its application to polymers will be surveyed briefly. We will then illustrate thermal FFF applicability to polymers by reporting a new high-speed programming technique and associated instrumentation capable of fractionating polymeric components from 10^4 to 5×10^6 molecular weight in a single run whose typical duration is 10–20 min.

In thermal FFF, a thin ribbonlike channel is clamped between two heat-conductive (copper) bars (Figure 1). By heating the top bar and cooling the bottom bar, a temperature drop ranging typically from 20 to 80 °C, and in exceptional cases from 5 to 150 °C, is established across the channel. Because the channel is thin, typically 75–125 μm thick in current work, temperature gradients on the order of 10^4 °C/cm are established across the channel. Such temperature gradients acting through the phenomenon of thermal diffusion are capable of creating strong driving forces on polymeric components (4). Polymers are usually driven to the cold wall of the channel by the thermal diffusive process (7), as shown in Figure 2.

When the flow of solvent is initiated in the channel, the polymeric materials of an injected sample are swept downstream. However, the velocity of downstream displacement of the polymer molecules depends upon where the polymer is located in the channel cross section because of the differential (parabolic) nature of the channel flow. In particular, components driven closest to the cold wall will be carried downstream only slowly because the flow velocity approaches zero at all solid walls and surfaces. Consequently, with the onset of flow, a fractionation process is initiated in which the polymer components driven closest to the cold wall are retarded in their passage through the channel more than the polymeric species with a greater average distance away from the cold wall. Because high-molecular-weight polymers are driven closer to the cold wall than are those of low molecular weight, the low-molecular-weight species emerge first, followed by fractions of successively higher molecular weight. Thus the elution order established in thermal FFF is opposite to that found in size exclusion chromatography (SEC) (5).

Thermal FFF was first applied to polymers in our laboratory early in the latter half of the 1960s, a time approximately coincident with the beginnings of SEC in the form of gel permeation chromatography (8). However, SEC was developed more rapidly into practical laboratory instrumentation than was thermal FFF, which was largely ignored except for ongoing work in our laboratory.

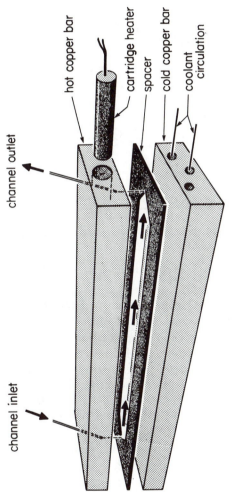

Figure 1. Basic thermal FFF channel system.

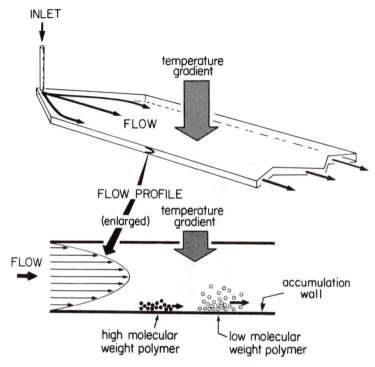

Figure 2. Flow and polymer separation in thermal FFF system.

A combination of three factors has substantially increased the interest of polymer scientists in thermal FFF in recent years. One factor is the continuous improvement in FFF instrumentation and methodology and the consequent introduction of a thermal FFF instrument into the commercial market. The second factor is the increasing need for new techniques in polymer analysis to cover an ever-expanding variety of new polymeric materials. The third factor is the heightened demand for improved characterization. In this context thermal FFF appears to be particularly advantageous in the analysis of very-high-molecular-weight polymers, species subject to shear degradation, copolymers, polymers that tend to interact with surfaces, polymers needing corrosive solvents, high-temperature polymer solutions, and narrow polymer samples requiring an accurate determination of polydispersity. At the same time, thermal FFF is now a workhorse technique flexibly applicable to routine polymer analysis problems.

Although the mechanism of thermal FFF bears very little resemblance to that of SEC, the two methods can in some cases be used for the same applications: the determination of molecular weight distributions for various industrial polymers. However, the unique mechanism of thermal FFF imparts some equally unique characteristics and advantages. These are summarized as follows.

First, as already noted, the elution order in thermal FFF is opposite to that observed for SEC: in thermal FFF the low-molecular-weight components emerge first and the high-molecular-weight species last. Despite this difference, the resolution and speed of separation of the two methods appear to be comparable as measured with present instrumentation (5).

Another unique feature of thermal FFF is its flexibility. A major element of this flexibility arises from the fact that the temperature drop ΔT across the channel can be varied rapidly and precisely to any desired level. Because the temperature drop controls retention, thermal FFF is a single system that can be quickly tuned (by means of ΔT) to accommodate almost any polymer analysis problem. For example, the temperature drop can be raised (to over 100 °C) to accommodate low-molecular-weight polymers (9), it can be lowered (to under 10 °C) to accommodate ultra-high-molecular-weight polymers (10), and it can be changed continuously (programmed) to handle a polymer sample having a wide molecular-weight range (11). Thus an enormous variety of polymers can be handled in a single system without special requirements.

In thermal FFF, greater flexibility arises through the control of flow velocity than in SEC. Resolution is more highly flow sensitive in thermal FFF than in SEC, and thermal FFF gives a better-defined tradeoff between polymer resolution and analysis speed (5).

The open unobstructed channel of a thermal FFF system is very different from the packed columns used in SEC. The absence of extensional shear in the FFF channel reduces shear degradation or shear-induced structural changes (10). The uniform channel structure also yields a far more predictable flow than a packed bed. Consequently, the theoretical predictability of retention, band broadening, and resolution is much better advanced in thermal FFF than in SEC (12). This predictability is a substantial advantage in the determination of optimum separation conditions. In addition, it has several special advantages, such as making it possible to exactly compensate for band broadening by using deconvolution techniques (13) and to measure the extremely low polydispersity values of narrow polymer standards in a straightforward fashion (14).

Finally, thermal FFF retention is sensitive to the chemical composition of a polymer as well as to its molecular weight. Thus thermal FFF has the potential to be used in the compositional analysis of copolymers and blends (15). In earlier work, thermal FFF was applied almost exclusively to polystyrene standards, but it has in more recent times been extended to polymers such as poly(methyl methacrylate), polyisoprene, poly-(α-methyl)styrene, polytetrahydrofuran, polyethylene, polycarbonate, polyurethane, and a variety of other polymers that suggest that the method is almost universal for polymers soluble in organic liquids. In addition, thermal FFF has been found applicable to polymers down to ~1000 molecular weight at one extreme (9) and up to >60 × 10^6 molecular weight at the other extreme (10).

Another field-flow fractionation method has been proven applicable to polymers; that method is flow FFF. The flow subtechnique of FFF has been used primarily with aqueous systems including water-soluble polymers (16), many of which cannot be analyzed by thermal FFF. However, flow FFF can also be applied to lipophilic polymers in organic solvents, providing the right membrane is chosen (17).

Power-Programmed Thermal FFF

As noted earlier, the ability to control the temperature drop ΔT in thermal FFF introduces great flexibility into thermal FFF operation and makes it possible to tune the system to accommodate almost any molecular-weight range of polymer. When the molecular weight extends over a range greater than about 25- to 50-fold, this flexibility is best exploited by using programmed operation. In this technique, first applied to thermal FFF in 1976 (11), the field strength ΔT begins at a high value and then drops continuously during the run. The high ΔT optimally separates the low-molecular-weight components; the high-molecular-weight components are separated as ΔT drops to lower values.

In the first experimental realization of programmed thermal FFF, ΔT was held constant at 70 °C for a period of time (the predecay time), after which ΔT was forced to decrease parabolically as a function of time (11). A linear program was also used. These experiments showed that polymers ranging in molecular weight from 4000 up to 7.1×10^{6} could be separated in a single run. An example of the original programmed separation is shown in Figure 3.

Although this early programming work exhibited excellent resolving

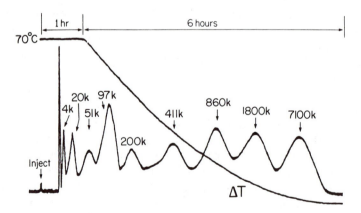

Figure 3. Slow but effective separation of polystyrene standards of indicated molecular weights by parabolic programming as reported in initial 1976 paper on programmed thermal FFF. (Reproduced from reference 11. Copyright 1976 American Chemical Society.)

power and broad-range molecular-weight applicability, it was flawed in one major respect: the run time of the experiment was over 6 h. Since that time extremely fast polymer fractionation has been achieved by thermal FFF. One study showed that several polymer standards could be resolved in only a few minutes (*18*). An example from that study is shown in Figure 4. Despite the success of the latter experiments, the conditions needed for high-speed polymer separation have not since been used in conjunction with programmed thermal FFF. The purpose of this work is to illustrate the successful marriage of programming methodology and high-speed instrumentation.

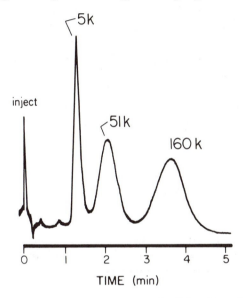

Figure 4. Fast separation of polystyrene standards by thermal FFF in 51-μm-thick channel. Numbers are molecular weights; k indicates thousand. (Reproduced with permission from reference 18. Copyright 1978.)

As shown in the high-speed study just noted, the key requirement needed to reduce separation time is a thermal FFF channel of reduced thickness. Whereas the slow programming run of Figure 3 was achieved in a channel having a thickness w of 254 μm, the separation in Figure 4 was realized with a channel thickness of 51 μm. Because the time of the run (other things being constant) is approximately proportional to the channel thickness squared (*18*), there is a built-in 25-fold difference in the intrinsic operating speeds of the two systems. However, programming runs for polydisperse samples generally require a longer time than do optimized isocratic runs achieving separation over a narrow molecular-weight range. Thus the time required for the run in Figure 3 was nearly 2 orders of magnitude greater than that required to reach separation in Figure 4. Although we have not found it practical to carry out programmed operation in a 51-μm-thick channel, we have been able to achieve effective programmed operation in a 76-μm-thick channel. The enhancement of speed relative to that of the 254-μm channel should be governed by the factor $(254/76)^2 = 11.2$, or approximately 1 order of magnitude.

In the earlier programming study, the mathematical form of the program (ΔT versus time) was parabolic in one case and linear in another (11). Any continuously decreasing function can be used. However, different mathematical programming functions display different elution spectra (molecular weight versus time) and different resolution levels in different parts of the resultant fractogram (19). In recent years special program functions have been developed to control these elution characteristics. Kirkland et al. (20), for example, have developed a time-delayed exponential programming function such that plots of the logarithm of molecular weight versus retention time are approximately linear. In our laboratory we have developed programming according to a power function (power programming) for the purpose of achieving uniform resolution or fractionating power over the full molecular-weight range of the experiment (21).

In power programming the field strength S is changed with time t according to the function (21)

$$S(t) = S_0 \left(\frac{t_1 - t_a}{t - t_a} \right)^p \tag{1}$$

where S_0 is the initial field strength, t_a is an arbitrary time constant, t_1 is the predecay time between the start of the run and the beginning of decay, and p is the decay power. These parameters are subject to the constraints $t \geq t_1 > t_a$ and $p > 0$.

For FFF, the retention ratio R is given by the equation

$$R = 6\lambda \left[\coth\left(\frac{1}{2\lambda}\right) - 2\lambda \right] \tag{2}$$

providing the assumption of a parabolic fluid velocity profile is valid. For thermal FFF, the retention parameter λ is given to a good approximation by (22)

$$\lambda = \frac{D}{D_T w (\mathrm{d}T/\mathrm{d}x)} \tag{3}$$

where D is the normal mass diffusion coefficient, D_T is the thermal diffusion coefficient, w is the channel thickness, and $\mathrm{d}T/\mathrm{d}x$ is the local temperature gradient. The temperature gradient $\mathrm{d}T/\mathrm{d}x$ differs only slightly from the overall temperature gradient $\Delta T/w$ even at high ΔT (23).

At high retention levels, where the polymer band is compressed closely to the accumulation wall, D and D_T in eq 3 may be assumed to correspond closely to their values at the temperature T_c of the accumulation (cold) wall. Hence we obtain the simple approximate equation

$$\lambda = \frac{D}{D_T \Delta T} \tag{4}$$

in which D and D_T values are those corresponding to temperature T_c. Coefficient D_T was shown (24) to be essentially independent of polymer molecular weight for any particular polymer–solvent system, and the molecular-weight dependence of λ therefore parallels that of D.

At high retention (corresponding to $R \ll 1$), eq 2 reduces to the form

$$R = 6\lambda = \frac{6D}{D_T \Delta T} \tag{5}$$

again with D and D_T taken at T_c. In practice, a plot of $-\ln R$ vs. $\ln M$, where M is the polymer molecular weight, is found to be a straight line, the slope of which is defined as the selectivity, generally taking a value between 0.5 and 0.65. Because D_T is independent of M, eq 5 makes it apparent that this dependence reflects principally the dependence of D on M. It follows that

$$\lambda = \frac{\phi}{\Delta T M^n} \tag{6}$$

where ϕ is a constant for a particular solute–solvent system at some fixed cold wall temperature and n in this equation is an exponent that becomes the limiting value of selectivity for small λ.

For significantly retained components eluted under conditions of a power-programmed field strength, the mass-based fractionating power F_M is given by (21)

$$F_M = \frac{n}{48w} \left[\frac{6(4p + 1)}{p + 1} Dt^0 \right]^{\frac{1}{2}} \times$$
$$\left[\frac{1}{\lambda_0} \left(\frac{6}{p + 1} \frac{t_1 - t_a}{t^0} \right)^p \right]^{\frac{3}{2(p + 1)}} \tag{7}$$

where t^0 is the system void time and λ_0 is the value of λ at the initial field strength. The mass-based fractionating power is simply defined as the resolution for two closely eluting components divided by their relative difference in molecular weight (19), that is,

$$F_M = \frac{R_s}{\delta M/M} = \frac{M}{4\sigma_t} \frac{\delta t_r}{\delta M} \tag{8}$$

where R_s is the resolution $\delta t_r/4\sigma_t$, in which δt_r is the difference in the two retention times and σ_t is the mean standard deviation in retention time. Quantity δM is the difference in the two molecular weights and M is the mean molecular weight. For the special case of $t_a = -pt_1$, eq 7 reduces to

$$F_M = \frac{n}{48w} \left[\frac{6(4p + 1)}{p + 1} Dt^0 \right]^{1/2} \times$$
$$\left[\frac{1}{\lambda_0} \left(\frac{6t_1}{t^0} \right)^p \right]^{\frac{3}{2(p + 1)}} \tag{9}$$

From the direct proportionality between λ and D expressed by eq 4, it is apparent that

$$F_M \propto D^{\left[\frac{1}{2} - \frac{3}{2(p + 1)} \right]} \tag{10}$$

A fractionating power independent of D, and therefore independent of M, is given for significantly retained components when the power of D in eq 10 is reduced to zero, that is, when $p = 2$. In this case

$$F_M = \frac{3n}{8w} \frac{t_1}{t^0} (2D_T \Delta T_0 t^0)^{1/2} \tag{11}$$

where a substitution using eq 4 was made in eq 9 for λ, ΔT_0 is the initial temperature drop across the channel, and it is assumed that $t_a = -2t_1$. A fractionating power independent of molecular weight is desirable when widely polydisperse samples are to be characterized.

The following expression was also derived (21) for elution time of significantly retained components

$$t_r = t^0 \left[\frac{p + 1}{6\lambda_0} \left(\frac{t_1 - t_a}{t^0} \right)^p \right]^{\frac{1}{p + 1}} + t_a \tag{12}$$

When $p = 2$ and $t_a = -2t_1$ eq 12 reduces to

$$t_r = t^0 \left[\frac{D_T \Delta T_0}{2D} \left(\frac{3t_1}{t^0} \right)^2 \right]^{1/3} - 2t_1 \tag{13}$$

The approach just described presupposes a parabolic fluid velocity profile. In thermal FFF the assumption of such a profile is not strictly correct. A deviation from the parabolic form occurs because of the temperature dependence of the solvent viscosity. Close to the cold wall where the viscosity is greatest, the velocity is lowered relative to that of the parabolic profile, while toward the hot wall the velocity is relatively greater (22, 25–27). This deviation, which has an influence on both band spreading and retention ratio, is a function of ΔT; thus for programmed thermal FFF the deviation will be time dependent. Therefore, constant fractionating power should be obtained for a power p close to, but not equal to, 2. Nevertheless, by setting p equal to 2 we likely have established conditions close to those needed for the optimum fractionation of polydisperse samples.

In order to achieve uniform fractionating power, the three constants (t_1, t_a, and p) in eq 1 describing the field program must be fixed at specified levels. For thermal FFF, $p = 2$ and t_a and t_1 are related by $t_a = -2t_1$. When the latter condition holds, the field strength—in this case the temperature drop ΔT—for thermal FFF is given by

$$\Delta T = \Delta T_0 \left(\frac{3t_1}{t + 2t_1}\right)^2 \tag{14}$$

For the general power-programmed decay described by eq 1, the time required for ΔT to decay to half of the initial value ΔT_0 is given by

$$t_{1/2} = 2^{1/p}(t_1 - t_a) + t_a \tag{15}$$

which reduces to $t_{1/2} = 2.243t_1$ for $p = 2$ and $t_a = -2t_1$.

Experimental Details

The channel system employed in these studies is similar to that used in the model T100 thermal FFF system from FFFractionation, Inc. (Salt Lake City, UT). It consists of two highly polished chrome-plated bars of electrolytic grade copper clamped together over a thin (76-μm) polyester (Mylar) spacer. The channel form was cut into the spacer. The resulting channel has dimensions of 76-μm thickness, 2.0-cm breadth, and 46.3-cm tip-to-tip length. The measured void volume V^0 was 0.685 mL.

Four cartridge heaters of 1500 W each were used to heat the upper bar. The cold bar had slots milled in it to provide efficient water circulation throughout the bar for cooling. The cold wall temperature was maintained at 32 ± 1 °C during programming by adjusting the flow of water. The temperature of the hot wall was controlled by programmed computer-activated solid-state relays. An 80-s time lag was used to account for the time required for the conduction of heat from the heating cartridges to the hot wall.

Small holes were drilled at two places on each bar to within 0.76 mm of the polished surfaces for the measurement of temperature by copper–constantan thermocouples with digital thermometers (Omega, Stamford, CT). The maximum variation in the temperature drop ΔT along the length of the channel relative to the setting was 2 °C. The variation became less as ΔT decreased. Two holes, one in the hot wall and one in the cold wall, were drilled from the smooth surfaces to the sides of the bars to form the inlet and the outlet of the channel, respectively.

The channel spacer was positioned in such a way that the two apices at the tapered ends aligned with the inlet and outlet holes. All connections were made with stainless steel tubing of 0.01 in. (0.0254 cm) internal diameter. The volume of the tubing used was 0.055 mL, 8% of the channel void volume of 0.685 mL.

Sample injections of 10 µL were made with a Valco (Houston, TX) valve. There was no stop-flow for relaxation following injection. A helium gas pressurized pump was used to generate the flow of carrier, which in all these experiments was ethylbenzene. The polystyrene samples were detected with a refractive index monitor (model 401, Waters Associates, Amherst, MA). The samples in ethylbenzene solution had typical concentrations of approximately 2 mg/mL. The samples of high molecular weight, of the order of millions, had concentrations of 3 to 4 mg/mL. The samples used in the study are polystyrene standards as described in Table I. Sample retention times were measured from the chart paper of an Omniscribe chart recorder (Houston Instruments, Austin, TX).

Table I. Polystyrene Standards Used in This Study

$\overline{M}_w{}^a$	Supplier	Cat. No.	Polydispersity[b]
9,000	Supelco, Inc.	4-5703	≤1.06
35,000	Supelco, Inc.	4-5705	≤1.06
90,000	Supelco, Inc.	4-5707	≤1.04
200,700	Supelco, Inc.	4-5708	≤1.05
575,000	Supelco, Inc.	4-5710	≤1.06
1,970,000	American Polymer Standard Corp.	PS 2000k	1.1
5,480,000	Polyscience, Inc.	PS 5000k	1.15

[a] Nominal weight averages provided by suppliers.
[b] As specified by suppliers.

Results and Discussion

In the work described we used a combination of power programming and thin-channel technology to demonstrate the rapid separation and characterization of broad polymer mixtures with molecular weights ranging from approximately 9000 to 5,500,000. Several polymer mixtures were prepared in ethylbenzene solvent with the polymer standards reported in Table I. A number of isocratic (constant temperature drop) runs were made with the four lowest-molecular-weight polymers at $\Delta T_0 = 80\ °C$ to establish the parameters of the separation. With the measured retention times from the isocratic runs, the values of n and ϕ were established by least squares as 0.656 and 7420, respectively. These values, although only moderately consistent with retention data obtained in earlier experimental work, are self consistent for the low end of the molecular-weight spectrum studied here and thus can be used effectively in this study.

Following the isocratic experiments, several replicate series of power-programmed runs were made with different flow rates and different values of the parameters t_1, t_a, and p in eq 1. Results for one linear and one exponential program are also reported. The parameters corresponding to the different series of runs are listed in Table II. Values of the observed and calculated retention times t_r are shown in Table III. Each observed t_r represents an average of values from two or three replicate runs. However, the individual runs of a replicate series were not identical because of slight variations in flow rate. The theoretical (calculated) values of t_r for the power-programmed runs were determined not from eq 12, but from a more rigorous iterative treatment (*see* eqs 23–26 in ref. 21). Retention, particularly for the lower-molecular-weight components, was generally not sufficient for eq 12 to yield satisfactorily accurate prediction.

Table II. Parameters Defining Eight Programmed Series of Runs

Programmed Series[a]	Flow Rate (mL/min)	ΔT_0 (°C)	p	t_1 (s)	t_a (s)	$t_{1/2}$ (s)	F_M (range)
1	0.748	80	1	151	−302	604	1.3–6.9
2	0.806	80	2	151	−302	339	1.2–3.0
3	0.460	80	2	187	−374	419	1.6–2.8
4	0.749	80	3	151	−302	269	1.3–2.3
5	0.790	80	4	151	−302	237	1.1–2.1
6	0.689	60	2	167	−334	375	0.9–2.6
7[b]	0.733	80	—	151	—	601	—
8[c]	0.711	80	—	151	—	256	0.8–2.1

[a] Series 1–6 are power programmed, 7 is linearly programmed, and 8 is exponentially programmed.
[b] Linear decay constant (ramp time) = 900 s.
[c] Exponential decay constant $\tau' = 151$ s.

By and large, the agreement between the observed and predicted retention times reported in Table III for the various programmed runs is very good. However, for the polymer of M 5,480,000, substantial departures are observed. These departures may be due in part to the fact that the calibration curve for retention parameters n and ϕ was obtained by using only the four

Table III. Retention Times (seconds) for Polymer Standards in Different Programmed Series

Polymer Standard		1	2	3	4	5	6	7	8
9,000	obs.	71	62	109	67	65	65	67	69
	calc.	70	65	113	70	66	69	71	73
35,000	obs.	101	94	164	100	97	85	100	105
	calc.	117	109	190	117	111	104	119	123
90,000	obs.	169	165	262	168	165	143	172	173
	calc.	194	180	298	191	182	166	199	199
200,700	obs.	285	254	382	249	242	220	298	257
	calc.	295	265	418	268	251	250	308	274
575,000	obs.	487	406	586	370	337	361	516	364
	calc.	490	407	615	383	346	394	521	376
1,970,000	obs.			890					
	calc.			913					
5,480,000	obs.	1010	592	1150	540	429	644	874	514
	calc.	1270	850	1230	697	581	847	1000	599

NOTE: Parameters for series 1–8 are defined in Table II.

lowest-molecular-weight standards. A factor contributing to the deviation might be the erratic elution profile of the highest-molecular-weight species.

A number of factors can contribute to the error in calculating programmed retention times. Among such factors are errors in the reported molecular weights of the standards, which can lead not only to erroneous calibration parameters but to retention anomalies for the affected polymers. The calibration parameters could likely be improved by using a larger set of polymer standards.

Some error in the calculated retention times is also introduced as a consequence of the high speed of the programmed runs. For such runs the temperature drop ΔT does not exactly follow the program selected because of time-lag effects. Therefore the ΔT-versus-time curve departs slightly from that used for calculation. Also, some error is introduced by small uncertainties in flow rate and retention time and by a slight unevenness in ΔT along the length of the channel.

Figure 5 shows a fractogram obtained by using a mixture of all the polymers shown in Table I. This run belongs to series 3 in Table III. (In this figure and those following, the peak labeled 200k corresponds to component of M 200,700.) The power-programming parameters are $t_1 = 187$ s, $t_a = -374$ s, and $p = 2$. According to eq 15, these parameters give a decay half-life $t_{1/2}$ of 419 s or 6.99 min. The flow rate is relatively slow for this series of runs: 0.460 mL/min average for the series and 0.429 mL/min for the fractogram in Figure 5. All the polymer component peaks in the fractogram are reasonably well resolved. Considerable noise is associated with the appearance of the polymer of M 5,480,000.

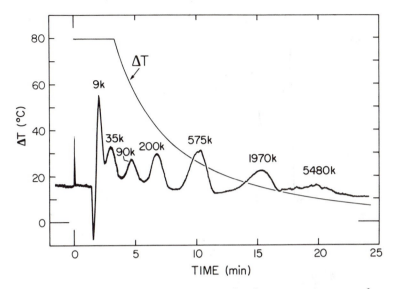

Figure 5. Fractogram and ΔT-versus-time plot for power-programmed run (series 3) of polystyrene standards of indicated molecular weights (in thousands, k) with parameters t_1 = 187 s, t_a = −374 s, p = 2, and $\hat{V}$ (flow rate) = 0.429 mL/min.

Because a good deal of the band broadening of polymer peaks is due to polymer polydispersity and not to underlying column processes, the experimental fractionating power calculated from the fractogram of Figure 5 (or from other fractograms) will be artificially small and should generally be exceeded by the theoretical values. (In theory, the band broadening due to polydispersity could be subtracted if polydispersity values were known precisely.) For example, in Figure 5 the observed F_M for the M 200,700 and 575,000 polymer peaks is 1.45, considerably lower than the calculated value, 2.78.

Experiments with sedimentation FFF (28) have shown that the calculated fractionating power is in reasonably good agreement with experiment, providing the experimental results are corrected for the small polydispersity effects characteristic of the latex bead standards used. Under the assumption that the theoretically calculated fractionating powers are reasonably accurate in thermal FFF, as they are in sedimentation FFF, the calculated range for this parameter is shown in Table II. For increased accuracy we used the more rigorous calculations of ref. 21 instead of the approximate equations given herein, which are valid only when the retention ratio at elution is fairly small.

If speed is more important than resolution, the power-programmed runs can be hastened to completion in several ways and by several combinations of parameters. For the fractogram in Figure 6 (from series 2) this increased

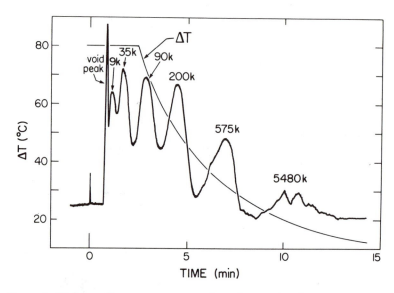

Figure 6. High-speed power-programmed run (series 2) with parameters t_1 =
151 s, t_a = –302 s, p = 2, and $\hat{V}$ = 0.773 mL/min.

speed is accomplished by nearly doubling (to 0.773 mL/min) the flow rate.
The time constants t_1 and t_a are reduced about 20% to 151 and –302 s,
respectively, to further enhance this effect. This leads to a program half-life
$t_{1/2}$ of 339 s or 5.64 min. The run is now completed in just over 10 min as
opposed to approximately 20 min for the previous run. However, the resolution has been somewhat degraded by the increase in speed. Because of
this degradation the standard of M 1,970,000 could not be resolved under
such conditions and was omitted from the run displayed in Figure 6 and
from the other series 2 runs summarized in Table III.

A further amplification of speed can be realized by increasing power p.
Figure 7, taken from series 5 in Table III, shows that the run can be completed in approximately 8 min by increasing p from 2 to 4. (There are no
changes in the time constants and only a minor change in the flow rate,
0.766 mL/min.)

Another important operating parameter is ΔT_0, the initial temperature
drop. A reduction in ΔT_0 (with other parameters constant) will also hasten
the separation process, but the lowest-molecular-weight polymers will suffer
a substantial resolution loss. Thus the run in Figure 8 with ΔT_0 = 60 °C,
instead of the 80 °C used previously, shows very poorly resolved M 9000
and 35,000 peaks. For higher molecular weights, however, the resolution
is comparable to that of most other runs, such as that shown in Figure 6.

As noted earlier in the text, many different programming forms can be
used to achieve the separation of a wide range of polymer molecular weights.
Figure 9 illustrates the use of a linear program (part of series 7) to resolve

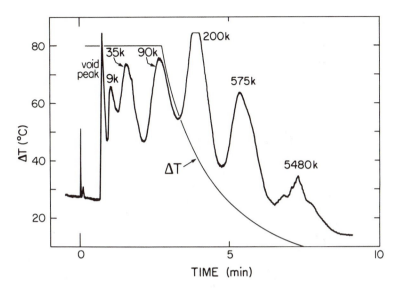

Figure 7. High-speed power-programmed run (series 5) with parameters $t_1 = 151$ s, $t_a = -302$ s, $p = 4$, and $\hat{V} = 0.766$ mL/min.

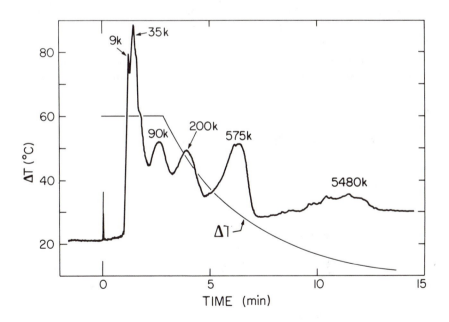

Figure 8. Low ΔT_0 (60 °C) power-programmed run (series 6) with parameters $t_1 = 167$ s, $t_a = -334$ s, $p = 2$, and $\hat{V} = 0.674$ mL/min.

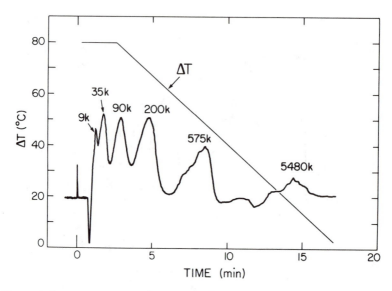

Figure 9. Linear-programmed run (series 7) with parameters t_1 = 151 s, ramp time = 900 s, and $\hat{V}$ = 0.750 mL/min.

the same polymer standards. Although the positions and resolution levels of the peaks are changed somewhat by going to the linear program, the molecular-weight range of our polymer standards is successfully covered by this alternative as well as by power programming. A linear program for thermal FFF was first used in our initial 1976 programming studies (11).

Acknowledgment

This work was supported by Grant CHE–8800675 from the National Science Foundation.

References

1. Giddings, J. C. *Anal. Chem.* **1981,** *53,* 1170A.
2. Giddings, J. C. *Sep. Sci. Technol.* **1984,** *19,* 831.
3. Caldwell, K. D. *Anal. Chem.* **1988,** *60,* 959A.
4. Giddings, J. C. *Chem. Eng. News* **1988,** *66,* 34.
5. Gunderson, J. J.; Giddings, J. C. *Anal. Chim. Acta* **1986,** *189,* 1.
6. Gunderson, J. J.; Giddings, J. C. In *Comprehensive Polymer Science, Vol. 1. Polymer Characterization and Properties;* Booth, C.; Price, C., Eds.; Pergamon: Oxford, 1989, Chapter 14, pp 279–291.
7. Giddings, J. C.; Martin, M.; Myers, M. N. *Sep. Sci. Technol.* **1979,** *14,* 611.
8. Moore, J. C. *J. Polym. Sci.* **1964,** A2, 835.
9. Giddings, J. C.; Smith, L. K.; Myers, M. N. *Anal. Chem.* **1975,** *47,* 2389.
10. Gao, Y. S.; Caldwell, K. D.; Myers, M. N.; Giddings, J. C. *Macromolecules* **1985,** *18,* 1272.

11. Giddings, J. C.; Smith, L. K.; Myers, M. N. *Anal. Chem.* **1976**, *48*, 1587.
12. Hovingh, M. E.; Thompson, G. H.; Giddings, J. C. *Anal. Chem.* **1970**, *42*, 195.
13. Schimpf, M. E.; Williams, P. S.; Giddings, J. C. *J. Appl. Polym. Sci.* **1989**, *37*, 2059.
14. Schimpf, M. E.; Myers, M. N.; Giddings, J. C. *J. Appl. Polym. Sci.* **1987**, *33*, 117.
15. Gunderson, J. J.; Giddings, J. C. *Macromolecules* **1986**, *19*, 2618.
16. Wahlund, K.-G.; Winegarner, H. S.; Caldwell, K. D.; Giddings, J. C. *Anal. Chem.* **1986**, *58*, 573.
17. Brimhall, S. L.; Myers, M. N.; Caldwell, K. D.; Giddings, J. C. *J. Polym. Sci. Polym. Lett. Ed.* **1984**, *22*, 339.
18. Giddings, J. C.; Martin, M.; Myers, M. N. *J. Chromatogr.* **1978**, *158*, 419.
19. Giddings, J. C.; Williams, P. S.; Beckett, R. *Anal. Chem.* **1987**, *59*, 28.
20. Kirkland, J. J.; Rementer, S. W.; Yau, W. W. *Anal. Chem.* **1988**, *60*, 610.
21. Williams, P. S.; Giddings, J. C. *Anal. Chem.* **1987**, *59*, 2038.
22. Myers, M. N.; Caldwell, K. D.; Giddings, J. C. *Sep. Sci.* **1974**, *9*, 47.
23. Martin, M.; Myers, M. N.; Giddings, J. C. *J. Liq. Chromatogr.* **1979**, *2*, 147.
24. Schimpf, M. E.; Giddings, J. C. *J. Polym. Sci. Polym. Phys. Ed.* **1989**, *27*, 1317.
25. Westerman-Clark, G. *Sep. Sci. Technol.* **1978**, *13*, 819.
26. Brimhall, S. L.; Myers, M. N.; Caldwell, K. D.; Giddings, J. C. *J. Polym. Sci. Polym. Phys. Ed.* **1985**, *23*, 2445.
27. Gunderson, J. J.; Caldwell, K. D.; Giddings, J. C. *Sep. Sci. Technol.* **1984**, *19*, 667.
28. Williams, P. S.; Kellner, L.; Beckett, R.; Giddings, J. C. *Analyst (London)* **1988**, *113*, 1253.

RECEIVED for review February 14, 1989. ACCEPTED revised manuscript September 1, 1989.

Determination of Long-Chain Branching Distributions of Polyethylenes

Francis M. Mirabella, Jr.,[1] and Leslie Wild[2]

[1] Quantum Chemical Corporation, Rolling Meadows, IL 60008–4070
[2] Quantum Chemical Corporation, Cincinnati, OH 45237

The long-chain branching distribution (LCBD) of polyethylene was determined from measurements obtained from an on-line viscometer detector (VD) combined with size exclusion chromatography (SEC). The SEC was calibrated according to the universal calibration method, and the VD output was used along with the concentration measurement obtained from the refractive index detector of the SEC to determine the intrinsic viscosity distribution of the eluting polymers. The SEC and VD data were then combined to yield the molecular-weight distribution (MWD) and LCBD of a variety of polyethylene resins. The SEC–VD technique was shown to be useful for determining the LCBD across the entire MWD, except at very low MW ($<3 \times 10^4$) because of the insensitivity of the VD to low-MW species. The LCBD of a large variety of commercial low-density polyethylene resins was constant across the MWD. This finding was confirmed by ^{13}C NMR spectroscopic measurement of the average LCB of narrow-MW fractions of one typical LDPE.

MEASUREMENT OF THE DEGREE OF LONG-CHAIN BRANCHING (LCB) in polymers continues to be an area of active investigation. The continuing interest is partially motivated by the difficulty of the available experimental methods. Further, the theoretical basis of the calculations used to extract the long-chain branching from the experimental data is virtually never rigorous for the polymer systems of interest. Thus, there are serious doubts about whether the calculated degree of LCB actually reflects that present in the polymer system. Efforts have concentrated on simplifying the experimental techniques for determining LCB and evolving a basis of comparison for confirming the accuracy of the methods. The work reported here demonstrates the use of an on-line viscometer detector (VD) combined with size exclusion chromatography (SEC). Because the SEC employs a refractive index detector (RID), this combination amounts to dual detection of the SEC eluent with VD and RID detectors. This system permits the molecular-weight distribution (MWD) corrected for LCB, the average degree of LCB, and the long-chain branch distribution (LCBD) to be obtained in a single SEC experiment (typically less than 1 h).

The RID yields the concentration of eluting species versus elution volume. The VD yields the intrinsic viscosity of eluting species versus elution volume. The elution volume is calibrated in terms of molecular weight by using narrow polymer standards. Thus, the RID output and the elution volumes yield the molecular-weight distribution; that is, the molecular weight versus the weight fraction. The VD output and the elution volumes directly yield the molecular weight versus the intrinsic viscosity; this is actually a Mark–Houwink plot and yields the Mark–Houwink constants from the slope and intercept. In practice, the RID and VD output are combined with the elution volumes to yield the molecular-weight distribution through the universal calibration procedure.

The usual data obtained from the SEC with the VD are (1) a combined plot of the RID output versus the elution volume and the VD output versus the elution volume; (2) the Mark–Houwink plot of intrinsic viscosity versus the molecular weight; (3) the normalized MWD; (4) the long-chain branching versus molecular weight (if applicable); and (5) the numerical data, including instrument conditions, data collection parameters, digitized data, MW averages, calculated intrinsic viscosity, and long-chain branching. The calculations of molecular weight and long-chain branching were done by the Viscotek Unical software package. These calculations were checked and found to be accurate. The following assumptions were made in these calculations: (1) the SEC separates molecules by hydrodynamic volume and not molecular weight; (2) the detector may contain a collection of species at any particular elution volume having varying LCB and MW, but the same hydrodynamic volume, and the average viscosity of these species is obtained and used to calculate LCB without any correction for mixing of linear and branched species in the detector; and (3) axial dispersion is considered to be negligible.

Polyethylene (PE) is a particularly important industrial polymer in which LCB plays an extremely significant role in determining the polymer properties. The work reported here was done on low-density polyethylene (LDPE) resins.

Experimental Details

A Waters 150C SEC was used at 140 °C (oven, RID, and injector temperature) with 1,2,4-trichlorobenzene (TCB), HPLC grade, twice filtered through a 1-m bed of activated silica and stabilized with 0.0009% w/v of Santanox-R antioxidant. The conditions were 1 mL/min, 250-mL injection volume, 0.1% w/v sample concentration, and Polymer Laboratories PL Gel (styrene–divinylbenzene) columns (10^6, mixed bed, 10^4, 500 Å, all 10-μm particle size with a 5-cm precolumn). Under normal operation, the solutions were prepared in an external oven at 160 °C for 1 h and equilibrated at 140 °C in the 150C SEC injector compartment for 15 min. The typical time between injections was 57 min, including polymer and small molecules elution time.

In special experiments, dissolution was achieved over 1 h at 130, 140, and 150 °C. Viscosity detection was achieved with a Viscotek model 100 differential viscometer detector connected in parallel with the RID using a 50:50 eluant splitter valve. Data handling was accomplished by using the Viscotek Unical software package on an IBM AT personal computer with a Hewlett–Packard (HP) Color Pro x–y plotter and a Citoh CI-3500 model 20 high-speed printer.

LDPE resins from Quantum Chemical Corporation were used: NPE 353, NPE 940, NBS 1476, USI 1016, USI NA205, USI NA102, USI 5602A, and USI 6009. The molecular-weight fractions were obtained by gradient elution fractionation of whole LDPE resins.

Preparative fractionation of polyethylene resins was carried out by using a gradient-elution column technique (1). Polymer (10 g) was loaded on the Chromosorb-P (diatomaceous earth) column packing by cooling from hot xylene solution. A continuous, exponential solvent gradient was employed at 115 °C with 70:30 and 20:80 mixtures of xylene and ethylene glycol monoethyl ether (Ethyl Cellosolve) as solvent and nonsolvent, respectively. The samples precipitated in acetone were suitably combined to give approximately 15 0.5–1.0-g fractions.

The average LCB of some of the PE resins was determined by ^{13}C NMR spectroscopy. The ^{13}C NMR spectra were obtained on a Varian 200-MHz spectrometer on each sample by using the following conditions: $\pi/2$ pulse width, 16.5-s recycle time, 125 °C, and overnight accumulation. Because of the poor signal-to-noise ratio in the spectra from the fractions, branching contents were determined by using peak heights rather than peak areas. This method will introduce some uncertainty into the absolute values calculated for LCB and for amyl and butyl branches, but the relative differences between the samples should be fairly indicative of overall branching. Ethyl branching in these samples is more problematic, as four of the five resonances used to calculate ethyl branching are too small to see in the fractions. As a result, only the highest intensity ethyl branch resonance (assigned to 1,3 ethyl pairs) was used in the analysis.

Four narrow-MW fractions of commercial LDPE resin NA205 (lot no. 12162) were analyzed by ^{13}C NMR spectroscopy. The fractions were selected to span the range of the whole resin's MWD. The whole resin was also analyzed by ^{13}C NMR spectroscopy. All samples were dissolved in a 4:1 mixture of trichlorobenzene–benzene-d_6 at 125 °C. The whole-resin sample was made up as a 15% w/v solution. The fraction samples were considerably more dilute because of the small amount of available sample.

Results and Discussion

The method of calculation of molecular weight and other parameters involves a combination of the universal calibration method and the intrinsic viscosity obtained from the VD. The SEC is first calibrated with a series of narrow-MWD polystyrene standards for which the intrinsic viscosity ([η]) and molecular weight are known. (Molecular weight $M \sim M_n \sim M_v \sim M_w$; that is, the number-average, viscosity-average, and weight-average molecular weights are all approximately equal.) A universal calibration line is obtained in which the product of individual values of $[\eta]M$ designated by $([\eta]M)_i$ is known as a function of the elution volume V_E. Thus, the product $([\eta]M)_i$ is readily obtained from the elution volumes for any unknown polymer when run on the SEC. Further, the VD and RID give the $[\eta]_i$ as a function of elution volume for an unknown polymer. The $[\eta]_i$ is obtained by calculation from the specific viscosity $(\eta_{sp})_i$ obtained from the VD corrected to infinite dilution by using the concentration from the RID at each V_E.

At this point the molecular weight M_i at each V_E is obtained by dividing the product $([\eta]M)_i$ obtained from the universal calibration by the $[\eta]_i$ obtained from the VD. The MWD is generated by plotting the concentration obtained from the RID of the species with molecular weight M_i at each elution volume. The typical moments of the MWD ($\overline{M}_n$, $\overline{M}_v$, $\overline{M}_w$, $\overline{M}_z$, and $\overline{M}_{z+1}$; the last two are the z-average and $z+1$-average molecular weights, respectively) are obtained in the usual way by insertion of the M_i and concentrations from the RID into the statistical formulas. The whole-polymer $[\eta]$ is obtained by summation of the $[\eta]_i$ at each elution volume using the concentration from the RID as a weighting factor.

Because the $[\eta]_i$ are actual intrinsic viscosities at each V_E, the M_i are correct molecular weights for long-chain branched polymers. Thus, for LDPE calculation of the "corrected" MWD is straightforward. The long-chain branching (LCB) is obtained by including the Mark–Houwink constants (k and a). These are used to calculate the intrinsic viscosity of a linear molecule $[\eta]_{i,L}$ for each M_i. The previously determined intrinsic viscosity will be lower than this value and can be redesignated as $[\eta]_{i,BR}$. The ratio $[\eta]_{i,BR}/[\eta]_{i,L}$ will yield the long-chain branching at each elution volume (LCB/1000C)$_i$ by the method of Drott and Mendelson (2). Thus, the long-chain branching per 1000 carbon atoms can be plotted as a function of molecular weight, and by summation of (LCB/1000C)$_i$, the whole-polymer LCB/1000C can be obtained.

The only theoretical derivation of consequence that purports to relate measurable parameters of polymers to the degree of LCB was published by Zimm and Stockmayer (3). This derivation arrives at an exceedingly simple relationship between easily measurable polymer parameters and LCB, but rests on several tenuous assumptions. The results of the derivation are simply that the ratio of the intrinsic viscosity of the branched polymer to the intrinsic viscosity of a linear polymer of equal MW ($[\eta]_{BR}/[\eta]_L$) is directly related to the degree of LCB (g^e, where the exponent e has a specific value for each polymer). This relationship is expressed mathematically in the following equation:

$$g^e = \frac{[\eta]_{BR}}{[\eta]_L} \qquad (1)$$

where the value of e was determined to be 0.5 for PE, as described in detail in a previous report (2). Thus, g^e is readily accessible from measurement of the intrinsic viscosity of the branched polymer $[\eta]_{BR}$ followed by calculation of the corresponding intrinsic viscosity for a polymer with equal MW from the Mark–Houwink equation:

$$[\eta]_L = kM^a \tag{2}$$

where M is readily obtained with the VD–SEC technique as just described and k and a are the Mark–Houwink constants.

To calculate the LCB, the value of e in equation 1 must be included also. Values of e ranging from 0.5 to 1.5 have been proposed for PE (4). Our work shows that the value should be closer to 0.5, and this value has been used previously. However, recent data (5) were used to produce the plot in Figure 1, which indicates that an average value of 0.65 in the LCB/1000C region of 0 to 10 is most reasonable. This value was used along with other values of e, and the best agreement of calculated and certificate values of

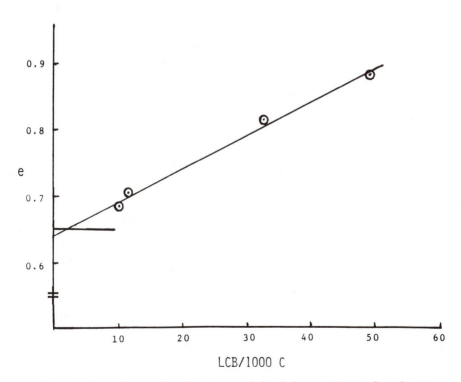

Figure 1. *Dependence of* e *value in* $g^e = [\eta]_{BR}/[\eta]_L$ *on LCB. A value of 0.65 appears to be the best average value in the 0 to 10 LCB/1000C region. Data are from Grinshpun et al.* (5).

LCB for a series of standards was obtained with $e = 0.65$. Thus, this value has been adopted for PE.

At this point, the value of g can be used to calculate the degree of LCB from simple equations. However, by assuming that branch points are trifunctional, the branch distribution has a uniform number of branches per molecule, and the branches have a random distribution of lengths, the following two equations can be derived (3). Here g is replaced by g_3 to signify trifunctional branching.

The equation for polydisperse polymers is

$$\langle g_3 \rangle w = \frac{6}{B_w} \left[\frac{1}{2} \left(\frac{2 + B_w}{B_w} \right)^{\frac{1}{2}} \ln \left(\frac{(2 + B_w)^{\frac{1}{2}} + B_w^{\frac{1}{2}}}{(2 + B_w)^{\frac{1}{2}} - B_w^{\frac{1}{2}}} \right) - 1 \right] \qquad (3)$$

where B_w is the weight-average number of long branches per molecule. The equation for polymer fractions monodisperse in MW is

$$\langle g_3 B_n \rangle = \left[\left(1 + \frac{B_n}{7} \right)^{\frac{1}{2}} + \frac{4 B_n}{9\pi} \right]^{-\frac{1}{2}} \qquad (4)$$

where B_n is the number-average number of long branches per molecule. Equations 3 and 4 are plotted in Figure 2. If g has been determined, B_n and B_w can be readily obtained from Figure 2. For example, if $g = 0.5$, then $B_n = 16.5$ and $B_w = 6.5$ from Figure 2. If the value of $M = 100,000$, then the LCB/1000C is obtained from

$$\frac{\text{LCB}}{1000\text{C}} = \frac{B_x}{M} (14,000) \qquad (5)$$

where $x = n$ or w.

Therefore, from the number-average equation LCB/1000C would be 2.31, and from the weight-average equation it would be 0.91. The question immediately arises as to which value is correct. The answer would presumably be 2.31 if the polymer were monodisperse in MW or 0.91 if the polymer were polydisperse in MW. At this point, a suspicion may arise that the equation involving B_n would never be appropriate for a "real" polymer because even a "narrow" polymer fraction, fractionated on the basis of the uniform hydrodynamic volume of the polymer chains, would be suspected to contain a collection of molecules varying from unbranched to some maximum branching level, all having approximately equal hydrodynamic volume.

Because the VD–SEC method measures intrinsic viscosity for small "slices" over an elution distribution for a polymer sample, it may be assumed that the number-average calculation (for monodisperse polymer) should be

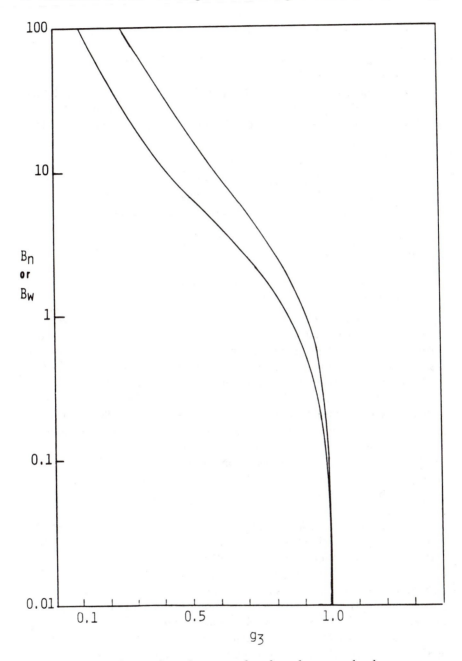

Figure 2. Number- and weight-average long branches per molecule versus g_3. B_n is the upper curve and B_w is the lower curve.

used. However, this work showed that the number-average (equation 4) calculation yields LCB/1000C values about 3 times higher than the weight-average (equation 3) calculation. For several polymers the LCB/1000C was determined by alternate means, such as ^{13}C NMR spectroscopy. Table I presents a comparison of LCB/1000C values for two polymers for which this parameter was determined by alternate means. These data show that the weight-average (equation 3) equation yielded values in agreement with the alternate techniques, but the number-average (equation 4) equation yielded values about 3 times too high.

Table I. LCB/1000C Calculated with Equation 3 (Weight-Average) and Equation 4 (Number-Average) Branching Equations Compared to Certificate and ^{13}C NMR Values

Polymer	Certificate	^{13}C NMR[a]	Number Average	Weight Average
NBS SRM 1476	1.2	1.3	3.0	1.3
NPE 350	–	4.5	9.4	3.3

[a]NMR spectroscopy overestimates LCB because it counts all branches greater than six carbons as long branches.

NMR values are always overestimates of LCB because all branches over six carbon atoms are counted as long branches. Thus, only the lower weight-average LCB values are in agreement with NMR values. The proposition that the number-average calculation yields fallaciously high values of LCB has been observed by others. Westerman and Clark (6) and Drott and Mendelson (7) discounted the use of the number-average equation because it yielded fallaciously high LCB values; their values were most clearly too high for some fractions for which the LCB values exceeded the short-chain branching values determined by IR spectroscopy, which is clearly impossible. On the basis of this analysis, the weight-average equation was judged to be appropriate and was used for the LCB calculations in this work. A reexamination of the derivation of these equations was beyond the scope of this study.

A compilation of data for "standard" and commercial polyethylene resins is presented in Table II. The agreement between the observed data and the certificate data is generally good. (Certificate values were determined by American Polymer Standards Corporation.) The observed data agree with past experience with low-density polyethylene (LDPE) and high-density polyethylene (HDPE) resins; that is, commercial LDPE resins typically exhibited LCB of 2 to 5 and HDPE resins of <1. For the HDPE resins, the values of LCB/1000C of several tenths may arise from other effects, such as short-chain branching, because LCB is assumed to be zero in HDPE resins. However, the low levels of LCB may be real. This method is probably the most direct method to obtain LCB values and, therefore, is expected to yield the most accurate values. The Drott and Mendelson approach suffers from the fact that the whole-polymer LCB value is obtained iteratively, but

Table II. MWD and LCB Data for HDPE and LDPE Resins

Sample	MI	Density	Observed Values × 10^{-3}								Certificate Values × 10^{-3}					
			$\bar{M}_n$	$\bar{M}_v$	$\bar{M}_w$	$\bar{M}_z$	$\bar{M}_{z+1}$	D	$[\eta]^a$	LCB/1000C	$\bar{M}_n$	$\bar{M}_w$	$\bar{M}_z$	D	$[\eta]^b$	LCB/1000C
NBS 1475 HDPE	2.07	0.978	17.3	52.4	58.3	122	204	3.4	1.01	0.4	18.3	53.1	138	2.9	1.01	0
NBS 1476 LDPE	1.19	0.931	28.2	79.8	101	248	420	3.6	0.91	1.3	23.7	102	—	4.3	0.90	1.2
Standard[c]	—	—	34.0	91.6	127	360	711	3.7	0.86	1.8	—	110	—	—	—	1.6
Standard[c]	—	—	24.2	90.1	123	346	631	5.1	0.89	1.6	—	83.9	—	—	—	1.7
Standard[c]	—	—	26.0	107	149	479	891	5.9	0.84	2.7	—	143	—	—	—	3.4
Standard[c]	—	—	19.7	124	187	683	1236	9.5	0.87	3.2	—	202	—	—	—	2.5
Standard[c]	—	—	14.1	122	190	744	1441	13.4	0.92	2.4	—	243	—	—	—	4.5
NPE353	2.0	0.923	28.5	120	162	473	845	5.7	0.83	3.3						
NPE940	0.25	0.918	34.6	181	273	855	1434	7.9	1.06	2.6						
USI 1016	3.8	0.923	26.9	140	275	1370	2450	10.2	0.93	2.0						
USI 5602A	0.25	0.955	20.6	103	128	467	898	6.2	1.55	0.3						
USI 6009	0.79	0.960	11.0	91.1	116	490	1043	10.5	1.43	0.3						

[a] In TCB at 140 °C from on-line Viscotek detector.
[b] In TCB at 130 °C.
[c] American Polymer Standards Corp.

the Viscotek detector makes the LCB of each molecular species over the MWD accessible. As indicated, the NMR method tends to overestimate LCB.

A novel result of this work is the characteristic shapes of the long-chain branching distributions (LCBD) of typical LDPE resins. Figures 3 and 4 show the LCBD for NPE 353 and NPE 940. The LCB decreases with increasing MW and remains constant to high MW. This behavior is typical of all commercial LDPE resins so far observed. The opposite behavior was reported by Wagner and McCrackin (8) as shown in Figure 5. However, more recently Rudin et al. (9) reported the same behavior as reported here, observed by on-line low-angle laser light-scattering detection. The LCBD for NBS 1476 determined by Rudin et al. is shown in Figure 6. There are two LCBD in Figure 6. Rudin et al. claimed that the dissolution procedure affected the detailed shape of the LCBD and that a lower dissolution temperature left polymer "aggregates" in solution that cause fallaciously high LCB at low MW. Therefore, higher dissolution temperatures would be required to totally dissociate these "aggregates" so that correct LCB is obtained.

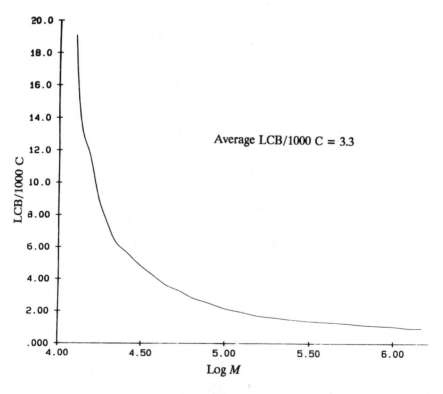

Figure 3. LCB/1000C versus log M for NPE 353 LDPE.

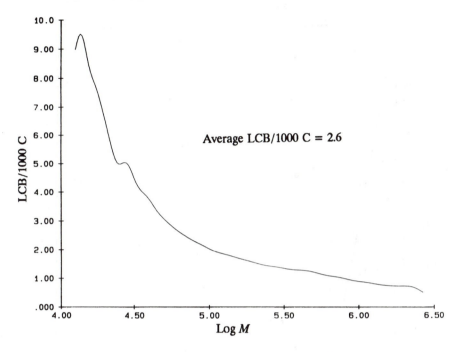

Figure 4. *LCB/1000C versus log* M *for NPE 940 LDPE.*

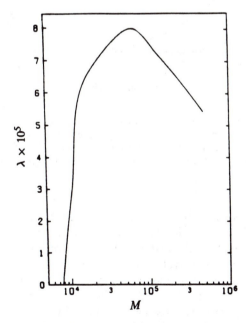

Figure 5. *Number of branch points per carbon atom for NBS 1476. (Reproduced with permission from ref. 8. Copyright 1977 Wiley.)*

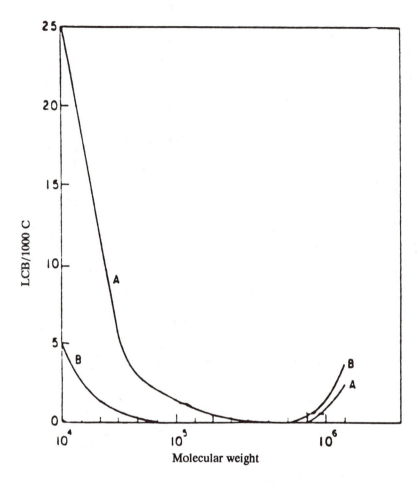

Figure 6. Long-chain branch frequency–molecular-weight relation for NBS 1476. Estimate was made with e = 0.5. A, polymer dissolved and analyzed in trichlorobenzene at 145 °C; B, polymer dissolved at 160 °C and analyzed at 145 °C. (Reproduced with permission from ref. 9. Copyright 1984, Marcel Dekker.)

This claim was tested by dissolving two LDPE resins over the temperatures range 130–160 °C. The data from this study are compiled in Table III. The LCBD for the NBS 1476 and USI 1016 resins studied are shown in Figures 7 and 8, respectively. These two figures show that there is no correlation between the dissolution temperature and the shape of the LCBD curve; that is, that the low-MW side of the curve does not become larger as dissolution temperature decreases, as claimed by Rudin et al. The source of this variability was suspected to be not a dissolution temperature effect, but a variability due to the base line chosen for the VD output data. Vari-

Table III. Comparison of LCB and MWD Data for LDPE Resins Dissolved at Different Temperatures

Dissolution Temperature (° C)	$\overline{M}_n$	$\overline{M}_v$	$\overline{M}_w$	$\overline{M}_z$	$\overline{M}_{z+1}$	D	LCB/ 1000C
NBS 1476							
160	28.2	79.8	101	248	420	3.6	1.3
160	28.3	80.1	100	239	396	3.6	1.3
150	18.9	76.5	94.4	237	426	5.0	1.3
140	25.8	78.7	101	253	442	3.9	1.1
130	25.6	76.7	93.9	208	318	3.7	1.4
USI 1016							
160	22.5	139	281	1440	2650	12.5	1.9
160	26.9	140	275	1370	2450	10.2	2.0
150	23.9	137	238	1020	1700	10.0	2.4
140	22.7	141	273	1200	1970	12.0	1.9
130	30.0	148	281	1330	2340	9.4	2.5

NOTE: In all cases, the dissolution time was 1 h. In all cases, the equilibrium time was 15 min at 140 °C. All MWD values are × 10^{-3}.

ability is often seen in MWD data as a result of the variability of drawing base lines for SEC peaks.

This suspicion was tested by varying the base line under the peak for the USI 1016 resin dissolved at 130 °C. Figure 9 shows a series of LCBD curves for this resin, with the base line being progressively moved further toward higher elution volume. The further the base line is moved toward higher elution volume (that is, the low-MW end) the smaller the M_n becomes, and this decrease results in a dramatically smaller low-MW side of the LCBD curve, as seen in Figure 9. Therefore, this base-line effect probably is the cause of the shape change in the LCBD curve and not the presence of undissolved polymer "aggregates".

The fact that the base-line drawing procedure results in major changes in the shape of the low-MW portion of the LCBD curve implies that this shape may be an artifact. This circumstance is considered to be likely because the VD is insensitive to low-MW species. The RID is more sensitive to low-MW species. Therefore, at low-MW the VD signal approaches zero while the RID signal remains finite. This behavior can be appreciated from Figure 10, which shows that the VD signal goes to zero at high elution volume (low MW) significantly before the RID signal reaches zero. The Viscotek software uses an extrapolation procedure in this vicinity and this procedure may give rise to the characteristic shape of the LCBD curves observed.

The question, however, may be asked as to the reason that Rudin et al. observed the same shape of the LCBD for branched polyethylenes by on-line low-angle laser light-scattering (LALLS) detection (9). The answer may be for the same reason as noted earlier because LALLS is even less sensitive at low MW than VD. Thus, both techniques may yield this rapidly decreasing

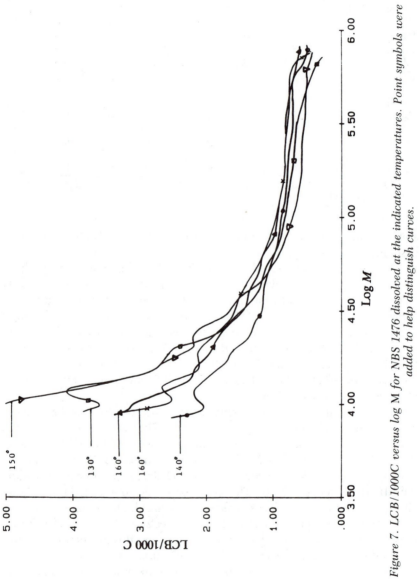

Figure 7. LCB/1000C versus log M for NBS 1476 dissolved at the indicated temperatures. Point symbols were added to help distinguish curves.

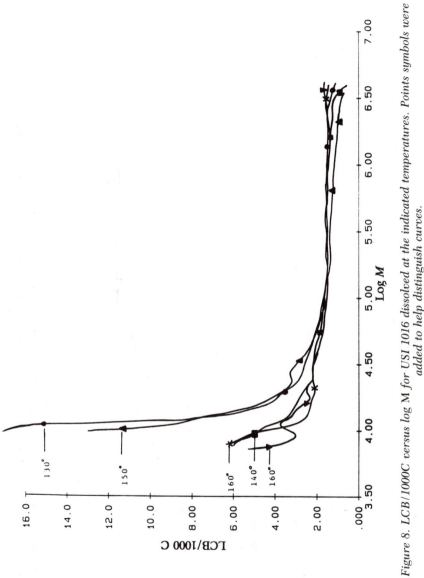

Figure 8. LCB/100C versus log M for USI 1016 dissolved at the indicated temperatures. Points symbols were added to help distinguish curves.

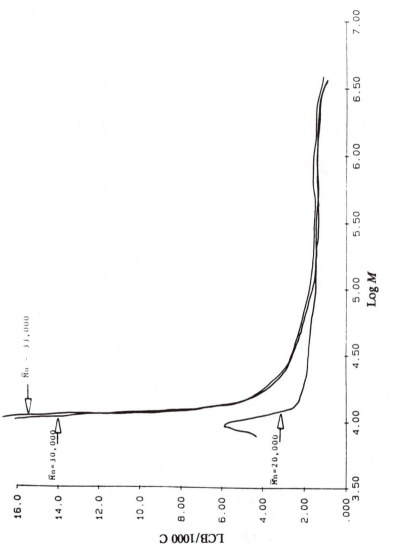

Figure 9. LCB/1000C versus log M for USI 1016 dissolved at 130 °C. Each curve was obtained with a different base line. As the base line was extended toward the lower-MW end of the curve, the $\overline{M}_n$ decreased as noted on each curve.

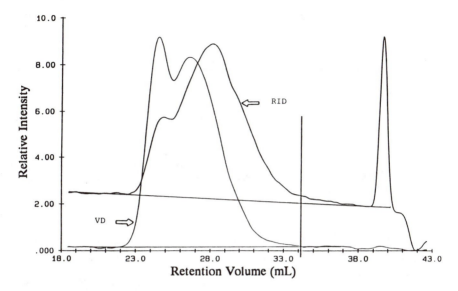

Figure 10. Raw data output from RID and VD showing that the VD signal goes to zero at high elution volume (low MW) while the RID signal remains finite to higher elution volume. Conditions: 130 °C for 1 h and 140 °C for 15 min.

LCB function at low MW for the same reason, that is, an artifact due to the insensitivity of both detectors at low MW.

To further investigate the accuracy of the shape of the LCBD of typical LDPE resins as produced by the VD–SEC technique, two LDPE resins were fractionated according to molecular weight and yielded a series of narrow-MW fractions. The molecular weight, long-chain branching, and ancillary data for these two LDPE resins are presented in Tables IV and V.

Table IV. Molecular Weights and Branching Parameters of Fractions of NA205

Fraction	Solution Viscosity (dL/g)	Melt Viscosity (poise)	$\overline{M}^w \times 10^{-4}$	LCB/ 1000C
4	0.345	—	2.0	4.2
5	0.439	8.82×10^2	2.7	2.7
8	0.646	1.97×10^3	6.2	2.5
10	0.873	1.97×10^4	9.9	1.6
12	1.095	2.37×10^5	18	1.6
14	1.351	2.05×10^6	31	1.4
15	1.561	1.02×10^7	51	1.4
16	1.785	7.10×10^7	81	1.4
Whole polymer	0.896	3.7×104	30.0	2.8

NOTE: The LDPE resin NA205 was lot 12162.

Table V. Molecular Weights and Branching Parameters of Fractions of NA102

Fraction	Solution Viscosity (dL/g)	Melt Viscosity (poise)	$\overline{M}^w \times 10^{-4}$	LCB/ 1000C
4	0.29	—	1.71	5.7
5	0.40	5.5×10^2	2.45	3.2
8	0.74	2.38×10^3	6.41	1.6
10	0.97	7.01×10^4	11.1	1.4
12	1.31	1.84×10^6	23.6	1.2
14	1.88	4.81×10^7	66.7	1.0
Whole polymer	0.896	6.35×10^4	21.3	1.3

NOTE: The LDPE resin NA102 was lot 12187.

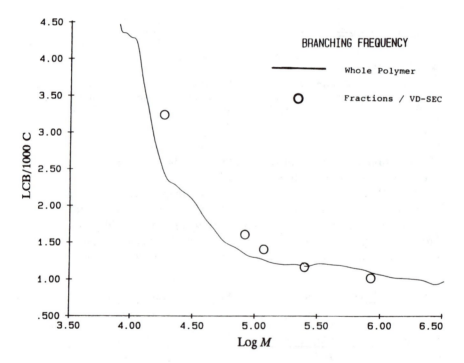

Figure 11. LCBD of USI NA205 and average LCBD of narrow-MW fractions.

The LCBD was determined by the VD–SEC technique for each whole LDPE resin, and then the average LCB was determined for each set of corresponding narrow-MW fractions. These values are plotted in Figures 11 and 12 for USI NA205 and USI NA102, respectively. In each case Figures 11 and 12 show that the average LCB of the fractions confirm the shape of the whole-polymer LCBD.

To further investigate this shape of the LCBD of typical LDPE resins,

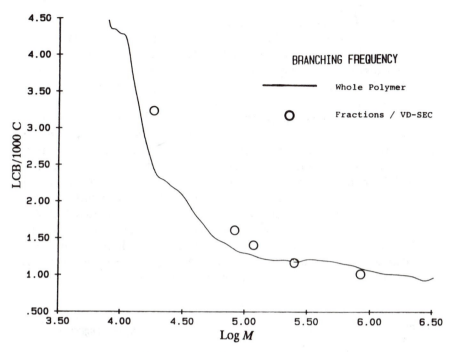

Figure 12. LCBD of USI NA102 and average LCBD of narrow-MW fractions.

Table VI. Molecular Weights and Branching Parameters
of Fractions of NA205

Fraction	$\overline{M}_n \times 10^{-4}$	LCB^a	Amyl	Butyl	Ethyl	Total
4	2.0	3.0	2.8	6.7	1.3	13.8
5	2.7	3.0	2.6	7.2	1.2	14.0
12	18	2.8	2.8	7.4	1.3	14.3
16	81	3.0	2.5	7.0	1.4	13.9
Whole resin		3.0	2.9	2.6	7.2	0.5	13.2

NOTE: The LDPE resin NA205 was lot 12162.
All values given are branches per 1000 carbons determined by ^{13}C
NMR spectroscopy.
^{a} Defined as any branch of six or more carbons long.

we sought an independent technique. Both short- and long-chain branching
may be simultaneously determined in LDPE resins by the use of ^{13}C NMR
(*10–12*). These experiments assume that the NMR resonance assigned to
hexyl branches is, in fact, due only to long-chain branches (*10*). Although
some controversy exists as to the validity of this assumption (*11*), the hexyl
resonance is generally used to determine LCB in LDPE resins.

Results of the NMR experiments are shown in Table VI. The whole-
polymer LCBD and the VD–SEC and NMR fraction-average LCB data are

plotted together in Figure 13. Few, if any, differences are noted in either the short-chain or the long-chain branching content across the MW range. If hexyl branches are actually LCBs, the NMR data indicate little variation in LCB across the MW range. The data reported here are consistent with the NMR work of Bugada and Rudin (12), who report LCBs on the order of 2.4–3.0 in a variety of LDPE resins.

These NMR results confirm the suspicion that the LCB determined at low MW by the VD–SEC technique is probably fallacious because of the low sensitivity of the viscosity detector to low-MW polymer species. However, the qualitative trend of a flat (zero slope) LCBD across the majority of the MW range (about 3×10^4 to 10^6 MW) of USI NA205 LDPE as determined by the VD–SEC technique (see Table IV) is supported by the ^{13}C NMR results in Table VI and Figure 13. Further, the values of the LCB of USI NA205 of about 3.0 across this same MW range as determined by NMR spectroscopy are in reasonable agreement with the values of about 1.5 to 2.0 as determined by the VD–SEC technique, especially when it is considered that the NMR technique overestimates LCB.

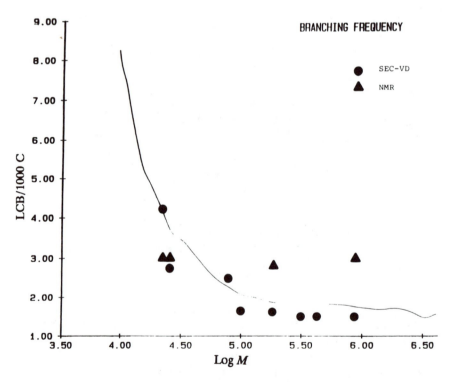

Figure 13. LCBD of USI NA205 whole-polymer (—) and molecular-weight (●, ▲) fractions.

Conclusions

The results of this study indicate the VD–SEC technique is suitable for the determination of the LCB and LCBD of polyethylene resins. However, these data were reliably determined only above a MW of about 3×10^4 by this technique, because of the low sensitivity of the viscosity detector to low-MW polymer species. An important observation of this work was that the LCB of a large number of typical commercial LDPE resins was constant across the majority of the MWD and does not increase at high MW. A common belief of workers studying polyethylene is that the often-observed high-MW "hump" (*see* Figure 14) on the MWD of these resins is an indication of highly long-chain-branched species. This type of species would yield a sharp increase of the frequency of LCB at high MW. Such an upturn in LCB at high MW was not observed in a large variety of typical commercial LDPE resins. Thus, the large "hump" on the molecular-weight distribution of typical LDPE resins, as seen in Figure 14, is apparently due not to an increased frequency of LCB relative to the rest of the resin, but rather to a simple bimodal molecular-weight distribution.

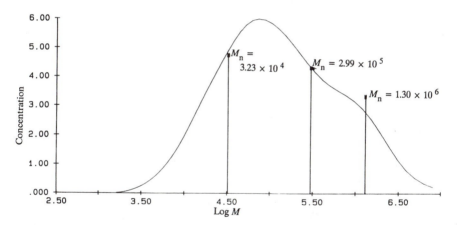

Figure 14. Molecular-weight distribution of a typical commercial LDPE resin (USI NA205).

Acknowledgments

The authors are grateful to David Bailey for making the ^{13}C NMR measurements, Kurt Klebe for the VD–SEC measurements, and Miriam A. Crandall for the typing of the manuscript.

References

1. Kamath, P. M.; Wild, L., paper presented at 22nd Annual Technical Conference of the Society of Plastics Engineers, Montreal, Canada, March 7–10, 1966; *SPE Tech. Papers* **1966**, *12*, XVII–6.

2. Drott, E. E.; Mendelson, R. A. *J. Polym. Sci.* **1970**, A–Z, 8, 1361, 1373.
3. Zimm, B. H.; Stockmayer, W. H. J. *Chem. Phys.* **1949**, *17(12)*, 1301.
4. Berry, G. C. *J. Polym. Sci.* **1971**, A–Z, 9, 687.
5. Grinshpun, V.; et al. *J. Polym. Sci. Polym. Phys. Ed.* **1986**, *24*, 1171; Table III.
6. Westerman, L.; Clark, J. C. *J. Polym. Sci. Polym. Phys. Ed.* **1973**, *11*, 559.
7. Drott, E. E.; Mendelson, R. A. *J. Polym. Sci.* **1970**, A–Z, 8, 1373.
8. Wagner, H. L.; McCrackin, F. L. *J. Appl. Polym. Sci.* **1977**, *21*, 2833.
9. Rudin, A.; Grinshpun, V.; O'Driscoll, K. F. *J. Liq. Chrom.* **1984**, *7*, 1809.
10. Usami, T.; Takayama, S. *Macromolecules* **1984**, *17*, 1756–1761.
11. Bugada, D.; Rudin, A. *Eur. Polym. J.* **1987**, *23*, 809–818.
12. Bugada, D.; Rudin, A. *J. Appl. Polym. Sci.* **1987**, *33*, 86–93.

RECEIVED for review February 14, 1989. ACCEPTED revised manuscript August 14, 1989.

Laboratory Robotics To Automate High-Temperature Gel Permeation Chromatography

Daniel G. Moldovan and Steve C. Polemenakos

Dow Chemical U.S.A., Polyethylene Research, B-3827, Freeport, TX 77541

This chapter presents the use of a commercial robotic system (Perkin-Elmer Masterlab) to automate the preparation of polyethylene samples for high-temperature gel permeation chromatography (GPC). The robotic system weighs the samples, calculates the appropriate volume of solvent, adds the solvent, caps the sample bottle, heats the sample for dissolution, and prepares the sample carrousel for the GPC autosampler. The advantages of this system include freeing technicians from tedious, repetitive tasks and from handling hot, hazardous materials.

INTEREST IN LABORATORY AUTOMATION has grown in the past few years because of the increasing complexity, diversity, and number of analytical techniques. "Hard" automation is used in areas where a single operation is performed. The use of "soft" automation (laboratory robotics) is implemented where flexibility and a range of operations is required.

The most repetitious and labor-intensive task for the chromatographer in high-temperature gel permeation chromatography (GPC) is sample preparation. The other steps in the analysis have been automated with a high-temperature (150 °C) GPC system (Waters) that will automatically analyze 16 samples and a computer system that will collect and reduce the data. The sample preparation step is the only procedure in the GPC analysis that is not automated and therefore is the bottleneck in the high-temperature GPC analysis. In this chapter, we report a procedure that has been developed

0065–2393/90/0227–0045$06.00/0

and uses a commercial robotic system (the Perkin-Elmer Masterlab) to automate the sample preparation procedure for the high-temperature GPC analysis.

The first step in developing this procedure or any robotic method is to define each of the laboratory unit operations (LUOs) (1). LUOs are common steps or building blocks of a laboratory procedure. Examples of some LUOs are weighing, manipulation, liquid handling, control, and documentation. These major LUOs can then be broken down into much smaller subclasses such as dispensing solvent and uncapping the bottle. After each of these smaller subclasses is defined, a flow diagram of the sequence of events is mapped out (Figure 1). At this point, individual robotic procedures can be written for each of the subclass LUOs. It is advantageous to make these robotic procedures as small as possible and to call them up individually in the main program or overall procedure. This approach allows one to access the individual procedures to make changes without having to overhaul the whole procedure.

Experimental Details

The Perkin-Elmer Masterlab robotic system used for the high-temperature GPC preparation consists of the Mitsubishi Move Master II model RM-501 robot equipped with a 0.7-m custom hand, a master syringe, symbol bar code reader, capper, Sartorius analytical balance (model A 200S), device interface, gas controller, custom racks, aluminum heat blocks, regripping station, and an IBM AT personal computer with a printer. The layout for our system is mounted on a 5- × 10-ft table and is shown in Figure 2. In an effort to retain the integrity of our sample preparation procedure, the 50-mL glass sample bottles that were used for the manual sample preparation procedure are used for the automated sample procedure. A set of custom racks that will hold 18 glass bottles was fabricated by the Perkin-Elmer Corporation. Also, two nine-hole anodized aluminum heat blocks were specially fabricated in-house to hold the bottles. The holes in the heat blocks were made deep enough to completely encompass the bottles and have a spherical clearance of 1 mm in order to provide good heat transfer to the bottles.

To begin the procedure, the operator fills the bottle rack with the appropriate number of 50-mL glass bottles. The operator then types the number of samples that need to be prepared. The robot tares each bottle. After all of the bottles have been tared on the Sartorius balance, the robot rings an alarm to let the operator know it has completed that part of the task. The robot then waits until the operator has added a representative amount of sample that consists of either pellets, film, or powder. The operator tells the robot to continue the task of sample preparation by hitting any key on the IBM AT keyboard. The robot reweighs each bottle and obtains the weight of the sample added.

The robot calculates the amount of solvent that needs to be dispensed to the bottle to obtain a concentration of 0.3% (w/w). The robot next takes the bottle to the regrip station, where it grips the middle of the bottle so that the bottle can be uncapped by the capper. While the robot is uncapping the bottle, the bar code reader reads the sample number from the bottle label that has been attached to the bottle. After removing the cap, the robot takes the sample bottle over to the dispensing station, where the syringe dispenses the required allotment of solvent to meet the concentration specification.

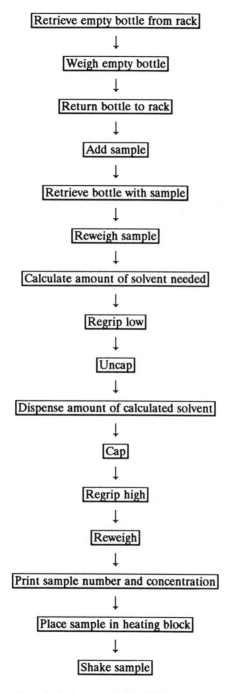

Sample is now ready for GPC analysis.

Figure 1. Flow diagram of sequence of events in sample preparation.

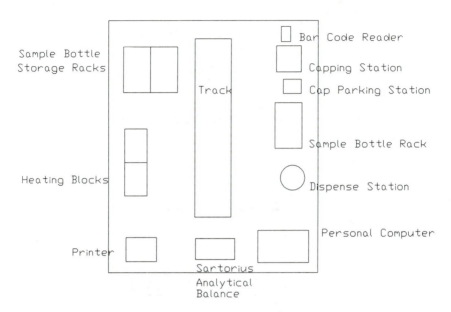

Sample Bottle
Storage Racks

Track

Bar Code Reader

Capping Station

Cap Parking Station

Sample Bottle Rack

Heating Blocks

Dispense Station

Personal Computer

Printer

Sartorius
Analytical
Balance

Figure 2. Robot table layout.

A 10-mL syringe is used for our application for dispensing the solvent because the solvent routinely used for high-temperature GPC is 1,2,4-trichlorobenzene. This solvent is very viscous at ambient temperature; therefore, it is difficult for the syringe mechanism to dispense large volumes accurately. The syringe must do repetitive fills for each bottle to obtain the correct concentration because of this limitation.

The robot takes the bottle back to the capper station, where it recaps the filled sample bottle. The capper turns a quarter of a turn counterclockwise to catch the lip of the bottle before turning clockwise to tighten the cap onto the bottle. The robot takes the bottle back to the balance, where it obtains the exact weight of solvent added; this step allows a more precise concentration calculation. The concentration, the sample weight, and the volume of solvent are printed for each sample prepared.

The robot again takes the bottle to the regrip station where it regrips the bottle by the cap so that it can place the bottle in the heat block, which is maintained at a temperature of 160 °C for 8 h for dissolution. The temperature in the heat block is monitored by a thermocouple that is inserted into the middle of the block. The temperature of the solution in the bottle is checked manually with a mercury thermometer during the setup of the system. (We found that the temperature of the solution in the bottle reached the target temperature of 160 °C after 35 min.) The robot shakes the sample bottle once during this period to help the dissolution process. The sample is now ready for high-temperature GPC analysis on a 150 °C GPC system.

Results and Discussion

The sample preparation in most analytical procedures is considered the weak link in the analysis for a number of reasons. Sample preparation is a major source of errors because it is subject to human variability. In our laboratory, a number of technicians perform the sample preparation, and each technician does it just a little bit differently from the others; the result is a variance in the concentration of the samples. Even an individual's laboratory performance changes from day to day, depending on the person's mood. Table I compares the weights and the volumes of the manual method versus the automated method.

In the manual method, the volume of the solvent is set by a repeating pipet (repipet); therefore, a target weight of 0.25 g is required for the sample to achieve a 0.3% (w/w) concentration. As shown in Table I, the range of the actual sample weight is wide; thus a greater error in the GPC analysis results.

Table I. Manual versus Automated High-Temperature GPC Method

Method	Sample Weight (g)	Solvent Volume (mL)	Conc. (% w/w)	Standard Deviation (%)
Manual	0.25 ± 0.0025	50 ± 0.5	0.03	10
Automated	0.15–0.25 ± 0.0001	30–50 ± 0.1	0.03	0.8

Another reason that sample preparation is considered the weak link is that the sample preparation procedure is labor intense and therefore expensive. In the manual GPC sample preparation, the laboratory personnel have to weigh out approximately 0.25 g of sample. Most of the sample is in the form of polyethylene pellets, which the technicians must cut into small pieces to get 0.25 g. After the 0.25-g sample is added to the 50-mL bottle, 50 mL of 1,2,4-trichlorobenzene is repipeted into the bottle. The manual method requires from 4 to 6 h each morning, depending on the sample load, in order to have samples ready for overnight analysis on the 150 °C GPC system. The typical sample load for the laboratory is from 16 to 48 samples per day. The automated method requires only 5–10 min of a technician's time to achieve the same task. This automated method therefore resulted in an immediate saving of half a worker-year.

Furthermore, sample preparation is time consuming, so sample turnaround time is slow. A person could prepare one, two, or more samples faster than the robot for the GPC sample preparation procedure, but on a daily basis the robot does not have to deal with meetings, answering the phone, and other tasks that distract the technicians from performing the sample preparation procedure.

Finally, sample preparation often exposes the personnel to a hazardous environment. The manual GPC sample preparation procedure requires the technician to wear both rubber and cotton gloves along with face shields and aprons to prevent exposure to hot 1,2,4-trichlorobenzene. With the automated method, only the robot is exposed to these hazards.

The automated system for high-temperature GPC preparation was successful in eliminating a very labor-intensive task requiring 4–6 h each day and replacing it with a 5–10-min computer interfacing task. In addition, the automated procedure has resulted in faster sample turnaround time and

more reproducible data for the high-temperature GPC analysis. The manual
GPC preparation procedure had a standard deviation of 10%, but with the
automated procedure, a standard deviation of 0.8% was attained. Besides
providing all of these advantages, the automated sample procedure has removed the solvent exposure and hot material handling hazards that were
part of the manual procedure.

Conclusions

The automation of the sample preparation procedure for high-temperature
GPC analysis using a Perkin-Elmer Masterlab robotic system has removed
the major bottleneck in the GPC analysis and therefore allowed essentially
complete automation of the high-temperature GPC analysis.

References

1. Hurst, W. J.; Martimer, J. W. *Laboratory Robotics: A Guide to Planning, Programming, and Applications;* VCH: New York, 1987; pp 15–23.

RECEIVED for review February 14, 1989. ACCEPTED revised manuscript August 16,
1989.

Characterization of Synthetic Charge-Containing Polymers by Gel Electrophoresis

David L. Smisek and David A. Hoagland*

Department of Polymer Science and Engineering and Department of Chemical Engineering, University of Massachusetts, Amherst, MA 01003

The effectiveness of gel electrophoresis as a tool for determining the molecular weight distributions of long-chain synthetic polyelectrolytes was evaluated extensively, primarily from results obtained with poly(styrene sulfonate) (PSS). We established that narrow distribution samples, when available, can be employed to construct chain length–mobility calibration curves. From these curves the full molecular weight distribution of an unknown can be calculated once its mobility distribution is measured. With PSS, this procedure was verified for $7 \times 10^3 < M < 15 \times 10^6$, a much broader molecular weight (M) range than has been achieved with size exclusion chromatography (SEC). The resolution of these fractionations is also significantly higher than normally observed in aqueous SEC. The quality and simplicity of gel electrophoresis, as documented here, provide strong motivation for wider application of the method in polymer science.

$\mathbf{E}$LECTROPHORESIS PLAYS A KEY ROLE in biopolymer research whenever complex protein or polynucleotide mixtures are to be analyzed (1). The first attempt to extend this traditional biopolymer technique to the realm of synthetic polymer science was by Chen and Morawetz (2), who described high-resolution electrophoretic separations of synthetic polymers in chem-

*Corresponding author.

0065–2393/90/0227–0051$06.00/0

ically cross-linked polyacrylamide gels. These researchers limited their experimental efforts to poly(styrene sulfonate) (PSS) and poly(acrylic acid) samples with molecular weights between 2×10^4 and 1×10^5.

We are principally interested in materials with molecular weights above this range; our objective is a characterization tool for the aqueous polymers used in processes such as water treatment and oil recovery. In these applications, high molecular weight is essential, and molecular weight averages above 1×10^6 are common. We are therefore conducting electrophoretic studies primarily in agarose gels, media appropriate to the separation of the largest macromolecules; agarose gels have a much "looser" pore structure than polyacrylamide gels and consequently permit entry of larger penetrant species. Variants of the techniques discussed here, such as those imposing pulsed electric fields, have been employed to isolate entire chromosomes of immense size (3). Such methods, as yet poorly understood, may prove crucial to the proposed effort to map the human genome.

Although size exclusion chromatography (SEC) is a valuable method for analyzing low- and medium-molecular-weight polyelectrolytes, fractions of higher molecular weight ($M > 1.0 \times 10^6$) have generally resisted this approach to molecular weight determination (4, 5). This failure can be attributed to polyelectrolyte expansion in the ionic solvent medium, to long-range electrostatic interactions with the chromatographic support, and to irreversible polymer adsorption on column surfaces. All three effects can be troublesome even when molecular weights are relatively small. Adsorption phenomena are particularly difficult to eliminate in aqueous polymer systems, and surface-active agents are often a necessary addition to the mobile phase.

Alternative analytical schemes to SEC have sometimes been employed to circumvent these difficulties, but the few successful options (band sedimentation, field-flow fractionation, and hydrodynamic chromatography) are generally too complex for routine use; they are certainly more difficult to operate than conventional gel electrophoresis. Given these conditions, commercial water-soluble polymer samples are commonly sold under the vague labels of "low", "medium", or "high" molecular weight. The application of gel electrophoresis to these samples would provide the same advantages cited for biopolymer characterization: reduced equipment cost (~$250 per device), more rapid sample turnover (up to 25 samples analyzed in a single run of several hours), and higher resolution.

Poly(styrene sulfonate) (PSS) was selected as a model compound in this study because of its well-understood and well-documented behavior in dilute ionic solutions (6, 7). Also, narrow distribution samples are commercially available or readily synthesized. Although gel electrophoresis is expected to have broad applicability for other highly charged polymers, quantitative molecular weight information can be obtained rigorously only when molecular size standards are available. Polymers other than PSS will be discussed only briefly. Hydrolyzed polyacrylamides will be specifically discussed because, in contrast to PSS, they possess a relatively low, easily adjusted charge density along the polymer backbone. In this case the separation of polymer fractions is a function of both molecular weight and charge density; whether unambiguous molecular weight data can be deduced from such separations is not yet certain.

Establishing a reliable procedure for detecting polymer in an aqueous gel constitutes a significant intermediate step in applying electrophoresis to a new polyelectrolyte species. High contrast of polymer bands against the gel regions not permeated by polymer is required; at the same time, the polymer concentration in the bands must be below the coil overlap concentration. For high-molecular-weight polymers, these conflicting requirements are not easily satisfied. An effort to develop a simple yet sensitive detection procedure for carboxyl-containing polymers in agarose gels is currently underway.

Experimental Details

Our electrophoresis technique is nearly identical to methods originally developed to isolate medium- to high-molecular-weight DNA fragments (8). Separation of samples loaded in a gel occurs as a steady electric field is applied over a time from 2 to 10 h. The gels are prepared by dissolving agarose powder at high temperature (95 °C) in an appropriate aqueous buffer. As the agarose solution is cooled to room temperature, it forms a mechanically stable gel possessing properties determined mainly by the agarose concentration. The gel is cast as a horizontal slab, with sample wells formed by the indentations of the teeth of a Plexiglas comb. The slab, typically measuring 15 cm on each horizontal edge, is about 1 cm thick. The different polymer fractions separate across the plane of the slab over distances on the order of several centimeters.

During the electrophoresis experiment, the weak agarose gel is supported on a tray bridging the two buffer reservoirs of a submarine cell (BioRad). Submersed platinum electrodes impose a voltage drop of 0.1 to 2 V/cm when connections have been made to a constant-voltage power supply (Ephortec 500 V). At the start of a run, polyelectrolyte solutions are pipetted into the sample wells, and the upper surface of the gel is left uncovered; after 15 min at constant voltage (a period designed to allow the polymer chains to migrate away from sample wells and into the gel matrix), the gel is submerged under a thin layer of buffer by carefully pouring additional solution into the buffer reservoirs. Throughout the remainder of the run, buffer solution is gently recirculated over the top of the gel by a peristaltic pump to ensure uniform electric field strength and solvent composition. Whenever desired, the field strength in the gel is measured with a hand-held voltmeter attached to platinum electrodes inserted in the gel at fixed separation; the magnitude of the measured field has always matched closely with the one that is applied.

When experiments with sulfonated polymer species are terminated, the transparent gel is transferred to a bath containing an aqueous dye solution (0.01% methylene blue, pH 4 acetate buffer). The dye attaches to the polymer, presumably by electrostatic interaction, over a period of 15 min. During this period the dye diffuses throughout the entire gel slab. Residual dye is then removed by placing the gel in a bath of distilled water. Complete destaining of agarose may take 12 to 15 h, although polyelectrolyte bands become visible after a few minutes. At this point, qualitative features of the molecular weight distribution are evident from even a cursory inspection of the stained gel. Quantitative analysis of the completely destained gels is accomplished with a densitometer (Isco model 1312) adjusted to measure absorbance at 580 nm.

PSS samples with a narrow molecular-weight distribution, derived from anionically polymerized polystyrene (PS) of polydispersity less than 1.25, constitute the largest class of materials so far examined. The PSS materials are either purchased from Pressure Chemical (for lower molecular weights) or prepared in our own laboratory from the PS molecular weight standards available from a variety of vendors (for $M > 1.2 \times 10^6$). The preparation and characterization of the linear PSS materials, as well as the electrophoresis apparatus, are described in detail in a separate publication (9). PSS "star" molecules are prepared from star polystyrenes (Polysciences) by the same methods employed to prepare linear PSS. In all cases the polyelectrolyte solutions that are pipetted into sample wells at the start of the experiment are dilute (<50 ppm). Under these conditions the optical density of the stained gel at 580 nm is linear in the local PSS concentration.

Poly(2-acrylamido-2-methylpropanesulfonic acid) was successfully fractionated under the same experimental conditions employed for PSS; the bands observed after staining with methylene blue reflect the broad molecular weight distribution of the samples studied (Scientific Polymer Products). These data cannot be quantitatively interpreted, however, because no molecular weight standards are available, and we have therefore chosen not to present specific results here. They are mentioned because of their importance in demonstrating the generality of the procedures outlined here for electrophoresis of sulfonated water-soluble polymers.

Efforts to stain hydrolyzed polyacrylamides in the gel were unsuccessful, so samples are labeled by a fluorescent dye prior to electrophoresis (10). After the run, the gel is placed on a UV transilluminator (360 nm) and photographed from above through a 450-nm cutoff filter. Mobilities are determined by measuring images on the resulting prints. The hydrolyzed polyacrylamide samples (gifts of R. Farinato of American Cyanamid) are labeled in the supplier's literature as SF210, SF212, and SF214. The nominal molecular weights of all three samples are the same (in the range 5–15 $\times$ 10^6), but the degrees of carboxyl substitution differ.

Results

The field-induced translational motion of a polyelectrolyte can be expressed in terms its electrophoretic mobility μ

$$\mu = \frac{x}{Et}$$

where x is the distance migrated in time t while under the influence of an electric field of magnitude E. A nominal value of x in our apparatus is 15 cm, E is of the order 2 V/cm, and t may range from 2 to 10 h. Mobility nearly always decreases with chain length, so the lower-molecular-weight fractions migrate to larger values of x.

A plot of the logarithm of PSS chain length versus mobility is displayed in Figure 1 for a set of typical run conditions (0.6% agarose; $E = 1.3$ V/cm;

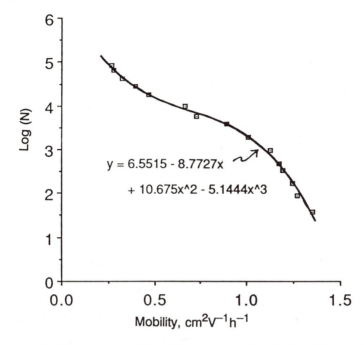

Figure 1. Calibration curve derived from narrow distribution PSS samples. N is the degree of polymerization. The highest value of N corresponds to a molecular weight of 15 × 10⁶. (Conditions: 0.6% agarose; 1.3 V/cm; 7.3 h; I = 0.03 M.)

t = 7.3 h; ionic strength (I) = 0.03 M). The mobility data can clearly be correlated with chain length over the entire span of chain lengths shown. Figure 1 is therefore a calibration curve that enables conversion of the mobility distribution of an unknown PSS sample to its molecular weight distribution. The standards from which the calibration curve is derived must be run simultaneously with the unknown if an accurate molecular weight determination is the goal. This side-by-side migration of unknowns and standards is desirable, as preparation of two identical gels is difficult.

Figure 2 shows chain-length distributions for two of the narrow distribution PSS samples purchased from Pressure Chemical. These samples, with average degrees of polymerization lying in the nearly linear central portion of the calibration curve of Figure 1, are essentially base-line resolved. The average molecular weights (in the 100% sulfonated forms) are 175,000 and 354,000. Such high resolution in the separation of high-molecular-weight, synthetic water-soluble polymers is unprecedented.

The polydispersities [weight-average molecular weight over number-average molecular weight (M_w/M_n)] calculated from the curves are both less than 1.02. This value is an upper bound, as the breadth of the polymer peaks mainly reflects the initial thickness of the sample wells; better characteriza-

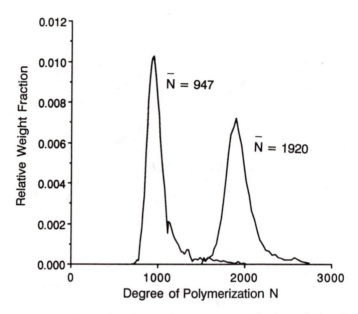

Figure 2. Chain-length distribution for two PSS standards. Calculated poly-dispersities from gel electrophoresis for both samples are less than 1.02. (Conditions: 0.6% agarose; 1.3 V/cm; 7.3 h; I = 0.03 M.)

tion of polydispersity could be obtained by using longer gels. The manufacturer merely states, on the basis of SEC of the polystyrene parents, that the polydispersities are less than 1.10. For each sample the degree of polymerization provided by the supplier agrees to within 1% with the value inferred from the calibration curve and the peak of the densitometry trace. The good agreement in peak values is expected, of course, because the same types of PSS standards were employed in developing the calibration curve.

Numerous analogies may be drawn between molecular weight determination by gel electrophoresis and determination by aqueous SEC. Although the separation mechanisms are quite different, both methods normally require calibration with a set of chain-length standards. In SEC, the polymer concentration, determined by refractive index difference or absorption, is measured as a function of retention volume; in gel electrophoresis, the polymer concentration, determined by the optical density of stained polymer in the gel, is measured as a function of position. Although not done at present, refractive index could provide a means of universal detection in electrophoresis just as it does in SEC. In fact, except in the conveyance of sample through the separation medium, gel electrophoresis and SEC manifest a parallel set of operating principles and obstacles.

A major distinction between SEC and gel electrophoresis is our comparative ignorance, in gel electrophoresis, of the molecular mechanisms that drive the separation. We are now confident that SEC fractionates flexible, nonpolar polymers according to hydrodynamic volume; in gel electrophoresis, however, we are not certain of the roles played by charge density, chain length, and molecular radius in controlling the fractionation. All three quantities are likely to be important in certain parameter ranges. For analytical purposes, however, a unique chain length–mobility relationship is sufficient knowledge for determining molecular weight by electrophoresis.

Several well-defined modes of macromolecular transport have been postulated for electrophoretic separations in highly swollen gels. The motion of proteins in polyacrylamide and agarose gels has been explained in terms of a hindered diffusion model, one in which spherical molecules of uniform segment density can pass through only those pores able to contain an entire, undeformed molecule (11). This mechanism is referred to as "sieving". More recently, the separation mechanism for large, flexible DNA fragments in highly confining gels has been explained in terms of the end-on motion of polymer chains by "reptation" (12, 13). In this depiction the size of the DNA molecules greatly exceeds the average pore dimension.

Finally, one can envisage an intermediate molecular size regime in which the polymer coils have a radius comparable to or slightly larger than the typical pore size. Polymer migration in this case is limited by the configurational barriers posed by the random gel matrix. Polymer chains pass through constraints only when the appropriate structural and translational Brownian fluctuations occur. The polymer thus travels by a series of jumps between more open regions in the gel matrix. All three transport modes produce a different dependence of mobility on chain length.

We will not compare our data to theoretical models here. When chain-length standards are available, molecular weight determination by electrophoresis does not require knowledge of the molecular transport mechanism in the gel. The third mechanism described, the one based on translational motion mediated by polymer configurational rearrangement, is probably the dominant transport mode for the polymer–gel systems described here. The radius of gyration of a 1×10^6 PSS chain in the buffer system of Figure 1 is nearly equal to the mean pore size of the 0.6% agarose gel; fluctuations in chain structure will therefore provide tremendous assistance to polymer transport during passage through the tortuous gel matrix. This picture of electrophoretic separation has developed only recently, and comparisons of the model predictions to the current data will be fully described in a later publication.

Chromatographic resolution is a function of both the magnitude of separation and the level of band broadening or dispersion. Although the magnitude of the separation produced in gel electrophoresis is comparable to that of SEC, band-broadening is significantly reduced. This attractive feature results primarily from the absence of convective flow in the separation; velocity gradients are the prime cause of band-broadening in most traditional chromatographic fractionations. The broadening of polymer peaks in gel electrophoresis occurs solely by diffusion (14), and polymer diffusion coefficients in gels are extremely small. For the smallest PSS chains ($M \leq 2 \times 10^4$), however, resolution in low-concentration agarose gels is correspondingly lowered because of large diffusion coefficients. This problem can be solved by using higher concentration gels, gels that might not allow larger polymers to migrate at reasonable rates. A good quality separation thus requires some tailoring of gel concentration to the molecular weight distribution of the sample.

Changing gel concentration also modifies the slope and position of the linear portion of the calibration curve. As illustrated in Figure 3, the linear section of the calibration curve can be extended to higher molecular weights by using lower concentration gels. The high molecular weight limit is achieved when the gels become too weak for handling. This breakdown occurs when the agarose concentration lies somewhere in the neighborhood of 0.05% to 0.10%. At these low concentrations agarose gels are easily able to fractionate polyelectrolytes with molecular weights up to the range of 1–2 $\times 10^7$. The cutoff is much above the limits encountered in aqueous SEC.

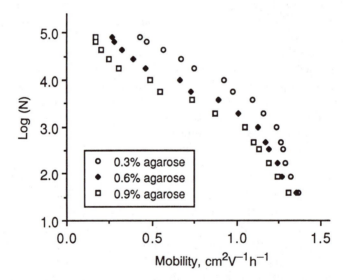

Figure 3. Mobility as a function of the degree of polymerization for three agarose concentrations. (Conditions: 1.3 V/cm; 7 h; I = 0.03 M.) (Adapted from ref. 9.)

Conversely, higher concentration agarose gels are able to resolve polymers with extremely low molecular weights; in Figure 1, for example, a 0.6% agarose gel easily separates PSS samples with molecular weights of 7,000 and 16,000 (the lowest two molecular weights shown). The ability to tailor the separation to a given sample's molecular weight distribution is a distinct advantage of gel electrophoresis, one that can be compared to the relative difficulty of switching SEC columns to obtain optimal resolution. Of course, it does take approximately 90 min to prepare each gel, a medium that is discarded after only a single run.

The buffer's ionic strength strongly affects the mobility of polyelectrolyte samples in agarose gels. For PSS separations these effects are complex and, surprisingly, not consistent with the analogous observations reported for DNA. The buffer used in all experiments cited here, 0.01 M Na$_2$HPO$_4$, maintains the ionic strength at 0.03 M. Good separations of PSS are observed in all phosphate buffer systems with ionic strengths above 0.01 M. At lower ionic strengths the quality of separation deteriorates, particularly for the very highest molecular weight samples ($M > 1 \times 10^6$); this effect is not yet understood. The upper ionic strength that can be achieved is limited by the current capacity of the power supply. For the standard conditions of this study, the upper ionic strength limit is about 0.3 M. A more complete discussion of ionic strength effects, particularly at low ionic strength, has been published elsewhere (9).

The "counterion condensation" theory of Manning (15) predicts that the effective linear charge density of a highly charged polymer is independent of the actual density of covalently attached ionic groups. For electrophoresis, this effective linear charge density is associated with the asymptotic long-range solution of the Poisson–Boltzmann equation around a highly charged cylinder (7). This result has significant implications for the present work: with any highly charged polymer species the distribution of charge within a sample should not affect its electrophoretic behavior. The separation is thus dependent only on the distribution of molecular weight. Charge effects should be significant, however, for polymers containing weakly ionizable groups because the linear charge density is correspondingly lower (assuming, in this case, that the charge density is below the onset of condensation). At neutral pH, for example, the separation of a carboxyl-containing polymer is likely to be highly sensitive to the actual density of carboxyl groups.

At the sulfonation levels of the PSS samples (>70%), effects of sulfonation level on mobility are not expected and have not been observed. In particular, a specially prepared 70% sulfonated sample (the degree of polymerization was 1550) has the mobility predicted by the calibration curve for a 100% sulfonated sample with the same degree of polymerization. This agreement is to within the experimental error in the mobility measurement ($\pm 1\%$).

Hydrolyzed polyacrylamides constitute a class of polymers for which the density of ionic groups along the polymer backbone is more readily varied. Table I lists the carboxyl substitutions and mobilities of the three hydrolyzed polyacrylamides studied. In contrast to the highly sulfonated PSS samples, the mobility displays a clear dependence on the polymer's charge density. According to Manning's theory, these low charge densities do not lead to condensation, a prediction consistent with the mobility results. In fact, the mobility roughly follows the level of hydrolysis, as expected from theory at

Table I. Mobility as a Function of the Degree of Carboxyl Substitution in Hydrolyzed Polyacrylamides

Sample	Carboxyl Substitution (%)	Mobility[a] (cm²/V h)
SF210	9.5	0.12
SF212	20	0.46
SF214	35	0.55

[a]At fluorescence maximum.

such low charge densities. Quantitative comparisons to theoretical models are to be avoided because the polydispersities of these samples are so large. Efforts to more accurately verify the range of charge densities over which counterion condensation compensates for charge variations are currently being conducted.

The application of gel electrophoresis to materials of broad molecular weight distribution can be illustrated by presenting data on two heterogeneous PSS samples. Figure 4 displays the distribution of chain length for a linear PSS as determined by both SEC and by gel electrophoresis. The SEC analysis is actually of the PS parent compound (Polysciences catalog no. 18544) in tetrahydrofuran (Polymer Laboratories PLgel columns in the pore-size series 10^5, 10^4, and 10^3 Å; flow rate = 1.0 mL/min). Chain-length distributions of the sulfonated and unsulfonated forms can be directly compared because the sulfonation reaction does not degrade or cross-link PS chains. The distributions plotted in Figure 4 show excellent agreement. Only minor discrepancies arise, mainly from noise in the detector signals and from uncertainties in selection of base-line levels. The PS parent sample did possess a significant fraction of monomer and short oligomeric material; these components are excised from the PS chain distribution displayed in the figure. Any such fractions would have been removed by the dialysis steps employed in preparing PSS samples for electrophoresis.

Figure 5 shows the mobility distribution of a sulfonated star polymer that nominally possesses six arms (Polysciences catalog number 18145) of well-controlled arm molecular weight (M_{arm} = 116,700; unsulfonated form). A series of discrete fractions is observed, each corresponding to a star topology with a different number of arms. Such a distribution in the number of arms within a single sample is expected if synthesis of the star topology is by the "nodule" method (16). The SEC of the PS parent in tetrahydrofuran (two Waters UltraStyragel (cross-linked styrene–divinylbenzene) linear columns in series, flow rate = 1.0 mL/min) displays a single broad peak (Figure 6), giving no hint of the actual complex distribution of molecular fractions. The high resolution of the gel electrophoresis characterization is obvious, extending over more than an order of magnitude in molecular weight.

The fractionation displayed in Figure 5 is superior to any previously obtained for a synthetic polymer at high molecular weight; fractions with molecular weight above 1×10^6 differing by an increment of only one arm

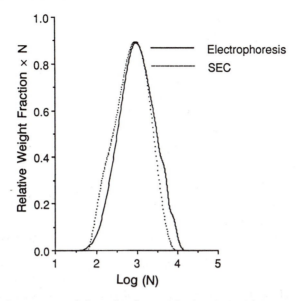

Figure 4. Comparison of the molecular weight distributions obtained by SEC and by agarose gel electrophoresis; N is the degree of polymerization. The SEC analysis is of the PS parent (nominal molecular weight 50,000), and the electrophoresis analysis is of the sulfonated product. By SEC: $N_w = 1100$; $N_w/N_n = 2.5$. By electrophoresis: $N_w = 1600$, $N_w/N_n = 2.6$.

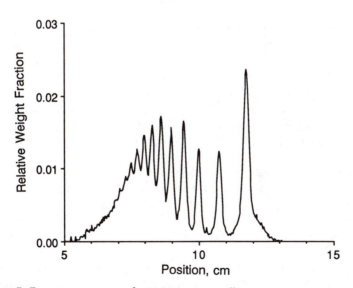

Figure 5. Densitometry scan of a PSS star nominally possessing six arms. The optical density is measured as a function of position from the sample well. (Conditions: 0.6% agarose; 1.3 V/cm; 7.3 h; I = 0.03 M.)

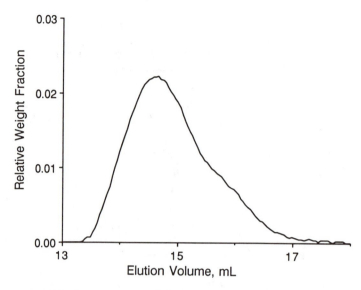

Figure 6. SEC chromatogram of the parent PS star for the PSS sample of Figure 5. Note the lack of detail as compared to the previous figure.

are resolved. Molecular mechanisms that could explain how these star polymer chains are being separated in the gel will be published elsewhere.

Discussion

The development of gel electrophoresis as a central method of biopolymer research has been largely overlooked by traditional polymer science, and one can foresee many new applications in synthetic polymer systems. In the mode described here the method should be regarded as a relative technique, one that provides only qualitative results in the absence of molecular weight standards. Qualitative information, however, may be sufficient in many cases. For example, we have studied polymer degradation in elongational flows by comparing the mobility distribution before and after flow; chain cleavage can be readily detected by electrophoresis.

Essentially, any highly sulfonated water-soluble polymer can be studied by the techniques we have described. Our efforts to generalize these techniques to other charged polymer species (those containing carboxyl groups, for example) have not yet been entirely successful, mainly because of detection problems. Staining appears to be the best technique for visualizing polyelectrolyte bands, and we are working to develop a dye treatment that will universally stain any negatively charged polymer while leaving the agarose matrix uncolored. The principle behind this approach is the electrostatic attraction of cationic dye molecules for oppositely charged polyelectrolyte chains. This attraction is weak for chains that contain weakly ionizable groups, so the pH and ionic strength conditions must be properly adjusted to produce a strong binding. The dye process that has been successful for PSS has not been effective when applied to such species as poly(acrylic acid). Fluorescent tagging constitutes a successful alternative in some situations, but caution must always be exercised when interpreting the behavior of polymer chains containing covalently attached labels.

Optimal conditions for the highest quality separations have not been completely determined yet. The number of experimental variables is substantial, and we have followed the precedent of DNA separations in selecting most parameter values. The ultimate resolution of the technique has therefore not been achieved. For example, we have not fully explored effects associated with modifying electric field strength, lengthening the gel, or varying temperature. Future work will address these issues. We believe that the resolution can be improved enough to explore problems as sensitive as the chain-length distributions from ideal anionic polymerization. In this case polydispersities below 1.005 must be accurately measured.

Summary

The major advantages of polyelectrolyte analysis by gel electrophoresis are high resolution and easy adjustment of the molecular weight range. The method appears to possess an upper chain-size limit that is well above the molecular weight of any synthetic polymer of commercial significance; the operational ease at high molecular weight stands in contrast to the well-known barriers encountered when attempting to operate aqueous SEC at molecular weight above 1×10^6. The resolution remains high throughout the entire molecular weight range if the gel concentration and ionic strength are adjusted properly. At this stage, the major difficulty in applying electrophoresis to unknown polyelectrolytes is the development of a sensitive and universal detection scheme for locating polymer bands in the gel. Overall, gel electrophoresis has the potential to lift the most ill-characterized class of synthetic polymers, polyelectrolytes of high molecular weight, to a position for which characterizations of equal or greater quality to those achieved with other synthetic polymers are possible.

Acknowledgments

We gratefully acknowledge the National Science Foundation Materials Research Laboratory at the University of Massachusetts for financial support; acknowledgement is also made to the Donors of The Petroleum Research Fund, administered by the American Chemical Society, for partial support of this research.

References

1. Jorgenson, J. W. *Anal. Chem.* **1986**, *58*, 743A.
2. Chen, J.-L.; Morawetz, H. *Macromolecules* **1982**, *15*, 1185.

3. Schwartz, D. C.; Cantor, C. R. *Cell* **1984**, *37*, 67.
4. Giddings, J. C. In *Advances in Chromatography;* Giddings, J. C.; Grushka, E.; Cazes, J.; Brown, P. R., Eds.; Marcel Dekker: Washington, DC, 1982; Vol. 20, pp 217–254.
5. Muller, G.; Yonnet, C. *Makromol. Chem. Rapid Commun.* **1984**, *5*, 197.
6. Davis, R. M.; Russel, W. B. *Macromolecules* **1987**, *20*, 518.
7. Davis, R. M.; Russel, W. B. *J. Polym. Sci. Polym. Phys. Ed.* **1986**, *24*, 511.
8. Hervet, H.; Bean, C. P. *Biopolymers* **1987**, *26*, 727.
9. Smisek, D. L.; Hoagland, D. A. *Macromolecules* **1989**, *22*, 2270.
10. Holzwarth, G. L. *Carbohydr. Res.* **1978**, *66*, 173.
11. Rodbard, D.; Chrambach, A. *Proc. Natl. Acad. Sci. U.S.A.* **1970**, *65*, 970.
12. Lumpkin, O. J.; Dejardin, P.; Zimm, B. H. *Biopolymers* **1985**, *24*, 1573.
13. Adolf, D. *Macromolecules* **1987**, *20*, 116.
14. Andrews, A. T. *Electrophoresis: Theory, Techniques, and Biochemical and Clinical Applications;* Oxford University Press: New York, 1986.
15. Manning, G. S. *J. Chem. Phys.* **1969**, *51*, 924.
16. Mays, J. W.; Hadjichristidis, N.; Fetters, L. J. *Polymer* **1988**, *29*, 680.

RECEIVED for review February 14, 1989. ACCEPTED revised manuscript July 31, 1989.

Particle Size Distribution by Zero-Angle Depolarized Light Scattering

L. Mark DeLong and Paul S. Russo*

Macromolecular Studies Group, Department of Chemistry, Louisiana State University, Baton Rouge, LA 70803–1804

Zero-angle depolarized dynamic light-scattering methods for particle size determinations are demonstrated by a study on aqueous colloidal suspensions of titanium dioxide. Previous results on a poly(tetrafluoroethylene) latex suspension are also briefly recounted and updated to include reanalysis by an additional independent Laplace inversion method, Provencher's CONTIN. Both of these colloidal suspensions depolarize light strongly, and the result is homodyne correlation functions with signal-to-noise ratios comparable to those of typical finite-angle quasi-elastic light-scattering measurements. Size distributions may be obtained by Laplace inversion of the correlation functions, with excellent agreement among the various Laplace inversion algorithms. Resolution is greatly improved compared to conventional quasi-elastic light-scattering measurements, because rotational motions, which are more sensitive to size than translational motions, are detected at zero angle. The size distribution from zero-angle depolarized light scattering agrees very well with that from electron microscopy in the case of the poly(tetrafluoroethylene) latex. There is reasonable agreement in the case of TiO_2, but significant differences do exist. It is not yet possible to determine which technique is more accurate in this case.

RECENT ADVANCES IN LAPLACE INVERSION OF NOISY DATA $(1–4)$ have made quasi-elastic light scattering (QLS) a powerful and versatile tool for particle sizing. The advantages of QLS for determining particle size distributions are

*Corresponding author.

0065–2393/90/0227–0065$06.00/0

many. QLS can be applied to polymers in corrosive solvents (5, 6) or systems that dissolve only at very high temperatures (7), and is applicable over a broad range of hydrodynamic radii (0.001 to more than 2 μm). Furthermore, QLS is a nonperturbing technique that relies only on very small spontaneous concentration fluctuations. This is in contrast to potentially more disruptive particle sizing methods such as size exclusion chromatography, which imposes bulk flow, shear, and the presence of a complex matrix.

A specialized variant of QLS, zero-angle depolarized light scattering (ZADS), was first reported by Wada et al. (8), who studied solutions of tobacco mosaic virus. The technique has been applied to a number of biopolymers (8–12). In most cases, relatively noisy heterodyne signals were recorded because of the presence of substantial stray depolarized light and the relatively weak depolarization of most biopolymers. The first homodyne measurement was made by Schurr and Schmitz (12). Light depolarized by the optics is not the only potential source of difficulty in ZADS. Hopman et al. (11) demonstrated the importance of double scattering effects in a study of bacteriophage T4 and T7. They were able to measure fairly quiet heterodyne correlation functions. After accounting for the double scattering effect, they obtained rotational diffusion coefficients that were in excellent agreement with electric birefringence results. Nevertheless, zero-angle depolarized light scattering has largely been supplanted by electric birefringence methods (13, 14) for the study of the rotational motions of biopolymers.

Han and Yu (15) reported the first ZADS measurements on synthetic polymers in a study of rotational diffusion of poly(hexylisocyanate) and internal motions of isotactic polystyrene. The zero-angle technique has also been extended to mineral colloids (16). Crosby et al. (6) were the first to attempt to use ZADS to obtain a size distribution. This study was partly successful, despite just moderate data quality in the heterodyne experiments and the very difficult nature of the system, which was poly(p-phenylene-benzobisthiazole) dissolved in an extremely aggressive solvent, chlorosulfonic acid.

Despite these several successful applications, the ZADS method has not enjoyed the overwhelming acceptance of conventional QLS. Perhaps this lack of acceptance is a result of its initial application to biopolymers and other systems that do not depolarize strongly enough to overcome imperfections of the optics or multiple scattering effects. This chapter concerns ZADS measurements of TiO_2 in suspension. Together with a previous article from this laboratory (17) on the sizing of colloidal poly(tetrafluoroethylene), this chapter demonstrates that when strongly depolarizing particles are measured in instruments designed to hold stray depolarized light to a bare minimum, the result can be very quiet homodyne correlation functions quite good enough for accurate Laplace inversion. Then high-resolution particle size distributions can be obtained simply, accurately, reproducibly, and with much better resolution than in conventional QLS.

Theory

The fundamental quantity of interest in any QLS experiment is the first-order (electric field) autocorrelation function, $g^{(1)}(\tau)$. In the usual homodyne measurements, this is obtained from the measured intensity–intensity autocorrelation function

$$G^{(2)}(\tau) = B(1 + f(A)|g^{(1)}(\tau)|^2) \tag{1}$$

where B is a base line and $f(A)$ is an instrumental parameter, $0 < f(A) < 1$, depending mostly on the number of coherence areas detected (*18*). In a polydisperse system, $g^{(1)}(\tau)$ consists of a weighted sum of discrete exponentials, but it can be closely approximated by a continuous distribution:

$$g^{(1)}(\tau) = \sum_i A_i \exp\left(-\Gamma_i\tau\right) \approx \int_0^\infty A\left(\Gamma\right) \exp\left(-\Gamma\tau\right) d\Gamma \tag{2}$$

The subscript i will be used to denote an individual value throughout; for example, A_i is a given scattering amplitude and Γ_i is a given decay rate. Popular Laplace inversion algorithms (*1–4*) yield a set of scattering amplitudes, $A\{\Gamma\}$, for a set of discrete values of the decay rates $\{\Gamma\}$. In the conventional QLS experiment, a vertically polarized incident beam is used, and either the unpolarized (Uv) or vertically polarized (Vv) scattered light is detected at some finite scattering angle, θ. The conversion from $A\{\Gamma\}$ space to concentration vs. size or concentration vs. molecular weight has been described in detail (*17, 19, 20*). In this chapter we need only consider the implications of the first steps of this process. Each decay rate Γ_i in a conventional experiment is directly proportional to the (mutual) diffusion coefficient D_i of the ith species with hydrodynamic radius, $R_{h,i}$:

$$\Gamma_i = q^2D_i = \frac{q^2kT}{6\pi\eta_oR_{h,i}} \tag{3}$$

Here, q is the magnitude of the scattering vector, equal to $4\pi n \cdot \sin(\theta/2)/\lambda_o$ where n is the refractive index, λ_o is the in vacuo wavelength of the incident light, kT is the thermal energy, and η_o is the solvent viscosity. The key feature is that, for two particles where one is twice as large as the other, the decay rates differ only by a factor of 2. As this factor is the approximate limit of resolution of Laplace inversion of imperfect data (*2, 3, 21*), two such particles can scarcely be resolved in conventional QLS.

Similarly, the determination of concentration in conventional QLS has its limitations. A given scattering amplitude, A_i, is proportional to the product of concentration expressed as weight of the ith species per unit volume,

c_i, and the molecular weight, M_i, of that species:

$$A_i \propto c_i M_i P(qR_{g,i}) \tag{4}$$

The "form factor" $P(qR_g)$ depends on size, usually expressed as radius of gyration, R_g, and also shape. It describes the reduction in intensity due to intramolecular interference, and it lies between zero and unity. In the limit $q = 0$, $P(qR_g)$ is unity for species of all sizes, as long as their refractive indices do not differ greatly from the solvent's refractive index. If one can successfully convert from $A\{\Gamma\}$ space to $A\{R_g\}$ space (see, for example, refs. 19 and 20), and if the particle shape is known, this term can usually be computed to sufficient accuracy from well-known theoretical expressions (22), and so poses no special problem. However, one would probably wish to make measurements at several angles and test for consistency, especially whenever any of the $P(q,R_{g,i})$ differ substantially from unity. These steps slow the analysis. What is worse, in the case that particle shape is not known, the form factor becomes a severe impediment to accurate sizing.

In zero-angle depolarized scattering (ZADS) the incident beam is again vertically polarized, but only the horizontal, or Hv, component scattered to $\theta = 0$ is detected. The particles must be optically anisotropic (18) for there to be any signal (apart from the multiple scattering signal (11)). As long as this condition is met, particles may have any geometrical shape—even spherical. In ZADS, as in conventional QLS, intensity fluctuations are observed. However, whereas in conventional QLS these fluctuations arise primarily from translational diffusion, in the ZADS experiment the intensity changes are due only to rotational diffusion. The principal advantage of this situation in a particle sizing application is that the rotational diffusivity, Ξ, depends on the cube of the particle size. The decay rates are now given by (18):

$$\Gamma_{\text{ZADS},i} = 6\Xi_i = \frac{6kT}{8\pi\eta_0 R_{h,i}^3} \tag{5}$$

When it can be successfully applied, the advantages of ZADS are significant. Considering again two particles where one is twice as large as the other, we see that their decay rates in ZADS would be separated by a factor of 8, and so could be resolved easily. Two particles differing in size by only 25% now define the resolution limit. Furthermore, form factor correction is unnecessary in ZADS. Finally, the slow number fluctuations (18) which plague conventional QLS measurements of large, strongly scattering particles that cannot be prepared at high concentration because of multiple scattering problems are usually negligible in ZADS as the scattering volume looking into the incident beam is many times larger than in conventional QLS.

These attributes do not come without a price. Aside from inapplicability to optically isotropic particles, the principal disadvantage of ZADS is that the Laplace inversion of eq 2 yields amplitudes that are not simply related to concentration. Instead, A_i is proportional to $N_i\beta_i^2$, where N_i is the number density of species in the scattering volume, and β_i is the optical anisotropy of the *i*th species (*18*). Thus, the distribution of $N\beta^2$ versus size is obtained. In some cases, the relationship between β and size is known (*see*, e.g., ref. 23), but in general it is not.

Two things may be said of this problem. First, it does not prevent studies of the stability of solutions—that is, changes in the size distribution will still be detectable. Second, it may be possible to empirically "calibrate" the dependence of β on size by making comparisons with another technique. Subsequent ZADS analyses of similar particles could be referenced to this calibration.

Experimental Materials and Methods

Conventional light-scattering and ZADS measurements and analyses were made prior to electron microscopic (EM) investigation, so as not to bias the results. A light-scattering spectrometer capable of both conventional and ZADS measurements has been described elsewhere (*17*), together with the ZADS alignment procedure and the methods used to prepare the poly(tetrafluoroethylene) (Fluon) samples. Anatase TiO_2 was kindly donated by Kemira, Inc., Savannah, GA.

Three TiO_2–water samples were prepared from a 3×10^{-5}-g/mL stock solution. Because of the high refractive index of TiO_2, even this dilute stock solution had a very faint blue tinge (in containers of 1-cm diameter) when held to direct light. To explore the importance of double scattering, experiments were conducted with polystyrene latex spheres of about the same size as the TiO_2. For a latex solution having about the same scattering power as the TiO_2 stock preparation, double scattering effects (*11*) were detectable, although the double scattering signal above base line was insignificant compared to the signal depolarized by TiO_2.

Nevertheless, in order to add an extra measure of certainty, dilutions with final concentrations of 4×10^{-6} and 2×10^{-6} g/mL were prepared by adding dust-free water from a water-filtration system (Millipore RQ, >2.5 MΩ-cm resistivity), supplemented by a 0.22-μm cartridge filter (Gelman), to aliquots of the stock TiO_2 solution in dedusted rectangular glass fluorimeter cells of 1-cm path length. A third sample consisted of the lower concentration sonicated for 1 h. These samples all appeared "water white"—that is, to the eye in natural light, they were as clear as pure water. Each sample was inspected for lack of "dust" by inserting the cuvette into the scattering device and observing the laser beam at about $100\times$ magnification, at a scattering angle of 22.1° (the unusual angle is the result of a Snell's law correction). Conventional dynamic scattering measurements in the Uv geometry were also made at this angle. At this low angle, $qR_h < 1$, so that the decay of the correlation functions is dominated by translational motion. All particle size distributions were obtained at 30.0 ± 0.1 °C and a wavelength of 632.8 nm. Because of the low concentrations, it was necessary to use a larger than normal scattering volume during the conventional QLS measurements of TiO_2 to prevent slow number fluctuations. This larger volume was accomplished by defocusing the beam and adjusting the apertures and pinholes in the detection system. A beam focused by a 16.5-cm lens was used for all ZADS experiments.

TiO$_2$ samples were viewed in a Jeol 100 CX electron microscope in the scanning mode at an acceleration of 80 kV and a magnification of 10,000×. In comparison to poly(tetrafluoroethylene), size distributions of TiO$_2$ were much more difficult to obtain by EM. Upon preparation for EM, the TiO$_2$ particles aggregated into large clusters (Figure 1). Some clustering of these particles in suspension is expected (24), but the aggregates seen by EM are much larger than any measured by QLS. Had they been present in solution, they would have caused bursts of high scattered intensity (similar to "dust") in the conventional measurements, but such bursts were not observed. Several preparative methods were tried to better approximate the true size distribution in suspension. Various concentrations of the colloidal TiO$_2$ were freeze-dried and air-dried on both protamine sulfate coated (25) and uncoated copper–glass slides. Many of these attempts failed to suppress the aggregation. Finally, an acceptable size distribution from EM was obtained by placing a 20-μL drop of an extremely dilute (2.0 × 10^{-7} g/mL) TiO$_2$ suspension on each of five different uncoated EM boats and allowing each to air-dry. Variously sized TiO$_2$ clusters of approximately spherical shape were randomly distributed on the grid (Figure 2). The distribution of radii (Figure 3) was measured from 266 particles on 75 separate EM fields and was determined as the average of the length and width.

The data analysis software has been described elsewhere (21). Briefly, each measurement consists of several short runs that are individually inspected for inten-

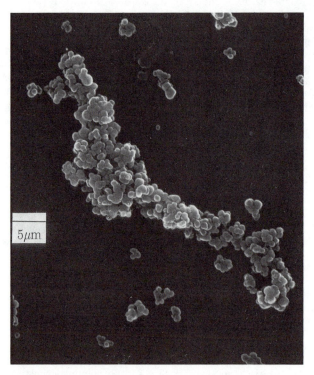

Figure 1. Image of aggregated TiO$_2$. A 20-μL drop of TiO$_2$ (4 × 10^{-6} g/mL) was placed on a protamine sulfate coated EM slide and freeze-dried.

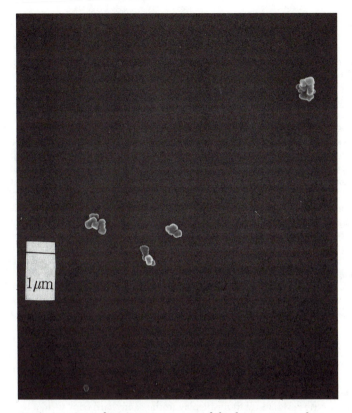

Figure 2. One of 75 SEM images used for histogram analysis.

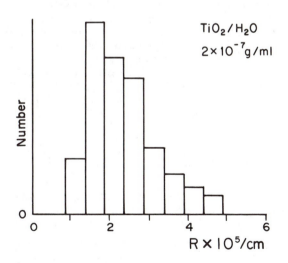

Figure 3. A histogram of TiO$_2$ radii, based on EM measurement of 266 particles.

sity anomalies and deviation of the last channels of the correlation function from the theoretical base line (described later). Each short run is also analyzed by second-order cumulants (26). These steps effectively identify "outliers" due to dust. The user has the option of deleting short runs, but in the present study, almost all were kept. The retained short runs are summed together and reanalyzed by cumulants, from first order to third order (3CUMU). The summed correlation functions are the gateway to all other analyses and plotting packages.

Decidedly nonexponential correlation functions with good signal-to-noise characteristics become candidates for the more complex fitting routines. First, a discrete multiple exponential analysis is performed (one to five exponentials, nonlinear least-squares program MARLIN (17, 21)) to determine the range (in decay rate space) and number of exponentials actually required for a good fit. These discrete exponential fits are often as good as those from the smoothed Laplace transform programs. However, although five or fewer exponentials will generally fit the data within noise, it is often known a priori that the true distribution is not discrete but continuous. Then Laplace inversion programs EXSAMP (17) and CONTIN (1) provide more realistic quasi-continuous distributions.

Program EXSAMP is generally used before CONTIN, because it is easier to vary the range of decay rate space over which solutions are sought, as well as the resolution, or number of exponentially decaying functions in the fit. Also, the sensitivity of the Laplace inversion to base-line error can be gauged easily. Thus, EXSAMP serves to suggest convenient operating parameters for CONTIN such that independent and completely impartial answers can be obtained in only one run of CONTIN, which produces 12 distributions of varying detail and provides a number of statistical selection parameters. CONTIN also automatically chooses the least detailed distribution that adequately fits the data. We routinely examine all 12 of CONTIN's outputs, and report them either as CHOSEN or by order of appearance in the output file (CONTIN varies its parameters in a consistent fashion). Screen-oriented software greatly facilitates the task of examining CONTIN's massive output and prepares the residuals of fit for display in the same hard copy format as our other analytical routines (Figure 4).

An ever-present problem in analysis of QLS data is the proper selection of base line. The cumulants fits herein used a theoretical base line obtained from acquisition time and intensity, $B_t = P(P - O)/\mathfrak{N}$, where P is the total number of photopulses (typically 10^8), O is the number of shift register overflows (usually 0), and $\mathfrak{N}$ is the acquisition time divided by the channel time (typically 10^8). Nonlinear and inverse Laplace transform routines occasionally employ fitted base lines, B_f, specific to a particular analysis. These fitted base lines sometimes vary perceptibly from B_t, in which case the difference is specified as the number of statistical uncertainties, $\sigma_B = B_t^{1/2}$, which were added to the base line (usually less than 10). The multiplicative factor relating B_f to B_t is also given (usually $B_f < 1.005 B_t$). Typical base-line adjustments appear in Figure 4.

Results of Light-Scattering Experiments and Discussion

Titanium Dioxide. From the outset, it was clear that the correlation functions were distinctly nonexponential and would not be well fit by single exponential or low-order cumulants (26) methods. Nonetheless, the average decay rates provided by such algorithms allow a quick check on the concentration dependence of the results. Such averages may be obtained in a

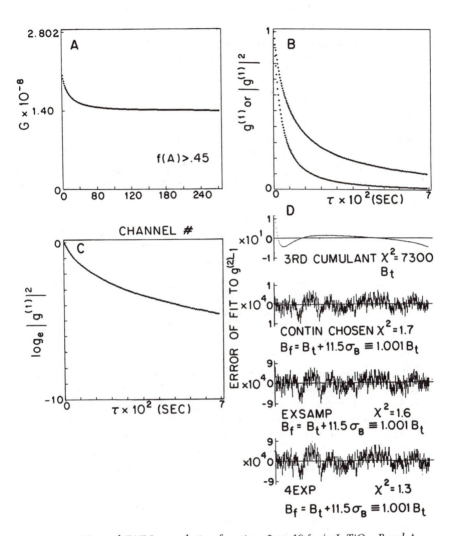

Figure 4. Typical ZADS correlation function; 2 × 10⁻⁶ g/mL TiO₂. Panel A: G⁽²⁾(τ), showing the base line and a large coherent signal above the base line. Panel B: normalized first-order correlation function and its square. Panel C: semilog representation. Panel D: error plots for various fits. The abscissa is the same as panel B. The height of each bar represents the uncertainty in |g⁽¹⁾(τ)|², and the center of each bar is plotted to show the difference |g⁽¹⁾(τ)|² − |g⁽¹⁾(τ)|²_fit, where g⁽¹⁾(τ) is calculated by using the theoretical base line Bₜ. The 3CUMU fit used base line Bₜ; all other fits used a fitted base line, Bf, related to Bₜ as shown, where σ_B is the base-line uncertainty, √Bₜ.

number of ways. Shown in Table I are the initial decay rates obtained from third-order cumulants analyses of conventional and ZADS experiments at various concentrations. No significant concentration dependence is seen, a result that attests to the unimportance of intermolecular effects, as expected of such dilute solutions in the absence of special efforts to emphasize long-range interactions by forcibly maintaining an extremely low ionic strength. These data also confirm the insignificance of any multiple scattering effects.

The average decay rates can also be converted to average hydrodynamic radii, whereupon a comparison with EM values can be made. In the conventional Uv geometry, and neglecting optical anisotropy and reductions in the scattered intensity caused by intramolecular interference at finite scattering angles, the average decay rate, or first cumulant (26), $\overline{\Gamma}_{Uv}$, is proportional to the z-average of the diffusion coefficients of the species in the sample:

$$\overline{\Gamma}_{Uv} = q^2 D_z = \frac{q^2 \sum_i c_i M_i D_i}{\sum_i c_i M_i} = \frac{q^2 \sum_i N_i M_i^2 D_i}{\sum_i N_i M_i^2} \tag{6a}$$

$$\overline{\Gamma}_{Uv} = \frac{q^2 kT \sum_i c_i M_i R_{h,i}{}^{-1}}{6\pi\eta_o \sum_i c_i M_i} = \frac{q^2 kT}{6\pi\eta_o \overline{R}_{h,Uv}} \tag{6b}$$

Thus, the apparent average hydrodynamic radius in the Uv geometry is the inverse of the z-average of the inverse hydrodynamic radius: $\overline{R}_{h,Uv} = [(1/R_h)_z]^{-1}$. The z-weighting arises because the intensity scattered by any given component is proportional to the product of its number concentration and the square of its molecular weight (18). In the ZADS experiment, the weighting goes according to the product of solute number density and squared optical anisotropy, leading to the following expressions:

$$\frac{\overline{\Gamma}_{ZADS}}{6} = \overline{\overline{\Xi}} = \frac{\sum_i N_i \beta_i^2 \Xi_i}{\sum_i N_i \beta_i^2} \tag{7a}$$

$$\frac{\overline{\Gamma}_{ZADS}}{6} = \frac{kT \sum_i N_i \beta_i^2 R_{h,i}{}^{-3}}{8\pi\eta_o \sum_i N_i \beta_i^2} = \frac{kT}{8\pi\eta_o (\overline{R}_{h,ZADS})^3} \tag{7b}$$

Only in special cases does this expression reduce to one of the conventional averages often associated with polymer chemistry. For example, for certain thin rods, β is adequately represented by the difference between the po-

Table I. Third Cumulants Analysis for TiO_2

Concentration (g/mL)	$\overline{\Gamma}_{ZADS}$ (Hz)	$\overline{\Gamma}_{Uv}$ (Hz)
4×10^{-6}	87.6 ± 20	38.2 ± 5
2×10^{-6}	86.1 ± 12	38.4 ± 4
2×10^{-6} (sonicated)	89.9 ± 15	40.6 ± 4

NOTE: This table corrects an error in our preprint (*28*), in which the two columns containing decay rates appear transposed.

larizability parallel to and perpendicular to the rod: $\beta = \alpha_{\parallel} - \alpha_{\perp}$. If $\alpha_{\parallel} \gg \alpha_{\perp}$, then $\beta \sim M$, and $\overline{\Xi}$ represents the z-average of the rotational diffusion coefficient.

Taking the average over the several concentrations as $\overline{\Xi} = 14 \pm 3$ Hz and solving eq 7 yields $R_{h,ZADS} = 244 \pm 20$ nm. This result is quite close to the number-average EM radius of 237 nm. Similarly good agreement was obtained in an earlier study of poly(tetrafluoroethylene) (*17*). Solving eq 6 and using an average value of 39 ± 4 Hz for the Uv decay rate yields $R_{h,Uv} = 183 \pm 20$ nm. This result is a bit lower than the either the EM average value or $R_{h,ZADS}$, reflecting either the different nature of the average or underrepresentation of the larger scatterers, due to intramolecular interference, despite the moderately low scattering angle.

Laplace inversion can provide more details about the distribution; however, the process is very sensitive to noise. Therefore, both the data quality itself and the quality of fit must be examined closely. In Figure 4 are typical data from ZADS measurements on TiO_2. An equivalent representation of the poly(tetrafluoroethylene) data appears in Figure 4 of ref. 17. Figure 4A shows the substantial signal above a large base line. The correlation function is undeniably homodyne and is easily as free of noise as many conventional QLS measurements. This finding is not surprising, considering our visual observation (using the eyepiece of the scattering instrument, with great care to prevent eye damage) that the depolarized scattering greatly outshines any stray light. Also, it was possible to force heterodyning by misaligning the polarizer, which allowed the horizontal components from the laser to reach the detector, and thus provided a local oscillator of the right polarization sense. The average decay rate of heterodyne correlation functions obtained in this way was precisely half that of the homodyne measurements obtained with the polarizer and analyzer exactly crossed.

If the correlation function contained information about only one particle size, the semilogarithmic plot (Figure 4C) would be a straight line. Clearly, it is not, so these relatively noise-free data are appropriate for Laplace inversion. The first of the error plots, Figure 4D, is for the third cumulants (3CUMU) method of obtaining average decay rates. The high value of the weighted mean square residual, χ^2 (*27*), and the high channel-to-channel correlation of errors indicate that 3CUMU inadequately fits the data. Application of the Laplace inversion algorithms or discrete analysis dramatically reduces χ^2, and these routines fit the data equally well and to within the noise of the experiment.

The final distributions from each fitting routine are in excellent agreement. Figure 5 compares distributions from the different fitting methods, all applied to the same sample of TiO_2. The abscissas are linear instead of

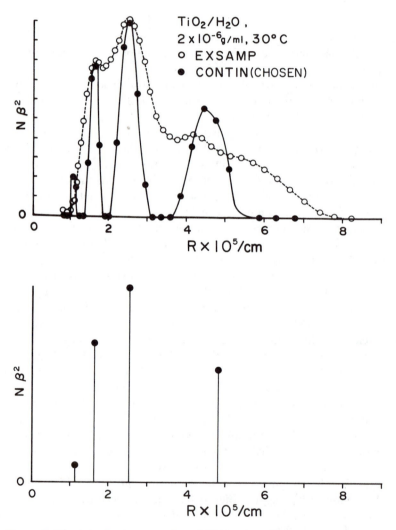

Figure 5. The top plot is an overlay of EXSAMP and CONTIN. The vertical error bars were smaller than the data points. The MARLIN fit (bottom plot) is to four exponentials.

the logarithmic scales usually associated with Laplace inversion of light-scattering data. For TiO_2, EXSAMP provides a somewhat smoother distribution than CONTIN, but the major features are similar. In our experience, it is unusual for CONTIN, which preferentially selects the smoothest solution that can fit the data, to return a tetramodal distribution. However, the discrete fit to four exponentials in $g^{(1)}$ is completely consistent with this highly detailed chosen CONTIN solution.

Sample-to-sample variation is shown in Figure 6 to be relatively minor. Although the tetramodal solution again appears, the peak locations are shifted slightly. We thus adopt the position that the best representation would be either to accept the smoother EXSAMP distribution or to blur the minor differences between CONTIN results. A smeared distribution was constructed graphically from the chosen CONTIN outputs of four repeat experiments (i.e., different TiO_2 samples, different acquisition times, etc.). It appears superimposed on the profile from electron microscopy in Figure 7. The major peaks from both EM and ZADS coincide at about 200 nm, with matching shoulders at about 100 nm.

The distribution from ZADS, however, contains a second peak at about 450 nm that is not found by EM. One possible explanation would be that a small number of large particles having a very high optical anisotropy exaggerates the importance of the second peak. Additional EM images might have revealed the large particles. But acquisition of additional images would be futile in the absence of any rigorous method for making a clear cut-off between aggregates that are actually present in solution and those that form

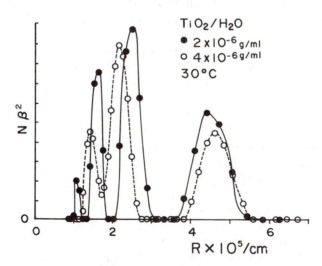

Figure 6. Comparisons of chosen CONTIN fits on 4 × 10⁻⁶ and 2 × 10⁻⁶
g/mL TiO₂, each with different acquisition times.

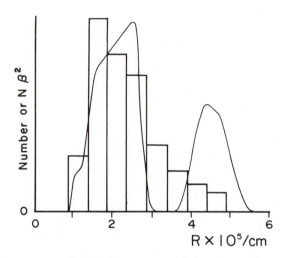

Figure 7. Comparison of the TiO₂ average Nβ² distribution from ZADS with the number profile from EM. The solid curve denotes a smoothed average of chosen CONTIN results from four separate TiO₂ measurements.

during preparation for EM. Probably, the EM size distribution contains a slight bias against larger particles, precisely because of this ambiguity. Apparently EM is innately limited as a sizing tool for particles in solution when those particles cluster during preparation.

The corresponding limitation for ZADS is that the size dependence of β must be known if true number distributions are to be obtained. Were it not for the uncertainties about the clustering of TiO_2 during preparation for EM, one might try converting amplitudes from $N\beta^2$ to N by dividing out trial functions $\beta^2(R)$ and seeking agreement between the ZADS and EM number distributions. For example, assuming $\beta(R) \sim R$ would greatly reduce the importance of the second peak. Unfortunately, this procedure is not possible in this particular case because of uncertainties in the distribution obtained from EM. Furthermore, a host of complexities to the ZADS experiment were delineated previously (*17*). For example, the correlation function from a monodisperse particle can actually contain more than a single exponential, because of coupling of the geometric and optical anisotropies (*18*). However, for particles that are not too aspheric, these modes are expected to have similar decay rates that could not be resolved, as was shown in detail for poly(tetrafluoroethylene) (*17*). Another potential artifact is that light scattered through 180° from the beam reflected by the cell window makes a small contribution. This possibility was considered for poly(tetrafluoroethylene) and found to be negligible (*17*) and, besides, this effect would lead to a rapidly decaying term, and would therefore result in extra peaks on the small side of the size distribution, not the large. Thus, with the aforementioned exception that β may be different at large sizes, the second peak in the ZADS distribution cannot be an artifact of the method, and the disagreement between EM and ZADS on the existence of the larger particles is unresolved.

In sum, it is entirely possible that a bimodal distribution better represents the true size profile of TiO_2 in solution than a unimodal one, but it cannot be proved. An additional method to measure the size profile in solution would be desirable. If there were a somewhat greater size difference between the two peaks, and if the particle sizes were not so large that accurate form factor corrections would be required, Laplace inversion of conventional Uv correlation functions could be used for this purpose.

Poly(tetrafluoroethylene) Latex. The distributions from Laplace inversion of poly(tetrafluoroethylene) data are superimposed on those obtained by electron microscopy in Figure 8. The chosen CONTIN distribution is new compared to our previous publication (*17*). It is in excellent agreement with the previous distributions from exponential sampling. The apparent hydrodynamic radii, shown in Figure 8, were calculated from eq 5. The length distribution from light scattering has also been obtained by applying Perrin's equations for a prolate ellipsoid of revolution (*17*). Whether the poly(tetrafluoroethylene) particles are treated as ellipses or apparent spheres, the size distributions from light scattering are in excellent agreement with those from EM.

Conclusion

The main conclusion from this work is that nonperturbing zero-angle depolarized light scattering can be simple and provide very high quality correlation functions for particles with large optical anisotropies, with the enhanced resolution that attends sizing based on rotational diffusion. Examples of systems that may be well suited to ZADS are catalysts and preceramic particles of mineral origin, polymeric suspensions of magnetic recording particles where the polymer matrix is only weakly optically anisotropic and, perhaps, soot particles in aerosol flames.

Although one may have to settle for the distribution of $N\beta^2$ versus size or calibrate β against another method, the ZADS technique remains a simple, useful, discriminating, and nonperturbing means of following changes in size, aggregation, or both. Additionally, the study of rotational diffusion of probes through polymeric matrices should be possible given appropriately monodisperse, optically anisotropic probes that do not aggregate.

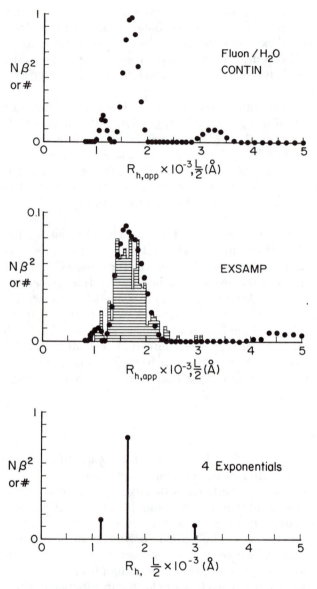

Figure 8. Comparisons of CONTIN, EXSAMP, and four-exponential MARLIN fits to data for poly(tetrafluoroethylene). Only three modes appear because the decay rates of two of the four exponentials were identical, so the amplitudes of these modes were summed. (EXSAMP and four-exponential data are reproduced with permission from ref. 17. Copyright 1986 Elsevier.)

Acknowledgments

This research was supported by National Science Foundation (NSF) Research Grant DMR–8520027 and by NSF Equipment Grant DMR–8413055. We thank Cindy Henk for invaluable assistance with the electron microscopy.

References

1. Provencher, S. W. *Comput. Phys. Commun.* **1982**, 27, 213; the latest information on CONTIN can be obtained from Dr. Stephen Provencher, Max-Planck Institut für Biophysikalische Chemie, Postfach 2841, D-3400 Gottingen, Federal Republic of Germany.
2. Ostrowski, N.; Sornette, D.; Parker, P.; Pike, E.R. *Optica Acta* **1981**, 28(8), 1059.
3. See, e.g., Bott, S. In *Measurement of Suspended Particles by Quasielastic Light Scattering;* Dahneke, B. E., Ed.; Wiley: New York, 1983.
4. See, e.g., Bertero, M.; Brianzi, P.; Pike, E. R.; deVilliers, G.; Lan, K. H.; Ostrowski, N. *J. Chem. Phys.* **1985**, 82(3), 1551.
5. Chu, B.; Wu, C.; Ford, J. R. *J. Coll. Int. Sci.* **1985**, 105, 473.
6. Crosby, C. R., III; Ford, N. C., Jr.; Karasz, F. E.; Langley, K. H. *J. Chem. Phys.* **1981**, 75, 4298.
7. Chu, B.; Wu, C.; Buck, W. *Macromolecules* **1988**, 21, 397.
8. Wada, A.; Suda, N.; Tsuda, T.; Soda, K. *J. Chem. Phys.* **1969**, 50, 31.
9. King, T. A.; Knox, A.; McAdam, J. D. G. *Biopolymers* **1973**, 12, 1917.
10. Thomas, J. C.; Fletcher, G. C. *Biopolymers* **1979**, 18, 1333.
11. Hopman, P. C.; Koopmans, G. ; Greve, J. *Biopolymers* **1980**, 19, 1241.
12. Schurr, J. M.; Schmitz K. S. *Biopolymers* **1973**, 12, 1021.
13. Lewis, R. J.; Pecora, R.; Eden, D. *Macromolecules* **1987**, 20, 2579.
14. Eden, D.; Elias, J. G.; In *Measurements of Suspended Particles by Quasi-elastic Light Scattering;* Dahneke, B. E., Ed.; Wiley: New York, 1983.
15. Han, C. C.; Yu, H. *J. Chem. Phys.* **1974**, 61, 2650.
16. Chu, B.; Xu, R.; DiNapoli, A. *J. Coll. Int. Sci.* **1987**, 116, 183.
17. Russo, P. S.; Saunders, M. J.; DeLong, L. M.; Kuehl, S. K.; Langley, K. H.; Detenbeck, R. W. *Anal. Chim. Acta* **1986**, 189, 69.
18. Berne, B.; Pecora, R. *Dynamic Light Scattering;* Wiley: New York, 1976.
19. Pope, J. W.; Chu, B. *Macromolecules* **1984**, 17, 2633.
20. Chu, B.; Gulari, E. *Macromolecules* **1979**, 12, 445.
21. Russo, P. S.; Guo, K.; DeLong, L. M. *46th Annual Conference Proceedings;* Society of Plastics Engineers: Fairfield, CT, 1988; p 983.
22. Kratochvil, P. In *Light Scattering from Polymer Solutions;* Huglin, M. B., Ed.; Academic Press: New York, 1972.
23. Berry, G. C.; Cotts, P. M.; Chu, S. G. *Br. Polym. J.* **1981**, 13, 47.
24. Private communication with Bruce Anderson of Kemira, Inc., Savannah, GA.
25. No special significance should be attached to the use of protamine sulfate coated slides; these are just routinely available.
26. Koppel, D. E. *J. Chem. Phys.* **1972**, 57(11), 4814.
27. Bevington, P. R. *Data Reduction in the Physical Sciences;* McGraw–Hill: New York, 1969.
28. DeLong, L. M.; Russo, P. S. *ACS Polym. Mat. Sci. Eng. Prepr.* **1988**, 59, 169.

Received for review February 14, 1989. Accepted revised manuscript August 1, 1989.

Latex Particle Size Distribution from Turbidimetric Measurements

Combining Regularization and Generalized Cross-Validation Techniques

Guillermo E. Eliçabe and Luis H. Garciá-Rubio

Chemical Engineering Department, College of Engineering, University of South Florida, Tampa, FL 33620

This chapter reports the recovery (deconvolution) of particle size distributions (PSDs) from turbidimetric measurements using a regularization technique (RT). Regularization techniques require the selection of a constraining parameter known as the regularization parameter. In this work the regularization parameter was calculated by using the generalized cross-validation (GCV) technique. The use of these complimentary techniques (RT and GCV) is demonstrated through the simulated recovery of PSDs of polystyrene latices. Unimodal and bimodal PSDs of varying breadth and mean particle diameters were investigated. The results demonstrate that the combination of these techniques yields adequate recoveries of the PSDs in almost every case. The cases where the techniques fail are identified, and strategies for subsequent recovery are discussed.

W HEN A SUSPENSION OF SPHERICAL PARTICLES is illuminated with light of different wavelengths, the resulting optical spectral extinction (turbidity) contains information that, in principle, can be used to estimate the particle size distribution (PSD) of the suspended particles. The recovery (deconvolution) of the PSD from turbidity measurements falls within the category of "inverse problems" to which several techniques have been applied with varying degrees of success (1–5). Recently (6), a regularization technique

0065–2393/90/0227–0083$06.25/0

was successfully applied to the estimation of the PSDs of polystyrene latices. Regularization techniques require the selection of a constraining parameter γ, known as the regularization parameter. The selection of the regularization parameter is critical for the adequate recovery of the PSD (6).

In this chapter some of the available methods for the estimation of γ are briefly introduced. Particular emphasis has been placed on the generalized cross-validation (GCV) technique. This technique, in our application, appears to be the most robust among the techniques available.

In the following section, the equations that relate the particle size and the turbidity are shown, and a discrete model for these equations is described in detail. Then, the regularized solution of the discrete model previously developed is introduced. A discussion about some of the techniques available for estimating the regularization parameter is given. The GCV technique is then revisited. Finally, the results of simulated examples are shown. In all the examples, the regularized solution is used along with the GCV technique to estimate a broad range of PSDs of polystyrene latices.

Absorption and Light Scattering of Spherical Suspended Particles

The loss of intensity experienced by a beam of electromagnetic radiation in passing through a sample of suspended particles, recorded as a function of the wavelength of the incident radiation, is known as the turbidity spectrum. The turbidity (τ) is related to the intensities at two points separated a distance l by

$$\tau = \frac{1}{l} \ln \frac{I^0}{I} \tag{1}$$

where I^0 is the intensity at the point where the electromagnetic radiation enters the sample, and it coincides with the intensity of the source; I is the intensity at the point where the electromagnetic radiation leaves the sample, and it coincides with the intensity at the detector. For a suspension of monodisperse isotropic spherical particles, the turbidity can be related to the wavelength of the incident radiation (λ_0), the particle diameter (D), and the optical properties of the suspension through Mie theory (7):

$$\tau(\lambda_0, D) = N_p \frac{\pi}{4} D^2 Q_{ext}[n_1(\lambda_0), k_1(\lambda_0), n_2(\lambda_0), \lambda_0, D] \tag{2}$$

where N_p is the total number of particles per unit volume in the sample and Q_{ext} is the extinction efficiency. Q_{ext} is a function of (1) the real and imaginary parts of the particle refractive index (n_1 and k_1, respectively); (2) the refractive index of the suspension medium, n_2; (3) the wavelength of the incident radiation in vacuo; and (4) the diameter of the spherical particles. The refractive indexes are, in general, functions of the wavelength. Equation 2 can be readily expressed in terms of the particle concentration (i.e., C is the weight of particles per unit volume of suspension):

$$\tau(\lambda_0, D) = \frac{3C}{2\rho D} Q_{ext}(\lambda_0, D) \tag{3}$$

where ρ is the density of the particle. (For simplicity, the refractive indexes have been omitted from the argument of Q_{ext} from eq 3 onward).

If the sample is a mixture with a distribution of particle diameters, and the PSD can be represented by a differential distribution, the turbidity can be rewritten as

$$\tau(\lambda_0) = \frac{\pi}{4} \int_0^\infty Q_{ext}(\lambda_0, D) D^2 f(D)\, dD \tag{4}$$

where $f(D)$ is such that

$$\int_0^\infty f(D)\, dD = N_p \tag{5}$$

If $f(D)$ is normalized with N_p, eq 4 becomes

$$\tau(\lambda_0) = N_p \frac{\pi}{4} \int_0^\infty Q_{ext}(\lambda_0, D) D^2 f'(D)\, dD \tag{6}$$

where now

$$\int_0^\infty f'(D)\, dD = 1 \tag{7}$$

Similarly, the turbidity of a polydisperse suspension, in terms of concentration C, can be written as

$$\tau(\lambda_0) = \frac{3C}{2\rho} \left[\frac{\displaystyle\int_0^\infty Q_{ext}(\lambda_0, D) D^2 f'(D)\, dD}{\displaystyle\int_0^\infty f'(D) D^3\, dD} \right] \tag{8}$$

By defining:

$$K(\lambda_0, D) = \frac{\pi}{4} Q_{ext}(\lambda_0, D)D^2 \tag{9}$$

eq 4 can be readily identified as a Fredholm integral equation of the first kind, in which $K(\lambda_0, D)$ is the corresponding kernel. The numerical solution to any of these equations (eq 4, 6, or 8) must be based on an appropriate discrete model. The solution to such a model will result in estimates of the number of particles and of the shape of the PSD. If the integrand in eq 4 is discretized into $n-1$ intervals, the integral can be approximated at a given wavelength $\lambda_{0,i}$ with a sum,

$$\tau_i \simeq \sum_{j=1}^{n} a_{ij} f_j \tag{10}$$

where $\tau_i \overset{\Delta}{=} \tau(\lambda_{0,i})$ and $f_j \overset{\Delta}{=} f(D_j)$. The details of the discretization procedure and the resulting coefficients a_{ij} are given in the Appendix.

If the turbidity is evaluated at m wavelengths $\lambda_{0,i}$, $i = 1, \ldots, m$, eq 4 can be written in matrix form,

$$\boldsymbol{\tau} \simeq \mathbf{Af} \tag{11}$$

where

$$\boldsymbol{\tau} = [\tau_1 \tau_2 \ldots \tau_m]^T \tag{12}$$

$$\mathbf{A} = \{a_{ij}\} \tag{13}$$

$$\mathbf{f} = [f_1 f_2 \ldots f_n]^T \tag{14}$$

T indicates the transpose.

Equation 11 can be written as an equality if the quadrature error ϵ_c introduced in the discretization is considered,

$$\boldsymbol{\tau} = \mathbf{Af} + \boldsymbol{\epsilon}_c \tag{15}$$

Finally, with the addition of the measurement error ϵ_m, the discrete equation for the representation of the experimental values of τ (i.e., τ_m), can be written as

$$\boldsymbol{\tau}_m = \mathbf{Af} + \boldsymbol{\epsilon}_c + \boldsymbol{\epsilon}_m = \mathbf{Af} + \boldsymbol{\epsilon} \tag{16}$$

Particle Size Distribution from Turbidity Measurements

Solution of Equation 16 Using Regularization Techniques.   The discrete model developed in the previous section (i.e., eq 16) transforms the problem of obtaining the PSD from turbidity measurements into a linear algebraic problem, where n points of the PSD can be estimated from m turbidity measurements (m has to be greater than or equal to n). If $m = n$, estimates of the PSD ($\hat{f}_d$) can, in principle, be obtained by the direct inversion of eq 16:

$$\hat{f}_d = A^{-1}\tau_m \tag{17}$$

Alternatively, if $m > n$, the least-squares solution of an overspecified system of linear equations yields

$$\hat{f}_{ls} = (A^T A)^{-1} A^T \tau_m \tag{18}$$

Although these solutions appear to be straightforward, it is well documented in the literature (8–12) that small errors (i.e., quadrature and experimental errors) result in large errors in $\hat{f}_d$ or $\hat{f}_{ls}$. The amplification of the errors occurs independently of the fact that the inverses of A and $(A^T A)$ can be calculated exactly, and it is a direct consequence of the near singularity of the matrix A (if $m = n$), or more generally (if $m > n$) of its near incomplete rank. This behavior can be explained by the near linear dependence between the functions $K(\lambda_{0,i}, D)$ from which the matrix A was obtained. In spite of the fact that these functions depend on the optical properties of the system under study, the preselected range of wavelengths and diameters, and the number of points that it is desired to recover, a certain amount of collinearity between some of the functions will be always present. Adding the fact that at least a small experimental error is also always present, it is possible to state that eqs 17 and 18 cannot give a solution to the problem under study.

However, by constraining the least-squares solution by means of a penalty function, approximate useful solutions can be obtained. This step can be achieved by using all of the prior information available regarding the PSD (i.e., the "true" f vector). For example, it is known that the values of f must be positive or zero, that there is an upper and a lower bound on the particle diameters, and that a certain amount of correlation exists between successive points on the distribution. Because eq 18 is the solution to the least-squares problem, then

$$\min_{\hat{f}} |A\hat{f} - \tau_m|^2 \tag{19}$$

where $| \bullet |$ indicates the modulus, and $\hat{f}_{ls}$ has been replaced by $\hat{f}$.

Prior information can be introduced by augmenting eq 19 with (8–12)

$$\min_{\hat{f}} \, [|\mathbf{A}\hat{\mathbf{f}} - \tau_m|^2 + \gamma q(\hat{\mathbf{f}})] \tag{20}$$

where $q(\hat{\mathbf{f}})$ is a scalar function that measures the correlation or smoothness of $\hat{\mathbf{f}}$, and γ is a nonnegative parameter that can be varied to emphasize more or less one of the terms of the objective functional given by eq 20. If γ is set to 0, eq 20 reduces to eq 19, a solution that generally exhibits large oscillations. On the other hand, when $\gamma \to \infty$ the minimization leads to a perfectly smooth solution judged by the measure of $q(\hat{\mathbf{f}})$ but totally independent of the τ_m values and, therefore, useless. Clearly, intermediate values of γ will produce acceptable solutions to the original problem (i.e., eq 16) and those solutions will have the smoothness or correlation characteristics imposed by the term $q(\hat{\mathbf{f}})$ in the functional. Also, it has been demonstrated that for bounded $(\mathbf{f}^T\mathbf{f})$, there exists a value of $\gamma > 0$ such that (13)

$$E[(\hat{\mathbf{f}} - \mathbf{f})^T(\hat{\mathbf{f}} - \mathbf{f})] < E[(\hat{\mathbf{f}}_{ls} - \mathbf{f})]^T(\hat{\mathbf{f}}_{ls} - \mathbf{f})] \tag{21}$$

where $E[\,]$ indicates expected value.

In other words, the error in the estimation of $\mathbf{f}$ associated with the solution of eq 20 will be smaller than that associated with the solution of eq 19. However, an appropriate form must be selected for the function $q(\hat{\mathbf{f}})$ and an adequate value for the parameter γ.

Several functions can be chosen to establish the desired correlation level or the smoothness of $\hat{\mathbf{f}}$. An interesting class of functions can be formulated by using a quadratic form of the vector $\hat{\mathbf{f}}$ because they yield an analytical solution to the minimization problem of eq 20. For example, if $q(\hat{\mathbf{f}}) = \hat{\mathbf{f}}^T\hat{\mathbf{f}}$, eq 20 can be readily identified with the well-known ridge regression (14). A more interesting example in which

$$q(\hat{\mathbf{f}}) = \hat{\mathbf{f}}^T\mathbf{K}^T\mathbf{K}\hat{\mathbf{f}}$$

with

$$\mathbf{K} = \begin{bmatrix} 0 & 0 & \bullet & \bullet & \bullet & \bullet & 0 \\ 1 & -1 & 0 & \bullet & \bullet & \bullet & 0 \\ 0 & 1 & -1 & 0 & \bullet & \bullet & 0 \\ \bullet & \bullet & \bullet & \bullet & \bullet & \bullet & \bullet \\ \bullet & \bullet & \bullet & \bullet & \bullet & \bullet & \bullet \\ 0 & \bullet & \bullet & 0 & 1 & -1 & 0 \\ 0 & \bullet & \bullet & \bullet & 0 & 1 & -1 \end{bmatrix} \tag{22}$$

gives the following $q(\hat{\mathbf{f}})$

$$q(\hat{\mathbf{f}}) = \sum_{j=2}^{n} (\hat{f}_j - \hat{f}_{j-1})^2 \tag{23}$$

which is a typical measure of smoothness. Another less restrictive $q(\hat{\mathbf{f}})$ is given by the sum of the squares of the second differences

$$q(\hat{\mathbf{f}}) = \sum_{j=2}^{n-1} (2\hat{f}_j - \hat{f}_{j-1} - \hat{f}_{j+1})^2 \tag{24}$$

In this case the matrix $\mathbf{H} = \mathbf{K}^T\mathbf{K}$ is given by

$$\mathbf{H} = \begin{bmatrix} 1 & -2 & 1 & 0 & \bullet & \bullet & 0 \\ -2 & 5 & -4 & 1 & 0 & \bullet & 0 \\ 1 & -4 & 6 & -4 & 1 & \bullet & 0 \\ \bullet & \bullet & \bullet & \bullet & \bullet & \bullet & \bullet \\ 0 & \bullet & 1 & -4 & 6 & -4 & 1 \\ 0 & \bullet & 0 & 1 & -4 & 5 & -2 \\ 0 & \bullet & \bullet & 0 & 1 & -2 & 1 \end{bmatrix} \tag{25}$$

It will be shown later that unimodal and bimodal latex distributions can be readily analyzed by using this last quadratic form with a slight modification that constrains the values of $\hat{f}_1$ and $\hat{f}_n$ to be 0, thus incorporating into the solution additional prior knowledge that had been imposed during the derivation of the discrete model. This last constraint can be implemented by summing β^2, with $\beta \gg 1$, to the $(1,1)$ and (n,n) elements of the $\mathbf{H}$ matrix shown in eq 25. In this form the final quadratic form of the $q(\hat{\mathbf{f}})$ function for the examples of the following sections will be

$$q(\hat{\mathbf{f}}) = \beta^2(\hat{f}_1^2 + \hat{f}_N^2) + \sum_{j=2}^{n-1} (2\hat{f}_j - \hat{f}_{j-1} - \hat{f}_{j+1})^2 \tag{26}$$

Having arrived at an explicit expression for $\mathbf{H}$, we can show that the solution to the constrained problem of eq 20 is given by (*11*):

$$\hat{\mathbf{f}} = (\mathbf{A}^T\mathbf{A} + \gamma H)^{-1}\mathbf{A}^T\tau_m \tag{27}$$

The value of $\hat{\mathbf{f}}$ obtained with eq 27 will be called the regularized solution of eq 16. If the matrix $\mathbf{H}$ of eq 25 is used with the modification that permits the constraint $\hat{f}_1 = \hat{f}_n = 0$, $(\mathbf{A}^T\mathbf{A} + \gamma H)$ is a positive definite symmetric matrix. Thus, efficient algorithms can be used to perform the inverse.

Selection of γ.   The regularized solution of eq 27 requires the selection of the regularization parameter γ. The existing methods for selecting γ can be roughly divided in two: those stemming from applications in physics and engineering, and those developed in statistics.

Among the first, Twomey's analysis of information content (11) has been applied mostly in atmospheric sciences. The idea of Twomey's method is to detect the number of independent pieces of information available in a set of experimental measurements. This analysis leads to a regularization parameter γ that also depends on an estimation of the root-mean-square value of the measurement noise. Using a completely different approach, Provencher (15) proposed a method for selecting γ in the problem of inverting the Fredholm integral equation that arises in the determination of molecular weight distributions (MWD) of polymers using photon correlation spectroscopy. Provencher's method is analogous to the standard procedure of constructing confidence regions for the sought solution. Although the method is rather arbitrary, the results obtained for the estimation of the MWD were satisfactory.

The methods for the selection of γ based on statistics theory were developed for the so-called ridge regression (RR). As stated previously, RR is a special case of eq 27 in which $\mathbf{H}$ is the identity matrix ($\mathbf{I}$). By defining

$$X = \mathbf{A}\mathbf{K}^{-1} \tag{28}$$

and

$$\hat{\mathbf{f}}' = \mathbf{K}\hat{\mathbf{f}} \tag{29}$$

then

$$\hat{\mathbf{f}}' = (X^T X + \gamma\mathbf{I})^{-1}X^T \tau_m \tag{30}$$

Therefore, the regularized solution of eq 27 can be seen as a RR, and the methods specifically developed for estimating γ in eq 30 can be directly applied to eq 27. The statistical methods for the estimation of γ may be divided into two: those that use a priori information, and those that use only the measured data.

Among the first, the method of Hoerl et al. (14, 16) and the closed form solution for the iterative method described in ref. 16 given by Hemmerle (17) can be cited. In these cases an estimation of the variance of the noise (σ^2) and an initial estimate of the solution are needed. Another method, which uses only an estimate of σ^2, is known as the "range risk" estimate, and it is briefly outlined in ref. 18. According to Golub et al. (18), the only methods available for estimating γ from the data are maximum likelihood, ordinary cross validation (OCV), and generalized cross validation (GCV).

These three methods do not require any a priori information, and therefore they can be used to completely automatize the PSD estimation process. Simulated studies (*18, 19*) have shown that the GCV technique is the most reliable and the best theoretically founded among those using only the measured data. This assertion justifies, in principle, its selection as a method for estimating γ in the context of PSD estimation from turbidimetric data.

The Generalized Cross-Validation Technique.   The GCV technique is a rotation-invariant version of OCV. The OCV technique may be derived as follows: define $\hat{f}'^{(k)}(\gamma)$ as the estimation of $f' = Kf$ using eq 30 with the kth value of τ_m omitted. The argument is that if γ is adequate, then $[X\hat{f}'^{(k)}(\gamma)]_k$ (the kth component of the $[X\hat{f}'^{(k)}(\gamma)]$ vector) should be a good predictor of τ_{m_k} (the kth component of τ_m). In order to obtain good predictors for all the measurements ($k = 1, \ldots, m$), γ should be chosen as the minimizer of

$$P(\gamma) = \frac{1}{m} \sum_{k=1}^{m} \{[X\hat{f}'^{(k)}(\gamma)]_k - \tau_{m_k}\}^2 \tag{31}$$

This function can be expressed as (*18*):

$$P(\gamma) = \frac{1}{m} |\mathbf{B}(\gamma)[I - \mathbf{Z}(\gamma)]\tau_m|^2 \tag{32}$$

where $\mathbf{B}(\gamma)$ is a diagonal matrix with entries $\{1/[1 - z_{jj}(\gamma)]\}$ where the element z_{jj} is the jj entry of

$$\mathbf{Z}(\gamma) = X(X^T X + \gamma I)^{-1} X^T$$

Although the idea developed as just described seems to be appealing, Golub et al. (*18*) pointed out that this method fails when the matrix $\mathbf{Z}(\gamma)$ is diagonal because $P(\gamma)$ does not have a unique minimizer. This behavior indicates that the OCV is not expected to perform successfully in the near diagonal case either. To circumvent this difficulty, the GCV technique was introduced as a rotation-invariant form of OCV (*18*). The GCV function of γ can be defined as the OCV function (eq 32) applied to the following transformed model

$$\tilde{\tau} = WU^T\tau_m = WDV^T f' + WU^T \epsilon = \tilde{X}f' + WU^T \epsilon \tag{33}$$

where U and V are the result of the singular value decomposition of X

$$X = UDV^T \tag{34}$$

and $\mathbf{W}$ is a complex matrix whose elements are

$$w_{ij} = \frac{1}{\sqrt{m}} \exp\left(\frac{2\pi ijk}{m}\right) \qquad j, k = 1, 2, \cdots m \qquad (35)$$

where $\sqrt{i} = -1$.

Therefore, using the transformed model in eq 32 results in

$$\tilde{P}(\gamma) = \frac{1}{m} |\tilde{\mathbf{B}}(\gamma)[I - \tilde{\mathbf{Z}}(\gamma)]\tilde{\tau}|^2 \qquad (36)$$

Using the fact that as a result of the transformation, $\tilde{\mathbf{Z}}(\gamma)$ is a circulant matrix and hence constant down the diagonals, the last equation can be expressed as

$$\tilde{P}(\gamma) = V(\gamma) = m \frac{|[I - \tilde{\mathbf{Z}}(\gamma)]\tilde{\tau}|^2}{\{\text{trace } [I - \tilde{\mathbf{Z}}(\gamma)]\}^2} \qquad (37)$$

It can also be shown that

$$V(\gamma) = m \frac{|[I - \mathbf{Z}(\gamma)]\tau_m|^2}{\{\text{trace } [I - \mathbf{Z}(\gamma)]\}^2} = m \frac{\sum_{i=1}^{m} \left(\frac{\gamma}{\lambda_i + \gamma}\right)^2 z_i^2}{\left(\sum_{i=1}^{n} \frac{\gamma}{\lambda_i + \gamma} + m - n\right)^2} \qquad (38)$$

where $z = [z_1, \ldots, z_m] = U^T \tau_m$ and λ_i ($i = 1, \ldots, n$) are the eigenvalues of $(X^T X)$.

Recovery of the Particle Size Distribution for Polystyrene Latices

In this section, eqs 27 and 38 will be used to estimate the PSDs of polystyrene latices. Using the measurements and the model, eq 38 will be minimized with respect to γ. The value of γ that minimizes eq 38 will be then used in eq 27 to estimate the PSDs. The turbidity spectra were simulated by using the results described under "Absorption and Light Scattering of Spherical Suspended Particles". For the simulated experiments, the refractive index of water was calculated from (20)

$$n_2 = 1.324 + \frac{3046}{\lambda_o^2} \qquad (39)$$

with λ_0 given in nanometers. The real and the imaginary parts of the complex refractive index for polystyrene were obtained from the data of Inagaki et al. (*21*). Figure 1 shows the optical properties of polystyrene and water as functions of the wavelength. Figure 2 shows the distributions analyzed. A broad range of possibilities is being considered, including bimodal and very narrow distributions. The mathematical expression for those distributions is

$$f(D) = N_p \left[\frac{C_1}{C_1 + C_2} \log N_1 + \frac{C_2}{C_1 + C_2} \log N_2 \right] \tag{40}$$

where

$$N_p = 8.49 \times 10^6 \text{ (particles per cubic centimeter)} \tag{41}$$

and

$$\log N_i = \frac{1}{\sqrt{2\pi}\sigma_i(D - D_s)} \exp \left[\frac{-[\ln (D - D_s) - \ln D_{g,i}]^2}{2\sigma_i^2} \right] \tag{42}$$

Table I shows the values for the leading parameters that characterize the distributions.

The simulated turbidity spectra were calculated by using a 51-point discretization of the distributions generated with eq 40 and the parameters shown in Table I. The range of wavelengths was chosen between 200 and

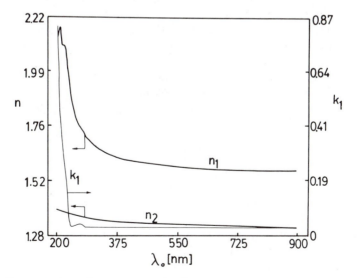

Figure 1. Optical parameters of polystyrene (n_1, k_1) and water (n_2).

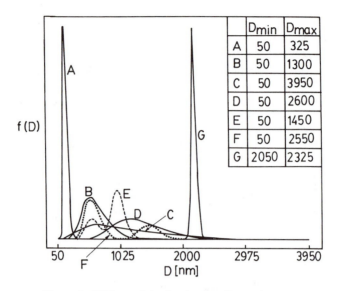

Figure 2. PSDs used in the simulated experiments.

900 nm with a resolution of 1 nm that results in a value of $m = 701$. A 3% maximum value random noise (relative to the maximum turbidity value) was added to the simulated spectra to represent extreme measurement conditions (in the instrument, noise is less than 0.01 absorption units). The use of an exaggerated noise level demonstrates that the technique would be robust for typical measurement errors. The simulated spectra with the added noise constitute the experimental data.

The number of recovered points on the distributions was $n = 51$ in all cases, and the ranges of diameters varied according to the case being analyzed (*see* Figure 2). The value of β was chosen as 1000.

To draw the most general conclusions, a Monte Carlo type experiment was carried out. Each spectrum was replicated five times, keeping the statistics of the noise constant. The objective function $r_f(\gamma)$ was defined as

$$r_f(\gamma) = \left\{ \frac{\sum_{i=1}^{n} [f_i - \hat{f}_i(\gamma)]^2}{\sum_{i=1}^{n} f_i^2} \right\}^{\frac{1}{2}} \tag{43}$$

The value of γ that minimizes eq 43 would give the best solution in the context of the regularization technique used in this work. Unfortunately, this objective function cannot be evaluated in a real situation because it depends on the unknown value of f. However, it permits us to examine, in a simulated experiment, the performance of eq 38 as estimator of γ.

Table I. Parameters That Characterize Particle Size Distributions Used in the Simulated Experiments

Case	D_{g1} (nm)	σ_1	D_{g2} (nm)	σ_2	C_1	C_2	D_s (nm)
A	175	0.2	—[a]	—	1	0	0
B	600	0.3	—	—	1	0	0
C	1000	0.65	—	—	1	0	0
D	1300	0.3	—	—	1	0	0
E	600	0.2	1000	0.1	1	2	0
F	600	0.2	1500	0.1	2	1	0
G	175	0.2	—	—	1	0	2000

NOTE: Symbols are the same as in eqs 40–42.

For each case, including replications, eqs 38 and 43 were minimized. The values of γ that minimize those equations, γ_{GCV} and γ_{opt}, respectively, are shown in Table II. The last two columns of that table show the value of the optimal objective function (eq 43) evaluated at γ_{opt} and γ_{GCV}. These data enable us to judge the accuracy of the solution obtained with the value of γ that minimizes the GCV objective function (eq 38) with respect to that obtained using γ_{opt}.

Figures 3 and 4 show eqs 38 and 43 as functions of γ in log–log plots for cases B and F, respectively. The values of γ range from the 37th to the 51st eigenvalue of the corresponding $(A^T A)$ matrix in both cases. For each case two replications are plotted to show closeness along the γ axis. Two replications of the same case should give values of γ_{opt} close together for eq 43, and thus indicate the validity of the γ_{GCV} values provided by eq 38 relative to the optimal values. In these figures the scales of the ordinate axis are different for each plot to clearly compare the locations of the minima attained for each function. (The use of the same scale does not give any additional information and prevents a clear comparison).

As can be seen in Table II, the results are very good in almost all cases. The method is able to distinguish between bimodal and unimodal distributions. For example, when the distribution is bimodal, as in case F, the optimum γ values are shifted to the left with respect to the unimodal cases (e.g., case B) to allow the inherent oscillations of a bimodal distribution (*see* Figures 3 and 4).

In case A, the values of γ provided by eq 38 do not give correct solutions in any replication. The distribution corresponding to this case is being very narrow and has a very small number-average particle diameter. To determine if the poor results are due to both characteristics or if they depend only on one of them, a simulation using the same distribution of case A but shifted to the large-particle diameters was analyzed. This simulation corresponds to case G and reveals that, although for this case the best solutions provided by the regularization technique are not as good as those for case A, the GCV technique gives values of γ very close to the optimal ones.

On the other hand, the results obtained for case C show that, even

Table II. Results for the Five Replications of Each Experiment (A to G) Using Equations 38 and 40

Case	Rep.[a]	γ_{opt}	γ_{GCV}	$r_f(\gamma_{opt})$	$r_f(\gamma_{GCV})$
A	1	1.32×10^{-15}	7.51×10^{-18}	0.1817	1.6838
	2	1.32×10^{-15}	5.86×10^{-17}	0.1626	0.4784
	3	1.32×10^{-15}	1.70×10^{-20}	0.2216	8.3119
	4	2.47×10^{-15}	1.61×10^{-16}	0.1665	0.5952
	5	1.61×10^{-16}	1.32×10^{-15}	0.1182	0.2014
B	1	2.05×10^{-12}	6.20×10^{-12}	0.0833	0.0919
	2	6.20×10^{-12}	6.20×10^{-12}	0.0575	0.0575
	3	7.83×10^{-13}	9.73×10^{-12}	0.0486	0.0805
	4	6.20×10^{-12}	9.73×10^{-12}	0.0671	0.0725
	5	9.73×10^{-12}	6.20×10^{-12}	0.0602	0.0681
C	1	2.50×10^{-9}	6.93×10^{-10}	0.0780	0.0955
	2	1.95×10^{-9}	1.78×10^{-10}	0.0788	0.1155
	3	1.39×10^{-9}	1.06×10^{-10}	0.0770	0.0937
	4	2.50×10^{-9}	1.39×10^{-9}	0.0770	0.0785
	5	2.05×10^{-10}	1.95×10^{-9}	0.0727	0.0779
D	1	9.05×10^{-10}	1.42×10^{-9}	0.0621	0.0624
	2	9.05×10^{-10}	9.05×10^{-10}	0.0450	0.0450
	3	1.94×10^{-9}	3.71×10^{-9}	0.0687	0.1745
	4	3.88×10^{-10}	1.94×10^{-9}	0.0412	0.0541
	5	9.27×10^{-11}	1.94×10^{-9}	0.0476	0.0593
E	1	1.43×10^{-12}	5.04×10^{-13}	0.1974	0.2177
	2	1.04×10^{-12}	4.65×10^{-12}	0.1196	0.1535
	3	1.43×10^{-12}	6.06×10^{-12}	0.1653	0.1924
	4	3.24×10^{-12}	3.24×10^{-12}	0.1117	0.1117
	5	1.38×10^{-11}	4.65×10^{-12}	0.1960	0.2062
F	1	9.31×10^{-13}	1.33×10^{-11}	0.2484	0.3169
	2	2.69×10^{-12}	3.96×10^{-12}	0.1736	0.1763
	3	2.25×10^{-12}	5.60×10^{-12}	0.2062	0.2174
	4	9.31×10^{-13}	3.96×10^{-12}	0.1417	0.1701
	5	4.78×10^{-12}	2.25×10^{-12}	0.1886	0.2085
G	1	1.41×10^{-11}	1.41×10^{-11}	0.2909	0.2909
	2	1.41×10^{-11}	3.94×10^{-9}	0.3829	0.4909
	3	1.41×10^{-11}	1.41×10^{-11}	0.2303	0.2303
	4	1.41×10^{-11}	1.41×10^{-11}	0.3049	0.3049
	5	9.54×10^{-12}	1.41×10^{-11}	0.1917	0.2049

[a]Replication.

though the corresponding distribution has a high number of small particles, the results obtained with the GCV technique are still good. Therefore, the poor behavior in case A may be attributed to a very narrow distribution in the small-diameters zone. Therefore, the GCV technique is expected to work poorly when the distributions are very narrow and have a small number-average particle diameter.

A clearer picture of the results can be seen in Figures 5–10 in which the estimated PSDs for cases B to G are shown, respectively. In these figures the true distribution and two replications are plotted for each case. Although

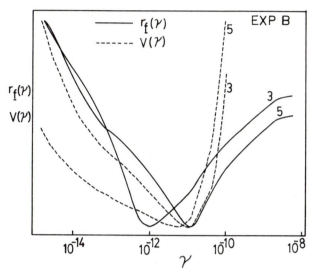

Figure 3. Equations 38 (---) and 43 (—) for case B and replications 3 and 5 as functions of γ.

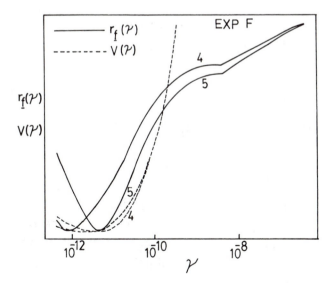

Figure 4. Equations 38 (---) and 43 (—) for case F and replications 4 and 5 as functions of γ.

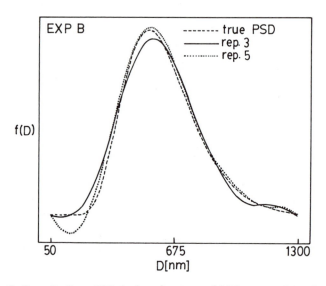

Figure 5. Case B. True PSD (---) and estimated PSDs using GCV for replications 3 (—) and 5 (• • •).

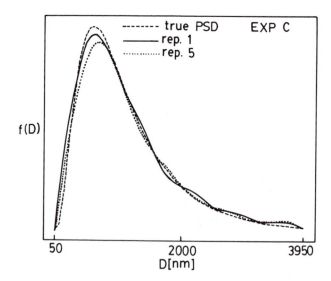

Figure 6. Case C. True PSD (---) and estimated PSDs using GCV for replications 1 (—) and 5 (• • •).

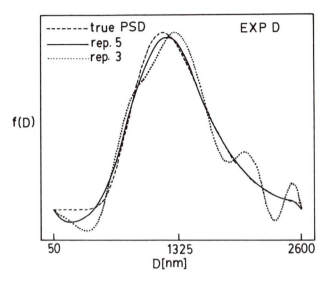

Figure 7. Case D. True PSD (---) and estimated PSDs using GCV for replications 5 (—) and 3 (• • •).

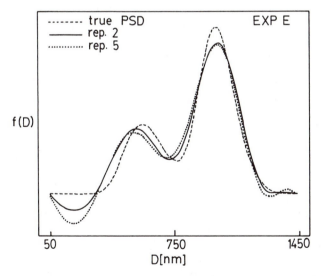

Figure 8. Case E. True PSD (---) and estimated PSDs using GCV for replications 2 (—) and 5 (• • •).

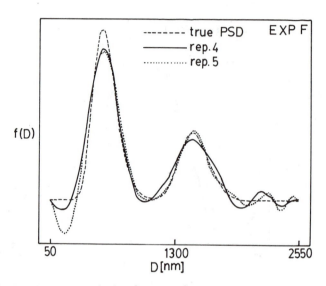

Figure 9. *Case F. True PSD (---) and estimated PSDs using GCV for replications*
4 (—) and 5 (• • •).

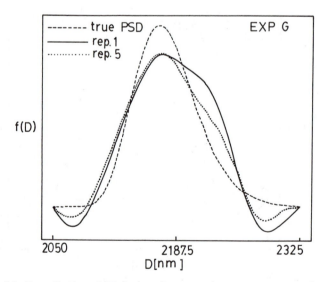

Figure 10. *Case G. True PSD (---) and estimated PSDs using GCV for repli-*
cations 1 (—) and 5 (• • •).

the optimal estimates of the PSDs are not plotted, they were very close to those obtained with the γ_{GCV} values. In case D (Figure 7) a mild oscillation is present in the solution obtained for replication 3. This result may be expected when replications are carried out. The possibilities of this kind of results are even higher when, as in this case, high levels of noise are present.

Summary and Conclusions

The results presented in this chapter verify the potential and versatility of the regularization technique when it is used along with the GCV technique. A completely different problem from those analyzed in refs. 18 and 22 using the same combination was solved with few and predictable limitations.

The use of these complimentary techniques was demonstrated through the recovery of the PSD of polystyrene latices. Unimodal and bimodal PSDs of varying breadth and mean particle diameters were investigated. The results were mostly satisfactory.

The GCV technique makes it possible to integrate the complete estimation process in a single step for the purpose of monitoring and controlling a variety of heterogeneous systems, among them emulsion polymerizations.

Appendix

For the discretization of eq 4, it is assumed that, for a given wavelength $\lambda_{0,i}$, the integrand can be approximated by the product of a linear interpolation between two successive points on $f(D)$,

$$f_j = A_j + B_j D_j \tag{A1}$$

$$f_{j+1} = A_j + B_j D_{j+1} \tag{A2}$$

with $f_j \stackrel{\Delta}{=} f(D_j)$, and $D_{j+1} - D = \Delta D$ for all j, and the kernel $K(\lambda_0, D)$ calculated at $\lambda_{0,i}$:

$$K(\lambda_{0,i}, D) \stackrel{\Delta}{=} K_i(D) = \frac{\pi}{4} Q_{ext}(\lambda_{0,i}, D) D^2 \tag{A3}$$

Dividing the integral in eq 4 into $n - 1$ sections and substituting in eqs A1–A3 show that τ_i can be expressed in the following form:

$$\tau_i \simeq \int_{D_1}^{D_2} K_i(D)(A_1 + B_1 D)\, dD + \int_{D_2}^{D_3} K_i(D)(A_2 + B_2 D)\, dD + \dots$$

$$+ \int_{D_j}^{D_{j+1}} K_i(D)(A_j + B_j D)\, dD + \dots + \int_{D_{N-1}}^{D_N} K_i(D)(A_{N-1} + B_{N-1} D)\, dD \tag{A4}$$

where it was assumed that $f(D) = 0$ for $D_1 > D > D_n$. The values for the parameters A_j and B_j can be obtained from eqs A1 and A2

$$A_j = \frac{D_{j+1}f_j - D_j f_{j+1}}{D_{j+1} - D_j} \tag{A5}$$

$$B_j = \frac{f_{j+1} - f_j}{D_{j+1} - D_j} \tag{A6}$$

Substituting eqs A5 and A6 into eq A4 yields

$$
\tau_i \approx \left[\frac{D_2}{\Delta D} \int_{D_1}^{D_2} K_i(D)\, dD - \frac{1}{\Delta D} \int_{D_1}^{D_2} K_i(D)D\, dD \right] f_1 + \left[\frac{1}{\Delta D} \int_{D_1}^{D_2} K_i(D)D\, dD \right.
$$
$$
\left. - \frac{D_1}{\Delta D} \int_{D_1}^{D_2} K_i(D)\, dD + \frac{D_3}{\Delta D} \int_{D_2}^{D_3} K_i(D)\, dD - \frac{1}{\Delta D} \int_{D_2}^{D_3} K_i(D)D\, dD \right] f_2 + \dots
$$
$$
+ \left[\frac{1}{\Delta D} \int_{D_{N-2}}^{D_{N-1}} K_i(D)D\, dD - \frac{D_{N-2}}{\Delta D} \int_{D_{N-2}}^{D_{N-1}} K_i(D)\, dD + \frac{D_N}{\Delta D} \int_{D_{N-1}}^{D_N} K_i(D)\, dD \right.
$$
$$
\left. - \frac{1}{\Delta D} \int_{D_{N-1}}^{D_N} K_i(D)D\, dD \right] f_{N-1}
$$
$$
+ \left[\frac{1}{\Delta D} \int_{D_{N-1}}^{D_N} K_i(D)D\, dD - \frac{D_{N-1}}{\Delta D} \int_{D_{N-1}}^{D_N} K_i(D)\, dD \right] f_N \tag{A7}
$$

Therefore a_{ij} is given by

$$
a_{ij} = \frac{1}{\Delta D} \int_{D_{j-1}}^{D_j} K_i(D)D\, dD - \frac{D_{j-1}}{\Delta D} \int_{D_{j-1}}^{D_j} K_i(D)\, dD
$$
$$
+ \frac{D_{j+1}}{\Delta D} \int_{D_j}^{D_{j+1}} K_i(D)\, dD - \frac{1}{\Delta D} \int_{D_j}^{D_{j+1}} K_i(D)D\, dD \tag{A8}
$$

for $j = 2, \dots, n-1$, and

$$
a_{i1} = \frac{D_2}{\Delta D} \int_{D_1}^{D_2} K_i(D)\, dD - \frac{1}{\Delta D} \int_{D_1}^{D_2} K_i(D)D\, dD \tag{A9}
$$

$$
a_{iN} = \frac{1}{\Delta D} \int_{D_{N-1}}^{D_N} K_i(D)D\, dD - \frac{D_{N-1}}{\Delta D} \int_{D_{N-1}}^{D_N} K_i(D)\, dD \tag{A10}
$$

For a small increment in the diameters, the integration of the functions $K_i(D)$ can be calculated by using a straight-line approximation and the same step ΔD used with $f(D)$. Thus, the integrals in eqs A8–A10 can be written as

$$\int_{D_k}^{D_{k+1}} K_i(D)\, dD \simeq D_{k+1} K_{ik} - D_k K_{i,k+1} \tag{A11}$$

$$+ \frac{K_{i,k+1} - K_{ik}}{2(D_{k+1} - K_k)} (D_{k+1}{}^2 - D_k{}^2)$$

$$\int_{D_k}^{D_{k+1}} K_i(D) D\, dD \simeq \frac{1}{2} K_{ik}(D_{k+1}{}^2 - D_k{}^2) \tag{A12}$$

$$- \frac{(K_{i,k+1} - K_{ik}) D_k}{2(D_{k+1} - D_k)} (D_{k+1}{}^2 - K_k{}^2)$$

$$+ \frac{(K_{i,k+1} - K_{ik})(D_{k+1}{}^3 - D_k{}^3)}{3(D_{k+1} - D_k)}$$

where $K_{ik} \overset{\Delta}{=} K_i(D_k)$.

By substituting eqs A11 and A12 into eqs A8–A10, the appropriate values for a_{ij} can be obtained.

Acknowledgments

This research was supported by National Science Foundation Grants RII 8507956 and INT–8602578. Guillermo Eliçabe holds a scholarship from Consejo Nacional de Investigaciones Científicas y Técnicas de la Republica Argentina.

References

1. Wallach, M. L.; Heller, W.; Stevenson, A. F. *J. Chem. Phys.* **1961**, *34*, 1796.
2. Wallach, M. L.; Heller, W. *J. Phys. Chem.* **1964**, *68*, 924.
3. Yang, K. C.; Hogg, R. *Anal.[31˜ Chem.* **1979**, *51*, 758.
4. Zollars, R. L. *J. Colloid Interface Sci.* **1980**, *74*, 163.
5. Melik, D. H.; Fogler, H. S. *J. Colloid Interface Sci.* **1983**, *92*, 161.
6. Eliçabe, G. E.; Garciá-Rubio, L. H. *J. Colloid Interface Sci.* **1989**, *129(1)*, 192–200.
7. Kerker, M. *The Scattering of Light and Other Electromagnetic Radiation;* Academic: New York, 1969.
8. Phillips, D. L. *J. Assoc. Comput. Mach.* **1962**, *9*, 84.
9. Twomey, S. *J. Assoc. Comput. Mach.* **1963**, *10*, 97.
10. Turchin, V. F.; Kozlov, V. P.; Malkevich, M. S. *Sov. Phys. Usp. Engl. Transl.* **1971**, *13*, 681.

11. Twomey, S. *Introduction to the Mathematics of Inversion in Remote Sensing and Indirect Measurements;* Elsevier: New York, 1977.
12. Bertero, M.; De Mol, C.; Viano, G. A. In *Inverse Scattering Problems in Optics;* Topics in Current Physics; Baltes, H. P., Ed.; Springer Verlag: New York, 1980; p 161.
13. Hoerl, A. E.; Kennard, R. W. *Chem. Eng. Progr.* **1962,** *55,* 54.
14. Hoerl, A. E.; Kennard, R. W.; Baldwin, K. F. *Commun. Stat.* **1975,** *4,* 105.
15. Provencher, S. W. *Makromol. Chem.* **1979,** *180,* 201.
16. Hoerl, A. E.; Kennard, R. W. *Commun. Stat.* **1976,** *A5,* 77.
17. Hemmerle, W. J. *Technometrics* **1975,** *17,* 309.
18. Golub, G. H.; Heath, M.; Wahba, G. *Technometrics* **1979,** *21,* 215.
19. Gibbons, D. I. General Motors Research Laboratories, Research Publication GMR 2659, Warren, MI, 1978.
20. Maron, S. H.; Pierce, P. E.; Ulevitch, I. N. *J. Colloid Sci.* **1963,** *18,* 470.
21. Inagaki, T.; Arakawa, E. T.; Hamm, R. N.; Williams, M. W. *Phys. Rev. B* **1977,** *15,* 3243.
22. Merz, P. H. *J. Comput. Phys.* **1980,** *38,* 64.

RECEIVED for review February 14, 1989. ACCEPTED revised manuscript August 1, 1989

On-Line Particle Size Determination during Latex Production Using Dynamic Light Scattering

Theodora Kourti[1], John F. MacGregor[1], Archie E. Hamielec[1], David F. Nicoli[2], and Virgil B. Elings[2]

[1]McMaster Institute for Polymer Production Technology, Department of Chemical Engineering, McMaster University, Hamilton, Ontario, Canada L8S 4L7
[2]Particle Sizing Systems, 6780 Cortona Drive, Santa Barbara, CA 93117

This chapter describes a system for automatic sample acquisition and dilution designed to interface with a particle-sizing instrument based on dynamic light scattering. Results are shown of the successful use of this technology to monitor on-line particle growth during the emulsion polymerization of vinyl acetate in a pilot-plant reactor. Automatic sampling every 10–15 min is achievable; therefore this system is a powerful new tool for on-line monitoring and control of latex production.

PARTICLE SIZE DISTRIBUTION IS A CRITICAL PARAMETER in emulsion polymerization because it influences the physical properties (and therefore the end use) of the latex product. The control of particle size is therefore of great importance in the production of latices. Even though the chemical recipe remains the same from run to run, the presence of impurities can significantly affect particle nucleation, and therefore particle size, during polymerization. When the latex is produced in continuous or semi-batch reactors, the particle size can be controlled during production by manipulating input variables such as emulsifier concentration and monomer feed rate. This task requires accurate and reliable on-line determination of particle size. Furthermore, the time required for particle size measurement must

0065–2393/90/0227–0105$06.00/0

be short enough to allow sufficient time for the appropriate control actions to be calculated and implemented. Various techniques have been developed for the determination of the particle size distribution in colloidal dispersions, but most of them are time consuming or unwieldy for on-line applications. Light-scattering techniques are fast, simple, sufficiently accurate and reproducible, and seem promising for on-line particle size measurements.

The technique of dynamic light scattering (DLS) has been evolved in recent years into a powerful research and quality control tool, able to effectively characterize simple submicrometer particle size distributions. Thus far, however, this technology has been confined almost exclusively to off-line quality control environments. It has yet to be integrated successfully into polymer production facilities to provide an automatic on-line sizing capability suitable for real-time process monitoring and control. The principal factor behind this obvious shortcoming is the requirement of significant operator intervention associated with sample acquisition, preparation, and introduction into the light-scattering instrument. The most critical requirement is the dispersion of the concentrated latex sample (30–50% solids) in a suitable diluent and dilution of the resulting suspension to a final concentration optimal for the light-scattering measurement.

With these needs in mind, we have developed a proprietary system for automatic sample acquisition and dilution (patents issued and pending), designed to interface with a DLS-based particle-sizing instrument. In the work presented here, this system was used in conjunction with a modified Nicomp 370 submicrometer particle sizer. Results are shown from the successful application of dynamic light scattering to monitor on-line the particle growth during the emulsion polymerization of vinyl acetate in a pilot-plant reactor.

Theoretical Background of Dynamic Light Scattering

Dynamic light scattering (also called quasi-elastic light scattering and photon correlation spectroscopy) is concerned with the time behavior of the scattered intensity obtained from a suspension of particles. This approach contrasts with traditional light-scattering techniques that measure the average scattered intensity. Submicrometer-sized particles in suspension exhibit significant random motion because of collisions with the molecules of the surrounding liquid medium (Brownian motion). As a result, when a colloidal dispersion is illuminated by a light source, the phases of each of the scattered waves (arriving at a detector at a fixed angle) fluctuate randomly in time because of the fluctuations in the positions of the particles that scatter the waves. Because these waves mutually interfere, the net intensity of the scattered light fluctuates randomly in time around a mean value. The DLS technique makes use of the fact that the time dependence of the intensity fluctuations (calculated from the autocorrelation function of the scattered intensity) can be related to the translational diffusion coefficient of the particles, which in turn is related to the particle size through the Stokes–Einstein equation. Details on the theory behind DLS and the experimental setup; examples from applications of the technique; and discussions of its advantages, problems, and difficulties can be found in a number of sources (1–4).

The autocorrelation function $G^{(2)}(t')$ of the scattered light intensity is given by:

$$G^{(2)}(t') = <I(t)I(t + t')>$$ (1)

where $I(t)$ is the intensity at time t, and t' is a time delay. The $<>$ symbol indicates a running sum of products taken at different times, t. For $t' \to \infty$, $G^{(2)}(\infty) = <I(t)>^2$, which is the square of the average scattered intensity, equal to the base line of the autocorrelation function. The normalized first-order autocorrelation function, $g^{(1)}(t')$, can be calculated from the measured function:

$$G^{(2)}(t') = B(1 + \beta|g^{(1)}(t')|^2)$$ (2)

where B is the base line and β $(0 < \beta < 1)$ is an instrument-related constant.

For systems of uniform particle size, $g^{(1)}(t')$ is a simple exponentially decaying function of t':

$$g^{(1)}(t') = \exp(-\Gamma t')$$ (3)

The decay constant Γ is related to the translational diffusion coefficient D_t by:

$$\Gamma = D_t K^2$$ (4)

where K is the scattering wave vector, which depends on the wavelength (in vacuum) of the light source (λ_0), the solvent refractive index (n), and the angle of detection, θ:

$$K = 4\pi n \sin\left(\frac{\theta}{2}\right)\lambda_0$$ (5)

For random diffusion of noninteracting particles, the single-particle diffusion coefficient (D_t) is obtained from equations 3–5; the hydrodynamic radius R is obtained from D_t via the Stokes–Einstein equation:

$$R = \frac{kT}{6\pi\eta D_t}$$ (6)

where k is the Boltzmann constant, T is the temperature (kelvins), and η is the shear viscosity of the liquid medium. Thus the particle size of a monodisperse (single-sized) suspension can be easily obtained from the measured autocorrelation function via equations 1–6.

For suspensions with broad unimodal or with multimodal distributions, the inversion of the autocorrelation data to obtain the particle size distribution is not an easy task and remains an area of active research (4–14). For a polydisperse suspension, $g^{(1)}(t')$ is a weighted sum of exponentially decaying functions, each of which corresponds to a different particle diameter D_i with decay constant Γ_i.

$$g^{(1)}(t') = \int_0^\infty F(\Gamma) \exp(-\Gamma t') \, d\Gamma \tag{7}$$

$F(\Gamma)$ is the normalized distribution of the decay constants of the scatterers in the suspension. The problem of obtaining the particle size distribution from the raw data, $g^{(1)}(t')$, in effect reduces to solving equation 7 for $F(\Gamma)$. A number of algorithms for inverting this equation (an ill-conditioned problem) have been presented (4–13). The approach used with significant success in the Nicomp 370 and elsewhere is based on a Laplace transform inversion of $g^{(1)}(t')$ using a nonlinear least-squares procedure (with a non-negative constraint). A review of most of the available algorithms for the determination of $F(\Gamma)$ and an evaluation of their performance for suspensions of unimodal and bimodal distributions can be found in ref. 15.

Fortunately, simple particle size distributions (smooth, unimodal populations) for which $F(\Gamma)$ is approximately Gaussian in shape are common. For these cases (including many synthetic polymer distributions), the much simpler method of cumulants analysis (16) usually provides a good fit to the autocorrelation function data, yielding moments of the distribution $F(\Gamma)$. In this approach, $\ln g^{(1)}(t')$ (which for a monodisperse sample is a straight line) is fitted to a low-order polynomial (quadratic or cubic). For a third-order cumulants fit:

$$\ln g^{(1)}(t') = -\overline{\Gamma}t' + \left(\frac{1}{2!}\right)\mu_2 t'^2 - \left(\frac{1}{3!}\right)\mu_3 t'^3 \tag{8}$$

where $\overline{\Gamma}$ is the mean value of the decay distribution, and μ_m is the mth central moment of $F(\Gamma)$, defined as:

$$\mu_m = \int_0^\infty F(\Gamma)(\Gamma - \overline{\Gamma})^m \, d\Gamma \tag{9}$$

The mean diffusivity of the suspension is calculated from $\overline{\Gamma}$ with equation 4. The standard deviation of the distribution of diffusion coefficients can be calculated from μ_2. An average diameter corresponding to the mean diffusion coefficient can be calculated from equation 6, and an indication of the spread of the particle size distribution is given by μ_2. The advantage of the cumulants analysis is that it is computationally fast and settles rapidly with improving statistical accuracy in the autocorrelation function. This method gives very accurate results for decay distributions with negligible high-order central moments (*15*), as, for example, smooth, nearly symmetric, single-peak distributions.

Commercially available DLS instruments usually employ two approaches to convert the autocorrelation data to particle size: (1) the method of cumulants; and (2) an algorithm that attempts to invert equation 7, solve for $F(\Gamma)$, and yield an estimate of the full particle size distribution. The Nicomp 370 particle sizer computes distributions by using both of these approaches and selects one of the computed distributions on the basis of goodness-of-fit criteria (*17, 18*).

Suitability of DLS for On-Line Applications: Importance of Autodilution

The DLS technique for particle sizing contains a number of inherent advantages over other methods (e.g., optical turbidity) that make it ideally suited to automated, on-line applications. First, it is an absolute technique. The scattering wave vector K (equation 5), which connects the time scale of the intensity fluctuations with the particle diffusivity D_t, depends on three parameters, all of which are constant (for a given choice of solvent). The conversion of the computed mean diffusivity into a particle of radius R (equation 6) depends on two additional parameters that either are known or can be held constant (temperature T and solvent viscosity η). Hence, any well-designed DLS instrument should yield consistent, reproducible results over extended periods of time, requiring no calibration. Second, the measured particle diffusivity (and hence the calculated radius) is essentially independent of the concentration of the measured suspension, provided it is sufficiently dilute that multiple scattering and interparticle interactions (i.e., electrostatic repulsions for charged colloids) have no appreciable effect on the autocorrelation function. Finally, the particle diffusivity depends only on its size and is independent of composition (density, molecular weight, index of refraction, etc.). Although these physical properties will certainly influence the average scattered intensity, they will not affect the particle diffusivity.

Clearly, these three characteristics of the DLS technique make it ideally suited to an on-line measurement, in which sample acquisition, dispersion, and dilution must be performed automatically. The crucial new ingredient required for successful on-line particle size analysis using DLS instrumentation is a computer-controlled mechanism capable of automatically acquiring a quantity of concentrated suspension from a process stream or reaction vessel and diluting it to a final concentration that is optimal for the DLS instrument. That is, the concentration must be sufficiently low to avoid multiple scattering and interparticle interactions but large enough to yield an acceptable signal-to-noise ratio in the autocorrelation function after a relatively short time of data acquisition (typically, just several minutes).

The difficulty associated with the use of conventional dilution schemes (i.e., those employing fixed dilution factors) is that the optimal dilution factor varies greatly with the properties of the starting concentrated suspension. The average scattered intensity from a suspension is a strong function of the particle size and the particle size distribution. For example, in the Rayleigh regime (diameters less than 100 nm using a HeNe light source), the single-particle scattered intensity is a function of the 6th power of the particle diameter. The scattered intensity is also a strong function of the ratio of the refractive index of the particles to that of the medium, and a linear function of the particle concentration in the suspension. In practice, the strong dependence of the scattered intensity on these characteristics of the particle suspension requires that any automatic dilution system possess a very wide dynamic range. That is, it must be capable of achieving dilution factors ranging from less than $100:1$ to greater than $100,000:1$. Because the particle concentration and size distribution are, at worst, completely unknown, there is no a priori knowledge of the correct dilution factor appropriate for the DLS measurement. A fixed dilution may not be acceptable even for a known recipe (routine analysis), because the dilution factor changes during the reaction as the particle size changes.

Motivated by these tradeoffs and requirements, Nicoli and Elings recently developed a proprietary method (and associated apparatus), known as Autodilution (19), which can automatically dilute any starting concentrated particle suspension for delivery to a flow-through scattering cell. A simplified block diagram of a DLS instrument with Autodilution is shown in Figure 1. A small, arbitrary quantity of a concentrated particle suspension is introduced by a valve (either manually or electrically operated) into a mixing chamber. Filtered diluent flows continuously into the chamber where the starting sample is continuously diluted. The diluted sample passes through the scattering cell and into the drain. The main system computer (Motorola 68000) monitors the light-scattering intensity produced by the continuously diluted suspension and stops the dilution process and flow when the appropriate, preset scattering level is reached. After a short time delay to reach temperature equilibrium, data acquisition commences.

Together with the apparatus of Figure 1, on-line analysis also requires a remote sampler–prediluter device attached to the process pipe, holding tank, or reaction vessel. After a series of experiments, we arrived at the

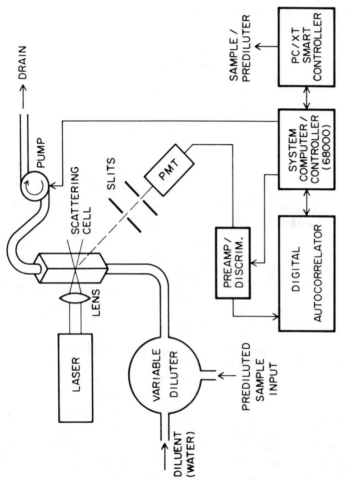

Figure 1. Simplified diagram of the DLS instrument with Autodilution.

configuration shown schematically in Figure 2. This simple apparatus consists of three one-way valves plus a check valve connected to a small central mixing chamber. The purpose of this accessory device is to capture an arbitrary small volume of fresh, concentrated-particle suspension (latex in our case) and predilute it to some arbitrary, but much lower, concentration suitable for delivery to the main Autodiluter in the DLS instrument (Figure 1). For this preliminary study we used a Nicomp model 370 submicrometer particle sizer with added Autodilution capability.

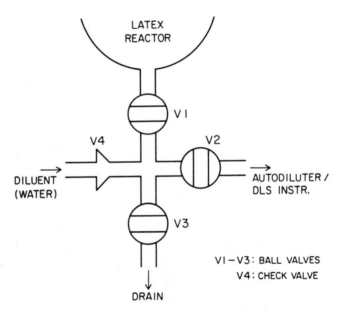

Figure 2. Configuration of the sampler–prediluter device used with the Autodilution–DLS instrument for on-line particle size measurements in a batch latex reactor.

The sampler–prediluter consisted of three pneumatically driven ball valves plus a check valve (1 psi) connected to a small manifold–mixing chamber (all parts are stainless steel). The valves were powered by compressed air (80 psi) and actuated remotely by electrically controlled solenoid valves. Filtered water (0.4-μm large-area Gelman filter) at 3–5 psi of pressure was the only other input requirement for the system, used for both predilution and final Autodilution of the captured latex emulsion sample. Air-driven valves were chosen because they are explosion proof and therefore meet the safety needs of a typical latex production facility. The valve assembly served to connect the three principal components of the system: (1) the pressurized pilot-scale batch reactor, (2) the DLS instrument, and (3) a source of filtered diluent.

The system was controlled by a PC XT computer, operating under MS DOS, with flexible software serving two main functions: (1) control of the external ball valves (via the electrical solenoid actuators) for the operations of sample acquisition, predilution, introduction into the Autodilution system of the DLS instrument, and flushing; and (2) an input–output device for serial communication with the DLS instrument (including data storage).

In fully automatic mode the sampling cycle commences by capturing a small quantity of concentrated sample from the latex reactor. Following a short predilution time, the partially diluted sample is then passed to the Autodiluter, where the dilution factor is allowed to increase continuously until the scattering intensity falls to a level appropriate for the digital autocorrelator and consistent with the considerations just discussed. After a predetermined delay to achieve temperature equilibration in the scattering cell, the diluted sample is analyzed by the DLS instrument. At a predetermined time the particle size distribution results are printed, the raw data are stored on a diskette, and the system is flushed with fresh diluent. The computer controller then awaits the preprogrammed start of the next measurement cycle. Cycle times of 15 min or less, where 7–8 min is allocated to data acquisition and analysis, are practical.

Off-Line Evaluation of the DLS Measurements: Potential for On-Line Applications

As should be evident from the preceding discussion, use of our DLS-based system with Autodilution for on-line latex particle size determination should yield results with accuracy and reproducibility comparable to those obtained in a "normal" off-line laboratory setting. That is, once a fresh latex sample has been captured and prediluted by the sampler–prediluter device of Figure 2, its treatment by the Autodilution–DLS instrument is identical to that occurring on a laboratory bench, where concentrated samples are introduced manually into the system.

Hence, it is useful to review our extensive experience in off-line analysis of a wide variety of latex samples using the Nicomp 370–Autodilution system. This review will serve to establish the potential of DLS in general, and our system in particular, for on-line latex particle size measurements. A brief presentation of this study is given here, and further details can be found in ref. 17. The method was evaluated for (1) ability to provide a reliable estimate of the particle size distribution in a short time, (2) consistency and ability to follow particle growth during the reaction, and (3) reproducibility.

A total of 650 samples were analyzed. These included a variety of latices [polystyrene, poly(styrene–butadiene), poly(methyl methacrylate), poly(butadiene–acrylonitrile), and poly(vinyl acetate)] with a variety of particle size distributions (bimodal, broad unimodal, and monodisperse) and covered a wide range of particle diameters (34 to 560 nm). Some of these latices were prepared with soap-free polymerizations and others with emulsifier present. Most of them were at conversions below 90% (and some at conversions as low as 10%). The presence of emulsifier and unreacted monomer in the particle suspension can potentially add undesired "noise" components to the DLS autocorrelation data and thus make deconvolution more difficult. These nonideal samples were chosen because they approximate the type of samples that would have to be analyzed by an on-line sizing instrument on a routine basis.

A Nicomp 370–Autodilution submicrometer particle sizer (Particle Sizing Systems, Santa Barbara, CA) was used for the off-line particle size determination. This instrument uses two different methods to convert the autocorrelation function data to particle size distribution: Gaussian analysis and Nicomp distribution analysis. The Gaussian analysis uses a second-order cumulants fit to the data, assuming a Gaussian distribution of decay constants, with higher order (≥ 3) central moments of the distribution equal to zero. A chi-squared (χ^2) fitting error parameter is used to test whether this assumption is reasonable (18). The analysis is a two-parameter fit, yielding a mean diffusivity and coefficient of variation (measure of the variance) of the distribution of the diffusion coefficients. The mean diffusivity is converted to an intensity-weighted mean diameter (D_{cum}). The distribution of diffusion coefficients is converted to a particle size distribution on an intensity, volume, or number basis, and the corresponding average diameters are calculated.

The Nicomp distribution analysis employs an algorithm based on a variation of Provencher's technique (7–9). This approach makes no assumption of the shape of the distribution; it is a nonlinear least-squares parameter estimation and requires longer times to settle because of its greater sensitivity to noise in the autocorrelation function.

A variety of polydisperse latices with known particle size distributions were analyzed with both of the techniques discussed. For unimodal distributions (broad and narrow), the Gaussian analysis gave a good estimate of the location of the main body of the true distribution on a weight (volume) basis, and a good estimate of the weight-average diameter (the estimated value was always within 8% of the true one). It should be kept in mind that the Gaussian assumption relates to the shape of the distribution of the decay constants (diffusion coefficients) on an intensity basis; depending on the particle size range covered by the distribution, the corresponding particle size distribution on a weight or number basis may be skewed. The mean diffusion coefficient estimated from the cumulants analysis was correct even for distributions where the Gaussian assumption does not hold (for example, bimodal distributions).

The weight-average diameter estimated from the Nicomp distribution analysis was sometimes very different from the true one (10–15% error). Furthermore, when the estimated distribution was overlaid with the true one, the larger particles were correctly estimated, but the small ones were not included. Detailed discussions and explanations for this behavior can be found in ref. 17. The Nicomp distribution analysis can detect some bimodal distributions (two populations of particles with significantly different diameters) in a short time, and this feature is useful when analyzing samples from processes where secondary nucleation may take place.

The results from the Gaussian analysis showed better reproducibility, and, as expected, settled faster than those from the Nicomp analysis. It was concluded that whenever the Gaussian assumption holds (indicated by a small χ^2 value), the Gaussian analysis can be used to obtain a reliable estimate of the weight-average diameter in a short time.

Finally, it was shown (17) that for routine analysis, D_{cum} (calculated from the mean diffusion coefficient), together with the coefficient of variation estimated from the Gaussian analysis, can be used successfully to monitor particle growth during latex production (both for monodisperse and polydisperse latex). In processes where secondary nucleation is likely to occur, the display form of the Nicomp analysis can be used, in parallel with the Gaussian analysis, to detect the presence of a second generation of particles.

The Nicomp 370 instrument collects scattered intensity data continuously, and a particle size distribution is estimated and displayed approximately every 45 s. To determine how fast the results from the Gaussian analysis settle, the intensity-average diameter (D_{cum}) obtained from this analysis was recorded as a function of time; this study was done for 217 samples, with a variety of distributions and values of D_{cum} ranging from 38 to 560 nm. The analysis time for each sample varied from 1 to 4 h. The estimate of the Gaussian analysis was considered to have settled when the fluctuations in the value of D_{cum} had a standard deviation less than or equal to 0.5% of their mean value. This mean value is referred to as settled D_{cum}.

Table I summarizes the settling history of these 217 samples; the samples are classified in columns according to the time that the value of D_{cum} was first recorded. Each column shows the percentage of samples (total number of samples is shown at the top) for which the value of D_{cum} recorded within the time range specified was within certain deviation of its final settled value.

Table I. Deviation of the Value of D_{cum} Recorded at Various Times with Respect to Its Settling Value for 217 Samples

Maximum Deviation from End Value	0.5–2 min, 78 samples	2–5 min, 50 samples	5–10 min, 47 samples	10–30 min, 42 samples
1.0	46	72	85	90
2.0	85	96	98	97
3.0	92	96	98	100
4.0	95	96	98	
5.0	97	98	100	
6.0	100	98		

NOTE: All values are given in percents.

For example, in the first column, for a total of 78 samples, a value of D_{cum} was first recorded at a time between 0.5 and 2 min; for 85% of those samples the diameter value recorded within that time slot deviated by less than 2% from the final settled value for that sample. Only the deviation of the first recorded value is listed. Clearly, D_{cum} settles very rapidly; after only 5 min of data acquisition, 96% of the samples measured yielded estimates of D_{cum} that were within 2% of the final settled values. After 10 or 15 min, no significant deviation of the measured value occurred with additional run time.

Second, 80 dispersions were tested for reproducibility. From each one of the original dispersions, two, three, or more diluted samples were prepared and run through the DLS instrument. The settled value of D_{cum} was recorded for each of the replicas. Table II gives the maximum difference observed between replicas of the same dispersion expressed as a percent of their mean value (the standard deviation for a set of replicas in each case is less than the maximum difference). In 91% of the cases, the maximum difference was 5%, and in 72% of the cases the maximum difference was less than 2%. The maximum difference observed between replicas of the same dispersion reflects the expected maximum experimental error due to preparation, dilution, and the presence of dirt particles in the suspension. The standard deviation for all the cases, expressed as percent of the mean diameter, was 1.08%. Tables I and II show that for a total data acquisition time of 5 min, the error in the estimated value of D_{cum} with respect to the final settled value is smaller than the reproducibility error.

Table II. Maximum Difference between Replicas of D_{cum} with Respect to Their Mean Value

Maximum Percent Deviation	Cumulative Number of Dispersions	Cumulative Percent
0.5	22	28
1.0	38	47
2.0	58	72
3.0	68	85
4.0	70	87
5.0	73	91
7.5	80	100

NOTE: 40 dispersions were duplicated; 27 were triplicated; and 13 had 4, 5, 6, or 7 repeats.

The latex samples used to study the behavior of D_{cum} included both monodisperse and polydisperse suspensions. The results just presented reflect the capability of the cumulants analysis to provide a reproducible estimate of a certain property of the suspension, namely D_{cum}, in a very short time. D_{cum} is the best estimate of particle size for monodisperse suspensions. For polydisperse suspensions, D_{cum} corresponds to an average of the particle size distribution. This average diameter depends on the size range covered by the distribution and the refractive index of the particles; for Rayleigh scatterers, D_{cum} corresponds to the D_{65} average (17), defined as the ratio of the sixth to the fifth moment of the particle size distribution.

As discussed earlier, a more meaningful average (the weight-average diameter) can be easily obtained when the Gaussian assumption holds. The behavior of the weight-average diameter estimated by the Gaussian analysis for broad distributions was also studied (17). For 90% of the suspensions studied, the maximum deviation observed between replicas was 5% of their mean value. The weight-average diameter also settles relatively fast; with a total of 5 min of data acquisition, the weight-average diameter is within 2 to 4% of its mean settled value, a deviation smaller than the reproducibility error.

The ability of the DLS method to follow latex particle growth during a reaction is shown in Figure 3. Particle diameters obtained from the Gaussian analysis (D_{cum}) are plotted as a function of the reaction time for two soap-free polymerizations of vinyl acetate (runs H21 and H22) and two polymerizations with low soap concentrations (runs H24 and H23). Latex samples were withdrawn from the reactor every 5 min; the off-line DLS system successfully detected the change in particle diameter that occurred within 2 h. These four runs have the same basic recipes and belong to a group where the effects of soap and impurities on particle nucleation were studied in a factorial design. The effects on particle size can be easily ascertained from the DLS results shown in Figure 3. The effect of impurities was studied for two pairs of runs: H21–H22 and H23–H24. The only difference between

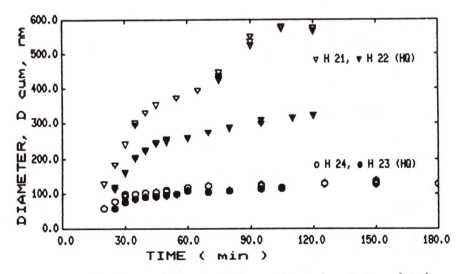

Figure 3. Particle growth histories for four emulsion polymerizations of vinyl acetate. Particle size was estimated with DLS.

the polymerization recipes of runs H21 and H22 is that very small amounts (~10 ppm) of water-soluble impurities were added in H22; these small amounts resulted in a significant decrease in particle size. [The results from off-line DLS were corroborated with results from electron microscopy (17)].

The impurities had the same effect in the runs with low soap concentration; the reacting mixture of run H23 had impurities in it, but run H24 with the same recipe was impurity free. Again, the presence of impurities caused a measurable decrease in particle size, but considerably smaller than in the soap-free cases. [The absolute decrease of 8–10 nm is significantly smaller than the 100–250 nm observed in the soap-free case. However, the relative decrease is significant (~10%) and indicates a 30% increase in the number of nucleated particles in H23 compared to H24 (17).] The effect of soap was studied for pairs H21–H24 and H22–H23. Latexes produced by soap-free polymerizations are expected to have much larger particles than those produced when soap is added to the reacting mixture. This result can be clearly seen in Figure 3. Runs H21 and H24 have the same polymerization recipe, but H21 is soap free. The particles produced in this run are much larger than those in H24. Similarly, the soap-free run H22 results in larger particles than run H23.

In these cases, the latex produced was nearly monodisperse, and the D_{cum} diameter was used to follow particle growth. Numerous examples of the use of weight-average diameter to follow particle growth in processes where polydisperse latex was produced are shown elsewhere (17); the results were obtained by using the Gaussian analysis of a Nicomp 370 for latices produced in a continuous stirred tank reactor or in a train of reactors.

The studies summarized showed that dynamic light scattering can provide reliable estimates of particle size in polydisperse and monodisperse latex samples in less than 5–10 min at any level of conversion. The method is consistent, reproducible, fast, needs no calibration, and therefore has an excellent potential for on-line applications.

On-Line Application of DLS: Results and Discussion

One of the primary concerns when dealing with latex measurements is whether the fluidics system of the instrument will become clogged after a certain time. Before being connected to a latex reactor, the sampler–prediluter apparatus was tested for reproducibility, stability, and ability to deliver samples to the Autodilution–DLS instrument over a prolonged time without plugging. A quantity of a poly(vinyl acetate) latex was placed in a holding tank pressurized with air, and the tank was connected to the automatic sampling–prediluter apparatus. The sampling system was then run in automatic mode to simulate an actual on-line measurement. Samples were automatically drawn from the tank every 15 min and analyzed with the system. The time for the on-line data acquisition was 5 min. The results obtained over 5 h of continuous sampling are shown in Figure 4. The reproducibility was excellent (maximum deviation observed between two extreme values, with respect to the mean value, was 3%), and no plugging occurred. The mean value of the on-line estimates of the intensity-average diameter was equal to the value obtained off-line with a data acquisition time of 40 min.

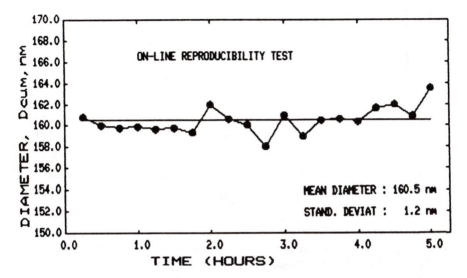

Figure 4. A preliminary on-line test with continuous sampling for 5 h. Latex samples were acquired automatically from a holding tank and analyzed in a DLS instrument.

The sampler–prediluter apparatus was then connected to a pilot-scale batch reactor [stainless steel jacketed, with temperature control (Chemineer, Dayton, OH)] to monitor particle growth during the emulsion polymerization of vinyl acetate. In these preliminary runs, the initiator was added to the reactor together with the other ingredients, and then the temperature of the reacting mixture was brought to 60 °C. The reaction time (shown on the abscissas of Figures 5 and 6) is equal to the total elapsed time following the charge of the initiator in the reactor. All the reactions were carried out under a nitrogen blanket (10 psi). Hydroquinone was added to the diluent to inhibit the reaction once the samples were acquired.

Figure 5 shows our first attempt to follow particle growth on-line during latex production. Intensity-weighted average diameters, obtained from the Gaussian analysis, are plotted as a function of the reaction time. Results are shown from both on-line and off-line measurements. The off-line measurements were performed on latex samples withdrawn at the end of the reaction. The time allocated for data acquisition was 7 min for the on-line measurements and more than 30 min for the off-line measurements. After a reaction time of 80 min the conversion had reached 100%. Therefore the last two

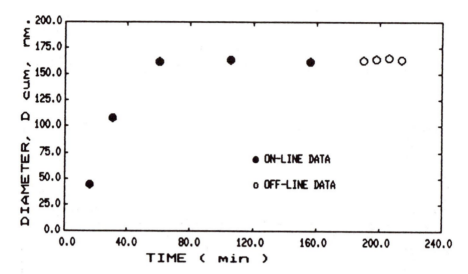

Figure 5. On-line particle size determination during an emulsion polymerization (Run A1) in a pilot-plant reactor. The sampling cycle was initiated manually.

on-line samples and the four off-line samples are essentially replicas of the same dispersion. There seem to be no significant differences between the on-line and the off-line estimates.

A complete cycle involves sampling, predilution, Autodilution, temperature equilibration, measurement, printout of the results, storage of the raw data, and flushing–cleaning the cell and the sampling valves. In this first run, the sampling cycle was initiated manually; parameters such as length of the predilution time, flushing time, and total cycle time were reset after each cycle to determine the optimal duration for these functions and decide on other parameters such as diluent flow rate.

Figure 6 shows results from a run where the complete cycle (sampling, predilution, Autodilution, measurement, flushing) was carried out automatically every 17 min for 4 h; 5 min of this cycle was devoted to collection of light-scattering data, and the remainder was dedicated to sample acquisition, Autodilution, temperature equilibration, and system flushing. For this run the reacting mixture was not degassed at the outset; consequently the presence of oxygen (an inhibitor in emulsion polymerization) resulted in a long induction period. On-line dynamic light scattering successfully followed particle growth during the reaction and detected the start and end of the reaction. More samples were analyzed at the end of the reaction both on-line and off-line to test the reproducibility and consistency of the results.

The results obtained after the reaction had reached 100% conversion are replotted in Figure 7, where the fluctuations in the particle size measurement can be seen in more detail. The off-line results showed excellent agreement with the on-line estimates, with no statistical difference between

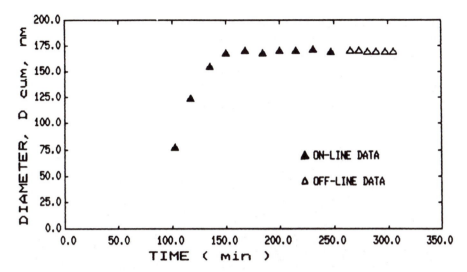

Figure 6. On-line measurements with automatic sampling during latex production (Run B2). The sample cycle was 17 min.

the means of these two groups. The maximum deviations observed in the
reproducibility test were less than 2% for the on-line estimates and 1% for
the off-line ones. As discussed earlier in this text, our off-line evaluation had
shown that an accurate estimate of D_{cum} can be obtained with 5–10 min of
data acquisition. We now show that on-line estimates obtained with 5 min

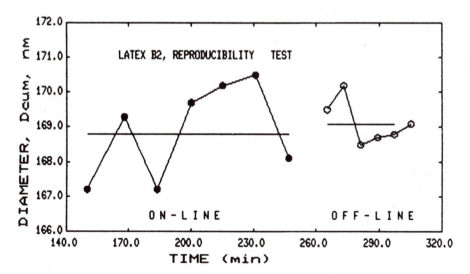

Figure 7. Reproducibility tests for the on-line and off-line results of latex B2
at 100% conversion.

of data acquisition were in excellent agreement with the off-line results of the same dispersion, obtained with longer times of data acquisition. It was originally expected that the on-line data would show a higher variation in the reproducibility tests because of the presence of noise sources (i.e., dust) that can be expected to be more significant in a pilot-plant environment. However, this variation did not seem to be significantly different from our off-line observations.

Conclusion

An on-line dynamic light-scattering instrument used in conjunction with a proper sampling–prediluting system is capable of producing stable, reliable estimates of the particle size distribution in a total elapsed time of 10–20 min. This time includes sample acquisition and dilution, temperature equilibration, autocorrelation buildup and analysis, and system flushing. The amount of time that must be allocated to data collection depends on the complexity of the particle size distribution and the degree of accuracy and the size resolution required. The results presented were obtained by using a total cycle time of 17 min, of which approximately 5 min was devoted to data acquisition. For this poly(vinyl acetate) latex the Gaussian analysis was sufficient to obtain an accurate reproducible characterization of the particle size distribution during all stages of the emulsion polymerization reaction.

In continuous polymerization reactions, more complex distributions may be present, and D_{cum} may not be sufficient to describe the particle size distribution or to follow the particle growth history; a more meaningful average may be required. A fast estimate of the weight-average diameter may be obtained by using the Gaussian assumption. Longer analysis times (20–30 min) are required for complex distributions where the Gaussian assumption is not valid, and the longer analysis times require a longer sampling cycle. Fortunately, the more difficult distributions come from continuous reactors. Reactors with long residence times allow longer times of data acquisition; therefore, any of the algorithms that provide the full particle size distribution may be used (in the Nicomp instrument, the Nicomp distribution analysis).

Another alternative would be to use the results from the Nicomp analysis in combination with results from the cumulants analysis. For example, for routine purposes, to monitor and control a known process, D_{cum} and the coefficient of variation (an indication of the spread of the distribution, given by the Gaussian analysis) may be followed with some type of calibration. The display from the Nicomp analysis can be used to detect any sudden changes, as, for example, a secondary nucleation. An example where the weight average from the Nicomp analysis together with D_{cum} and the coefficient of variation from the Gaussian analysis could be used to follow the changes in an unknown process is discussed in ref. 17 for a polystyrene emulsion polymerization in a continuous reactor.

References

1. Chu, B. *Laser Light Scattering;* Academic: New York, 1974.
2. *Dynamic Light Scattering: Applications of Photon Correlation Spectroscopy;* Pecora, R., Ed.; Plenum: New York, 1985.
3. Berne, B. J.; Pecora, R. *Dynamic Light Scattering;* Wiley: New York, 1976.
4. *Measurement of Suspended Particles by Quasi-Elastic Light Scattering;* Dahneke, B. E., Ed.; Wiley Interscience: New York, 1983.
5. Pike, E. R. In *Scattering Techniques Applied to Supramolecular and Non-Equilibrium Systems;* Chen, S. H.; Nossal, R.; Chu, B., Eds.; Plenum: New York, 1981.
6. Ostrowsky, N.; Sornette, D.; Parker, P.; Pike, E. R. *Opt. Acta* **1981**, *28*, 1059.
7. Provencher, S. W.; Hendrix, J.; De Maeyer, L. *J. Chem. Phys.* **1978**, *69*, 4273.
8. Provencher, S. W. *Makromol. Chem.* **1979**, *180*, 201.
9. Provencher, S. W. *Comput. Phys. Commun.* **1982**, *27*, 213–227, 229–242.
10. Morrison, I. D.; Grabowski, E. F.; Herb, C. A. *Langmuir* **1985**, *1*, 496–501.
11. Herb, C. A.; Berger, E. J.; Chang, K.; Morrison, I. D.; Grabowski, E. F. In *Particle Size Distribution: Assessment and Characterization;* Provder, T., Ed.; ACS Symposium Series No. 332; American Chemical Society: Washington, DC, 1987; pp 89–104.
12. Gulari, E.; Gulari, E.; Tsunashima, Y.; Chu, B. *J. Chem. Phys.* **1979**, *70*, 3965.
13. Gulari, E.; Gulari, E.; Tsunashima, Y.; Chu, B. *Polymer* **1979**, *20*, 347.
14. Bott, S. E. In *Particle Size Distribution: Assessment and Characterization;* Provder, T., Ed.; ACS Symposium Series No. 332; American Chemical Society: Washington, DC, 1987; pp 74–88.
15. Stock, R. S.; Ray, W. H. *J. Polym. Sci. Polym. Phys. Ed.* **1985**, *23*, 1393.
16. Koppel, D. E. *J. Chem. Phys.* **1972**, *57*, 4814.
17. Kourti, T., Ph.D. Thesis, "Polymer Latexes: Production by Homogeneous Nucleation and Methods for Particle Size Determination", McMaster University, Hamilton, Ontario, Canada, 1989.
18. Nicoli, D. F. In "Nicomp 370 Submicron Particle Sizer, Version 5.0, Instruction Manual"; Pacific Scientific, Instruments Division, Silver Spring, MD, 1987.
19. Nicoli, D. F.; Elings, U. B., U.S. Patent 4 794 806, January 3, 1989, "Automatic dilution system"; foreign patents pending.

RECEIVED at ACS for review February 14, 1989. Submitted November 10, 1988. ACCEPTED revised manuscript November 6, 1989.

Pressure Bonding and Coherency of Plastic and Fragmentary Materials

N. G. Stanley-Wood and A. M. Abdelkarim

Department of Chemical Engineering, University of Bradford, Bradford BD7 IDP, West Yorkshire, United Kingdom

The manufacture of plastics, now in particulate or granulated forms, for subsequent densification into geometrical shapes by pressure has necessitated the exploration of the physicomechanical behavioral effects of compression. Two categories of powders (plastic and fragmentary) were uniaxially compressed in a cylindrical mold, and the mechanisms of pressure bonding and volume reduction were continuously monitored in terms of volume change with axial stress. Two plastic powders (polypropylene and polystyrene) under high axial stress (450 MPa) did not produce coherent compacts. Polyethylene and sodium chloride (which has plastic properties) readily formed compacts in the stress range 30–250 MPa. The variation of stress with time and the degree of compact coherency of these plastics were compared with the physicomechanical properties, surface hardness, and diametral strength of fragmentary materials (dicalcium phosphate and sugar) to predict the pressure bonding and strength of various densified plastic polymers.

T HE RHEOLOGICAL BEHAVIOR OF POLYMER PARTICLES and powders has received scant attention in the scientific literature, although much work has been done on the behavior of solid polymers and polymer melts (1, 2). The properties of polymer resins depend mainly upon the characteristics of individual regularly or irregularly shaped particles. Schwaegerle (3) showed that the particle size, particle size distribution, and particle shape of poly(vinyl chloride) (PVC) resins are affected by the manufacturing process and influence the bulk properties, flow, and compactibility of polymers. The

0065–2393/90/0227–0125$06.00/0
© 1990 American Chemical Society

degree of compactibility achieved by these PVC resins was obtained at low stresses and expressed as the density difference between a tap density and an apparent bulk density.

Cleereman et al. (4) stated that general-purpose unoriented atactic polystyrene is brittle, but anisotropic uniaxially oriented polystyrene has a high tensile strength and ductility in the direction of orientation. The ductility and strength of anisotropic polystyrene changes, however, when stress is applied in the transverse direction, which can ultimately lead to more "crazing" than that seen with atactic polystyrene. Uncontrolled orientation has a beneficial effect only by serendipity (5), atactically oriented polymers normally being a source of weakness and failure (6). Multilayered and multiplastic plastics have been used to achieve either reinforced mechanical strength or flexibility. Nowadays multilayered composites are being superseded by designed particulate and multiparticulate compacted composites to promote toughness, chemical resistance, and predictive surface and matrix characteristics (2).

Bhateja et al. (7) illustrated the application of pressure compaction fusing, as opposed to heat fusing, of toner powders in the image copying or printing process. They developed two tests to provide basic rheological information about pressure fusing of plastic toner powders. A compaction test was used to ascertain the resistance of plastics or polymers to plastic pressure deformation; after uniaxial stress application, a relaxation analysis gave a measure of the recovery of particles to regain both their initial shape and elastic energy. A monitored compaction–relaxation procedure was used in the investigation reported in this chapter to characterize the rheological behavior of a number of powders that could or could not form coherent compacts when uniaxially compressed in an instrumented mold.

Experimental Details

Powders.   Two categories of powders were chosen, one that had a rheological plastic behavior and a second that was fragmentary in nature when uniaxially compacted. Powders within the plastic behavior category had an additional subdivision into those plastic polymers that could form coherent uniaxially compressed compacts and those that were noncompactible.

Plastic Rheological Behavior.   *Noncompactible Polymers.*   The two noncompactible polymer powders investigated were a spherical low-density-grade polypropylene (homopolymer) supplied by I.C.I. Plastics Division, Welwyn Garden City, Herfordshire, U.K., which had an ASTM D 695 compressive strength in the range 59–69 MPa, and a conventional beaded polystyrene supplied by Shell Chemicals U.K. Ltd. (Styrocell), which has an ASTM D 695 compressive strength in the range 80–110 MPa. The polypropylene had a mean particle size (by sieve analysis) of 950 μm and a density (by air pycnometry, Beckman model 930, Glenrothes, Fife) of 0.922×10^3 kg/m^3.

Compactible Powders.   Coherent compacts that could subsequently be tested for hardness and fracture strength were produced from a low-density general-purpose polyethylene with no additives (I.C.I. Plastics Division 18-003GA), which had an ASTM 1248 yield stress of 10 MPa and an ultimate stress of 18 MPa. Sodium chloride that was a white cubic crystalline powder (Reynolds and Branson, Leeds, U.K.), which has plastic rheological behavior when compressed (*8, 9*), also produced compacts. The density of sodium chloride as determined by air pycnometry was 2.16×10^3 kg/m^3, and the particle size was 250–300 μm when sieved in accord with British Standard (B.S.) 1791 (1976).

Fragmentary Powders.   Coherent compacts could be prepared from dicalcium phosphate dihydrate and granulated sugar when uniaxially compressed. Dicalcium phosphate dihydrate (DCP) is a white crystalline water-insoluble powder with a mean particle size, as determined by photosedimentation, of 12.8 μm and a size range of 5.3–27.5 μm. The density of uncompacted dicalcium phosphate measured by an air compression pycnometer was 2.40×10^3 kg/m^3. Dicalcium phosphate is a brittle, fragmenting material when compacted (*8, 9*). Granulated sugar was supplied by Tate and Lyle.

Instrumented Punch-and-Die Assembly.   A specially designed punch-and-die assembly, which consisted essentially of a stainless steel tool 20-mm diameter die and upper and lower punches hardened after fabrication, was manufactured by A.B. Hobley Ltd., Bradford, U.K. The die rested on a Kistler piezoelectric force transducer placed within the die support body, and measured the axial force (F_d) on the die. Around the circumference of the die, a horizontally split collar was fastened. Onto this collar a piezoelectric force transducer was fastened; it measured the force transmitted radially (F_R) to the die wall. The lower punch was mounted directly onto another force transducer to measure the lower punch force (F_L). The upper punch made a sliding fit into the die. The broader head of the upper punch contained a force transducer to measure the applied uniaxial force (F_a) and a base onto which to clamp a linear variable dispacement transducer (LVDT). Force was applied uniaxially by positioning the whole assembly between the platens of a hydraulically operated Denison T 42B3 tensile test machine (Denison Press Ltd., Leeds, U.K.). The axial mean stress (σ_A) was calculated from

$$\sigma_A = \frac{F_a}{\text{cross-sectional area of die}}$$

and the radial mean stress (σ_R) was calculated from the relationship

$$\sigma_R = \frac{F_R}{\text{perimeter of die} \times \text{compact height}}$$

The piezoelectric force transducers were connected to calibrated charge amplifiers (Kistler Ltd., Winterthur, Switzerland), which converted the electrostatic charge to a dc (direct current) voltage that was subsequently recorded on a Digitronix data logger.

Compacted Powders.   Individual compactions were prepared by placing twice the powder density weight into a lubricated stainless steel 20.0-mm i.d. die. The lubrication used for the punches and die prior to each compaction was a 0.1% (m/m) solution of magnesium stearate in acetone. Time was allowed for the acetone to evaporate from the die surfaces. As uniaxial force was continuously applied to the upper punch, the millivolt output from the force transducers (Kistler Ltd., types 906B and 903A), located on the upper and lower punches and on the radial collar around the instrument die, was recorded onto a cassette tape in a Digitronix super 8 data logger (Digitronix Ltd., U.K., Milton Keynes). Simultaneously with the millivolt force recordings, the height of the upper punch and compact was measured by a linear variable displacement transducer (type D401, Electro-Mechanisms Ltd., Slough, U.K.) and recorded onto the cassette tape. From previous calibration curves of the Kistler force transducers and LVDT, the radial stress σ_R, normal stress σ_A, and compact volume were computed from a FORTRAN program run on an ICL 1900S (10) together with a function of shear $(\sigma_A - \sigma_R)$ and mean normal stress $(\sigma_A + \sigma_R)$.

Mechanical Strength of Compacts.   To ascertain the strength of the compacts produced from polyethylene, sugar, DCP, and sodium chloride, three or four compacts from each powder and each compaction pressure were subjected to a diametral crushing test and to hardness tests.

Diametral Fracture Test.   The measured compact was placed diametrically between the upper and lower platens of a Denison press (Denison Ltd., Leeds, U.K.). The upper punch moved at a constant strain rate (0.4 in reciprocal minutes) to increase the applied force on the circumference of the compact. The fracture force was recorded when the compact underwent catastrophic cleavage. The fracture stress, σ_f, was calculated from B.S. 1881, Part 4 (1970)

$$\sigma_f = \frac{2P}{\pi dL} \quad \text{in N/mm}^2$$

where P is the recorded fracture force (N), d is the compact diameter (mm), and L is the compact height or thickness (mm).

Hardness Tests.   A Vickers hardness test (H_V) and a Brinell hardness test (H_B) were performed on compacts in accord with B.S. 427, Part I (1961) and B.S. 240, Part I (1962), respectively.

Results and Discussion

Compaction Profiles—V versus ln σ_c.   Figures 1–3 show the relationship between the volume (V) of a mass of polymer or powder in a mold and a uniaxial stress, expressed as the logarithmic applied stress $(\ln \sigma_c)$ as it is applied and subsequently released, allowing the compressed or compacted material to expand or relax axially within the die.

Materials that did not form coherent compacts (polypropylene and polystyrene) showed a unified compression and relaxation curve at high compression stresses (200–400 MPa). This curve is a characteristic of plastic deforming material and is also seen with coherent-compact-forming plastic behavioral materials like polyethylene and sodium chloride. The relaxation curves of

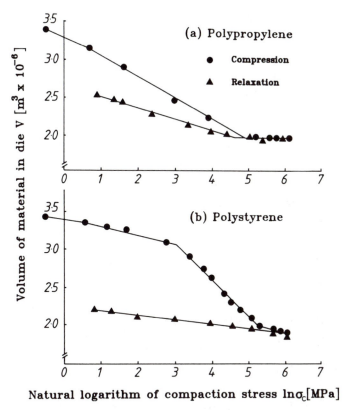

Figure 1. *Volume of powders undergoing compression (V) versus natural logarithm of compression stress (ln σ_c) for noncoherent compacts.*

polypropylene and polystyrene remain relatively close to their respective compression curves, a result indicating that a certain amount of elastic behavior remains in these particulate materials.

In engineering terms, the area beneath a pressure–volume curve is a work function. In these compaction profiles the range of applied uniaxial stress has to be expressed on a logarithmic scale. The area between the compression–compaction line and the relaxation line may therefore be regarded and defined as "logarithmic work". This area gives an indication of the amount of work retained in a coherent or noncoherent compressed mass.

The compaction profiles of polyethylene and sodium chloride, which are of a plastically deforming nature (8, 9), are intermediate between polypropylene and polystyrene, although atactic general-purpose polystyrene has brittle behavior (4).

Particles of dicalcium phosphate and sugar are, when compacted, fragmentary in nature because of the increase in powder surface area when

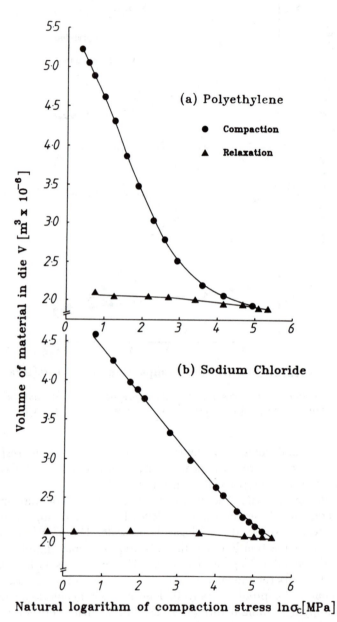

Figure 2. Volume of powders undergoing compression (V) versus natural logarithm of compression stress (ln σ$_c$) for plastic material producing coherent
compacts.

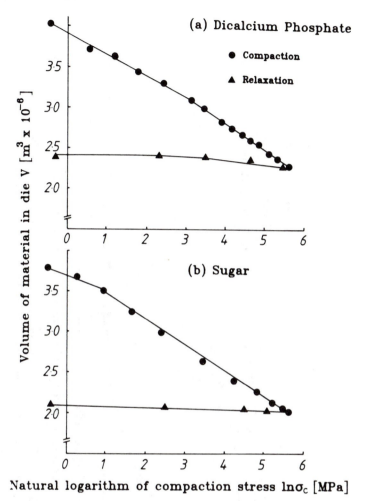

Figure 3. Volume of powders undergoing compression (V) versus natural logarithm of compression stress (ln σ_c) for fragmentary material producing coherent compacts.

subjected to uniaxial compression (*10–13*). These materials show no curvature in the applied compression stress line (Figure 3), and they undergo little relaxation within the die when the applied stress is released.

Differentiation of these various compaction profiles is shown in Figure 4, which summarizes both the uniaxial compression and relaxation behavior of these, plastic, and fragmentary materials.

Compaction Profiles—σ_c versus Time.   Figures 5–7 show the compression–compaction profiles and relaxation rate of applied stress on the six materials investigated. The analysis of a time–displacement profile in

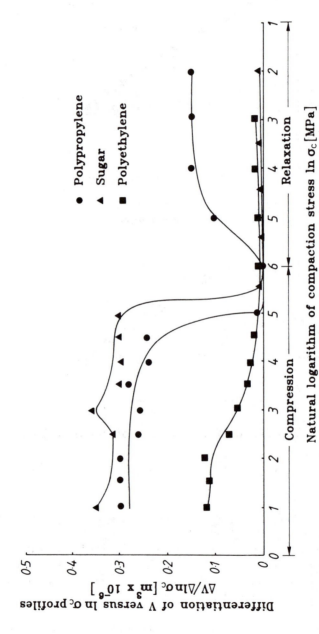

Figure 4. Differentiation $(\Delta V / \Delta \ln \sigma_c)$ of the compaction V versus $\ln \sigma_c$ curves for polypropylene, polyethylene, and sugar for compaction and relaxation.

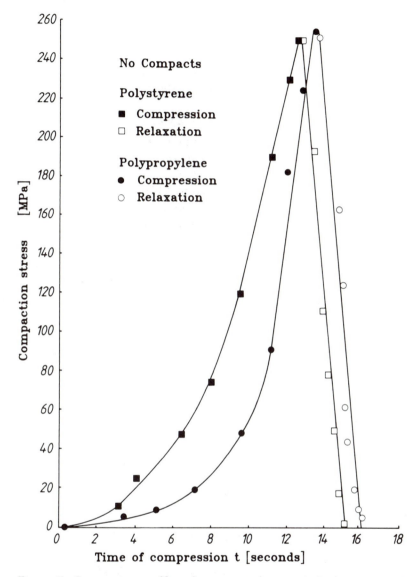

Figure 5. Compaction profiles of compression stress (σ_c) versus time of compression (t) for polypropylene and polystyrene.

pharmaceutical tableting cycles was used to identify the lamination of compacts due to formulation differences (*14*). Sophisticated computer-controlled compaction simulators can be used to evaluate stress rates, relaxation times, and ejection rates of compacted materials (*15*).

Heistand and co-workers (*16*) showed that shear deformation occurred during the decompression or relaxation stage, and this shear deformation

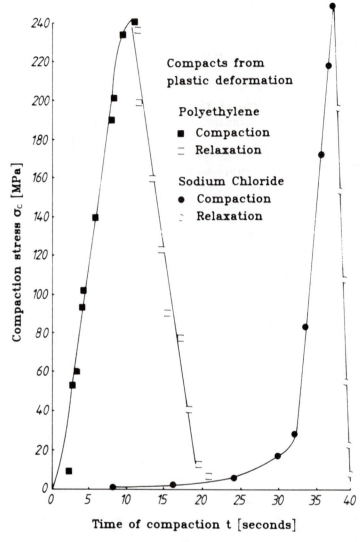

Figure 6. Compaction profiles of compaction stress (σ_c) versus time of compaction (t) for polyethylene and sodium chloride.

was responsible for the coherency of the material. Compactibility and coherency of compacts could thus be related to the stress relief of compacted materials by plastic deformation.

Rees and Shotton (17) observed a doubling of diametral strength and increase in interparticle bonding of sodium chloride in the first hour after ejection. This result was attributed to a continuing deformation and stress relief of materials. Measurement of the relaxation rate in terms of displacement or stress relief within the die can then be used to correlate the physicomechanical behavior of ejected compacts (*see* the following section). In

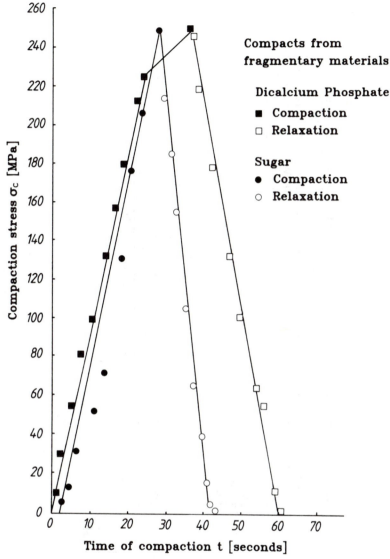

Figure 7. Compaction profiles of compaction stress (σ_c) versus time of compaction (t) for dicalcium phosphate and sugar.

Table I are given the mechanical properties of compacts, the compaction process logarithmic work (W_L), and the relaxation rates (RR) of all six materials investigated.

Explanation and Correlation of Compactibility and Compact Diametral Fracture Stress with Plastic and Fragmentary Material. The rheological data, in terms of uniaxial volume change with axial stress and the rate of both axial compression and relaxation, can be obtained

Table I. Mechanical Properties of Compacts, Logarithmic Work, and Relaxation Rates of Materials Investigated

Powder	Compaction Stress (MPa)	Logarithmic Work[a] (m³)	Relaxation Rate (MPa s⁻¹)	Diametral Fractural Stress (MPa)	Ratio RR/W_L[b] (MPa s⁻¹ m⁻³)	Power of Compaction[c] (J s⁻¹)
Polypropylene						
Homopolymer	249.6	84.5	—[d]	0	—	—
	206.0	70.5	—	0	—	—
Copolymer	256.0	100.5	107.1	0	1.06	1.07
	442.8	102.0	—	0	—	—
Polystyrene	405.5	—	101.4	0	0.618	1.66
Polyethylene	243.5	207	90.0	0.9–1.2	0.434	1.86
Sodium chloride	243.8	464	26.6	2.97–3.40	0.057	1.23
Dicalcium phosphate	248.9	249	10.3	2.89–3.20	0.041	0.256
Granulated sugar	250.8	323	17.6	1.06[e]	0.055	0.568

[a] Logarithmic work (W_L) is the area between the compaction and relaxation curves. All values are $\times 10^{-6}$.

[b] All values are $\times 10^{6}$.

[c] Power of compaction is $W_L \times RR$. All values are $\times 10^{-2}$.

[d] — indicates not available.

[e] Laminated.

directly from the uniaxial compression of powders. Therefore, an assessment of the behavioral strength of interparticle binding was obtained. Observation and differentiation of the volume V versus ln σ_c compaction profiles could possibly predict the degree of binding possible with various polymeric particles. Figure 8 shows the relationship between the interparticulate binding strength (σ_f) and the relaxation rate (RR) of both plastic polymers (the plastic-deforming and fragmentary-producing materials). The slower the relaxation rate, the greater the possibility of forming a coherent compacted mass regardless of the type of material. Figure 9 and Table I show that the ratio of relaxation rate to logarithmic work divides the six materials investigated into two groups: one group has a high relaxation rate and low amount of logarithmic work, and the other group has a low relaxation rate and a high amount of logarithmic work. Plastic polymers (polypropylene, polystyrene, and polyethylene) are in the first group, and crystalline material (sodium chloride and dicalcium phosphate) are in the second group.

Variation in the degree of crystallinity of polymers can therefore have a profound influence on the binding and compactibility of polymers. Heat fusing can induce the risk of recrystallization and create a disadvantageous effect, but pressure fusing would have little or no such risk. A measurement of the power of a uniaxial compaction process (Table I) in which logarithmic work is multiplied by the relaxation rate (megapascals per second multiplied

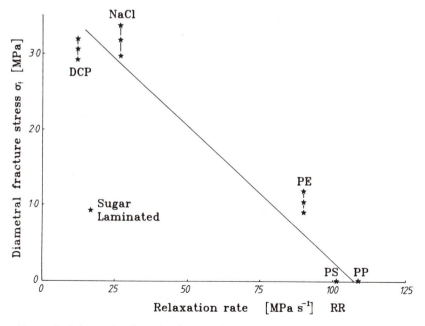

Figure 8. Relationship between diametral fracture stress of coherent and non-coherent compacts (σ_f) versus relaxation rate (RR) obtained from Figures 5–7.

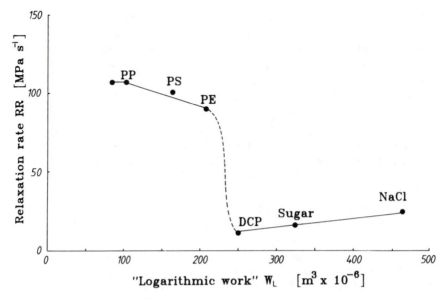

Figure 9. Correlation between relaxation rate of coherent and noncoherent compacts and logarithmic work (W_L) obtained from the V versus ln σ_c compaction profiles.

by cubic meters is identical with Joules per second) shows that for plastic-deforming material the numerical value is in the range 1–2×10^4 J/s, and for crystalline material the power is less than 1.0×10^4 J/s.

Conclusion

The characterization of particles and particulate polymers in the complex field of uniaxial compaction has shown that when the compression and relaxation uniaxial profiles of a V versus ln σ_c relationship coincide at high stresses, the material undergoes deformation and volume change, but no bonding between particles occurs. The rate of relaxation within a pressure die gives, with the materials investigated, a linear relationship with diametral fracture strength. The uniaxial relaxation rate together with the logarithmic work from the V versus ln σ_c profiles discriminates between plastic polymeric materials and crystalline materials.

References

1. Hegazy, A. A.; Blayinznski, T. Z. *J. Mat. Sci.* **1987**, *22*, 3321–3327.
2. Alfrey, T. *Technological Aspects of the Mechanical Behavior of Polymers;* American Chemical Society Applied Polymer Symposium No. 24, Detroit, MI, May 1973; Boyer, R. F., Ed.; Wiley–Interscience: New York, 1974.
3. Schwaegerle, P. R. *J. Vinyl Technol.* **1985**, *7(1)*, 16–21.

4. Cleereman, K. J.; Karam, H. J.; Williams, J. L. *Mod. Plast.* **1987**, *30* (*153*), 119.
5. Broutman, L. J.; McGarry, F. J. *J. Appl. Polym. Sci.* **1965**, *9*, 609.
6. Kresser, T. O. *Polypropylene;* Reinhold: New York, 1960.
7. Bhateja, S. K.; Gilbert, J. R.; Andrews, E. H. *J. Appl. Polym. Sci.* **1987**, *33*, 2305–2316.
8. Khan, K. A.; Rhodes, C. T. *J. Pharm. Sci.* **1965**, *64*, 445.
9. de Boer, A. H.; Bokhuis, G. K.; Lerk, C. F. *Powder Technol.* **1978**, *20*, 75.
10. Stanley-Wood, N. G.; Abdelkarim, A. M. *J. Powder Metall. Int.* **1982**, *14*, 135.
11. Stanley-Wood, N. G.; Abdelkarim, A. M. *Proceedings of the RILEM/CNR International Symposium on Principles and Applications of Pore Structural Characterization, Milan 1983;* Haynes, J. M.; Rossi-Doria, P., Eds.; Arrowsmith: Bristol.
12. Stanley-Wood, N. G.; Johansson, M. E. *Analyst,* **1980**, *105*, 1104.
13. Stanley-Wood, N. G. *Size Enlargement and Compaction of Particulate Solids;* Butterworths: London, 1983.
14. Mann, S. C. *Acta Pharma. Succia.* **1987**, *24*, 54–55.
15. Drew, P.; Sisson, M. L.; Atkin, G. J. *Symposium on Tablet Technology, Stockholm, February 1987;* Swedish Academy of Pharmaceutical Sciences, Apotekarsocieteten: Stockholm, Sweden.
16. Heistand, E. N.; Wells, J. E.; Peot, C. B.; Ochs, J. F. *J. Pharm. Sci.* **1977**, *66*, 511.
17. Rees, J. E.; Shotton, E. *J. Pharm. Pharmacol.* **1970**, *22*, 17S.

RECEIVED for review February 14, 1989. ACCEPTED revised manuscript December 19, 1989.

DYNAMIC MECHANICAL ANALYSIS AND RHEOLOGY

9

The Glass Transition Temperature as a Parameter for Monitoring the Isothermal Cure of an Amine-Cured Epoxy System

Guy Wisanrakkit and John K. Gillham*

Polymer Materials Program, Department of Chemical Engineering, Princeton University, Princeton, NJ 08544

Torsional braid analysis (TBA) and differential scanning calorimetry (DSC) were used to investigate the feasibility of using the glass transition temperature (T_g) as the prime parameter for monitoring the cure process. The objective is attractive for systems in which a one-to-one relationship, independent of the cure temperature, exists between the T_g of the material and its chemical conversion. It follows for such systems, for example, that the chemical activation energy can be determined equivalently from both the time to reach a fixed conversion and the time to reach a fixed T_g. In addition, for such systems, a simple methodology permits calculation of the time for the material to achieve any attainable T_g (i.e., conversion) at different cure temperatures, and the time for its T_g to reach any cure temperature below $T_{g\infty}$ (i.e., vitrification). Results are presented in the form of a time–temperature–transformation (TTT) cure diagram. Preliminary results are presented for a tetrafunctional aromatic diamine–diepoxy system. The results show, in particular, that the reactions before vitrification are kinetically controlled.

T HE GLASS TRANSITION TEMPERATURE, T_g, is a sensitive and practical parameter for following the cure of reactive thermosetting systems. A wide

*Corresponding author.

0065–2393/90/0227–0143$06.75/0
© 1990 American Chemical Society

range of values of T_g is encountered during cure, and it can be measured easily throughout the entire range of cure. The fact that T_g increases non-linearly with conversion in cross-linking systems makes it even more sensitive in the later stages of reaction. Sensitivity is especially needed when the reaction rate is low, for example, at high conversion and after vitrification (solidification).

A convenient summary of the changes that occur during cure of a thermosetting system is the isothermal time–temperature–transformation (TTT) cure diagram. The diagram, schematically shown in Figure 1, displays the states of the material and characterizes the changes in the material during isothermal cure as a function of time (1–3). Material states include liquid,

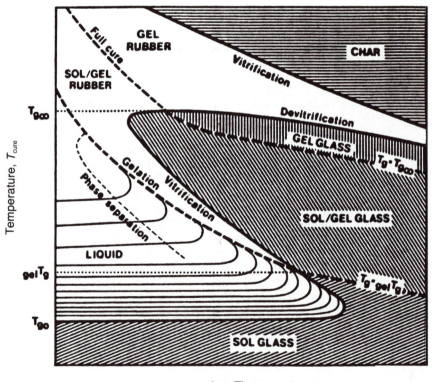

Log Time

Figure 1. A generalized isothermal time–temperature–transformation (TTT) cure diagram for a thermosetting system, showing critical temperatures (i.e., T_{g0}, $T_{g\infty}$, and $_{gel}T_g$), states of the material and contours characterizing the setting process. The full cure contour corresponds to $T_g = T_{g\infty}$. Isoviscous contours in the liquid region correspond to fixed viscosity levels (1, 3). Phase separation occurs before gelation for systems that undergo phase separation.

sol glass, sol–gel rubber, gel rubber, sol–gel glass, gel glass, and char. The various changes occurring in the material during isothermal cure can be characterized by contours of the times to reach those events. Relevant contours include molecular gelation (corresponding in simple systems to the unique conversion at the molecular gel point), vitrification (corresponding to T_g rising to the cure temperature, T_{cure}), devitrification (corresponding to T_g decreasing to T_{cure} because of thermal degradation), and char formation (corresponding to T_g increasing to T_{cure} because of thermal degradation). The progress of the isothermal cure process and the state of the material can be clearly summarized in terms of these contours in the TTT diagram.

T_g is an important and useful parameter in the TTT cure diagram. For example, for vitrification, devitrification, and char formation, T_g equals T_{cure}. The T_g of the unreacted material is T_{g0}, and that of the fully reacted material is $T_{g\infty}$. Molecular gelation corresponds to a definite chemical composition for which $T_g = {}_{gel}T_g$. The extrapolated temperature for which macroscopic gelation (corresponding to an isoviscous state) occurs simultaneously with vitrification is designated ${}_{gel}T_g{}'$ (discussed later).

The basic parameter governing the state of the material is the chemical conversion. Therefore, more details of the cure process can be summarized by incorporating isoconversion contours in the TTT diagram. The purposes of the work described in this chapter were to study the feasibility of using T_g as a direct one-to-one measure of chemical conversion and to demonstrate a simple methodology for constructing iso-T_g contours and the vitrification curve in the isothermal TTT diagram from T_g-vs.-time data at different values of T_{cure}, provided that a one-to-one relationship between T_g and conversion exists (4).

Chemical System and Experimental Procedure

The system chosen for this study is a diglycidyl ether of bisphenol A (DER 337 from Dow, epoxide equivalent weight = 230 g/eq) cured with a stoichiometric amount of a tetrafunctional aromatic diamine, trimethylene glycol di-*p*-aminobenzoate (TMAB from Polaroid, amine equivalent weight = 157 g/eq) (*see* structures). TMAB

Diglycidyl ether of bisphenol A (DER 337)

Trimethylene glycol di-*p*-aminobenzoate (TMAB)

is a highly crystalline solid with a melting point of 125 °C; DER 337 is a highly viscous liquid at room temperature. A solution of the reactants (~1 g/mL) was formed by adding the amine to a solution of the epoxy in methyl ethyl ketone (MEK) at room temperature. MEK was later removed prior to each experiment by holding each small specimen (ca. 10–15 mg of solid) at 70 °C for 2 h in a flowing inert atmosphere.

The initial conversion using this procedure was estimated from the kinetics of the conversion data at higher cure temperatures (which was developed in the course of the present work) to be less than 4%. The mixture, after the removal of MEK, was taken as the initial (uncured) state of the system ($T_{g,\text{initial}} < 25$ °C). This procedure was adopted to avoid the necessity of mixing the reactants at a higher temperature. An alternative procedure involves melting the amine crystals before adding to the liquid epoxy above 125 °C to prevent recrystallization. However, this step usually results in a substantial conversion of the initial mixture; the initial mixture is then a solid at room temperature (i.e., $T_{g,\text{initial}} > 25$ °C).

Two complementary techniques were used in this study: differential scanning calorimetry (DSC) and torsional braid analysis (TBA). A Perkin-Elmer calorimeter (DSC-4) was used to measure T_g and the heat evolution of the reaction, from which T_g vs. the fractional extent of chemical conversion was determined. Samples were partially cured in sealed DSC aluminum pans in an oven at three different temperatures (136, 150, and 160 °C) under N_2 flow. After cure times ranging from 15 min to 6 h, samples were removed from the oven, allowed to cool to room temperature, quenched to –20 °C in the DSC unit, and then scanned from –20 to 350 °C at 5 °C/min to determine T_g and the residual exotherm (ΔH_R) to complete the reaction. The scan rate of 5 °C/min was used because at higher scan rates, the sample began to degrade at high temperatures before the reaction was fully completed.

T_g appears as an endothermic step change over an interval of temperature in a DSC scan. In this study, T_g was taken as the midpoint of the transition interval. The total heat of reaction (ΔH_T) was determined by scanning uncured samples. The extent of reaction, α, was quantitatively calculated as $\alpha = (\Delta H_T - \Delta H_R)/\Delta H_T$ (5). Figure 2 shows the DSC trace of an uncured sample. The apparent total heat of reaction was 90 cal/g of mixture (or 27.8 kcal/mol of epoxide), and the initial glass transition temperature, T_{g0}, was 0 °C. The glass transition temperature of the fully reacted sample, $T_{g\infty}$, taken as the highest value of T_g obtained at 150 and 160 °C after extended cure times, was 145 °C.

In this study, TBA (1) was used to directly measure the times to macroscopic gelation and vitrification during isothermal cure at different temperatures and the T_g of the material vs. time at different temperatures. In the TBA experiment, the specimen was intermittently activated into free torsional oscillation to generate a series of damped waves, from which two mechanical parameters, namely, relative rigidity and logarithmic decrement, were determined. After impregnating a heat-cleaned glass braid with the solution of the reaction mixture, mounting the specimen at room temperature, and heating at 70 °C for 2 h, the temperature was raised at 10 °C/min to the cure temperature and held at the cure temperature for a definite time. The TBA spectrum during prolonged isothermal cure yielded two important parameters, the time to macroscopic gelation and the time to vitrification. They appeared as two successive maxima in the logarithmic decrement vs. time of cure (Figure 3A).

After the isothermal cure, the TBA specimen was control-quenched at 10 °C/min to –50 °C and then scanned from –50 to 250 °C at 1 °C/min to obtain the T_g attained by the particular isothermal cure. A subsequent scan from 250 to –50 °C at 1 °C/min gave the maximum glass transition temperature, $T_{g\infty}$, of the material. The data shown in Figure 3B are to a lower temperature limit (–180 °C)

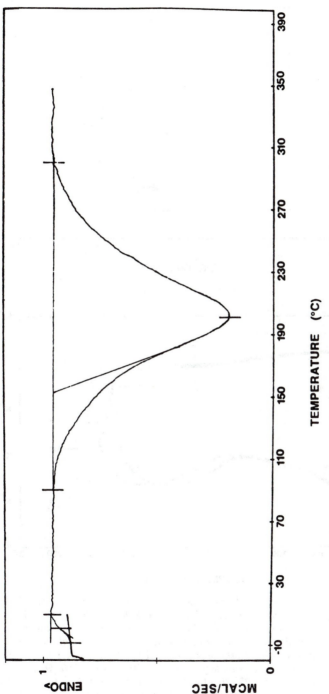

Figure 2. *DSC scan of the initial uncured DER 337–TMAB system from –20 to 350 °C at 5 °C/min to determine the initial glass transition temperature (T_g) and the total heat of reaction (ΔH_T). Prehistory: heated at 70 °C for 2 h, cooled to room temperature, and quenched in the DSC unit to –20 °C.*

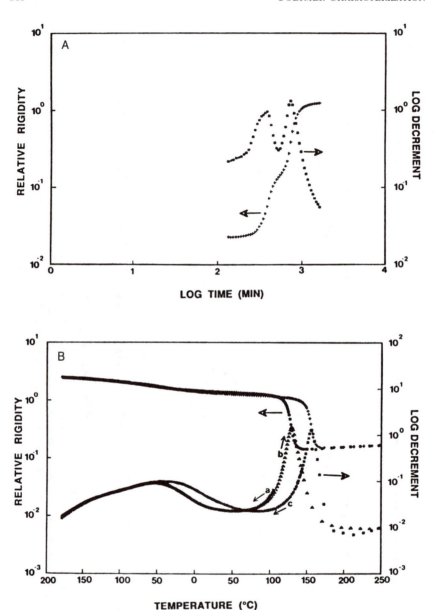

Figure 3. TBA dynamic mechanical spectrum of the system (DER 337–TMAB)
(A) during isothermal cure at 110 °C in He for 26 h; key: +, relative rigidity;
and *, logarithmic decrement; the two successive maxima in the logarithmic
decrement are assigned as macroscopic gelation and isothermal vitrification,
respectively; and (B) after isothermal cure at 110 °C for 26 h during sequential
temperature scans at 1 °C/min; key: a (+, *), 110 to –180 °C; b (●, ▲), –180
to 250 °C; and c (+, *), 250 to –180 °C. (For other TBA data in this work,
the lower temperature limit was –50 °C).

so as to include cryogenic secondary transitions. All T_g points were identified by the maxima of the peaks of the logarithmic decrement spectra (0.9 Hz). With this definition, T_g could be specified easily and unambiguously.

The TBA technique provides a useful means of monitoring the cure in terms of these measured parameters, (i.e., T_g, macroscopic gel times, and vitrification times), and provides a convenient way of directly constructing an experimental isothermal TTT cure diagram for the system from the times to gelation and to vitrification vs. isothermal cure temperature.

T_{g0} and $T_{g\infty}$ determined by the TBA technique for this system were 10 °C (0.9 Hz) and 156 °C (0.9 Hz), respectively. There was a difference of about 10 °C between the value of T_g determined by the DSC and TBA techniques (TBA values being higher). This difference was expected because of the difference in the operational definitions of T_g, the difference in the measuring time scale, the differences in the procedures of initial heating to and subsequent cooling from the isothermal temperature, and the difference in the temperature scanning rates during measurement.

The T_g-vs.-time data obtained by the TBA technique were used in the calculation of the iso-T_g and vitrification contours, although the calculation can also be performed entirely with the DSC results.

The advantages, however, of using TBA vs. DSC for measuring T_g are as follows:

1. For TBA, T_g is a prominent relaxation and can be measured accurately because the maximum of the logarithmic decrement can be located easily and unambiguously. For DSC, T_g is a weak relaxation and is measured from an endothermic shift in the specific heat-vs.-temperature curve.

2. The slow temperature scan rate (1 °C/min) in the TBA technique provides good temperature resolution of the resulting spectra, such that T_g can be determined to within 1 °C. In comparison, the DSC technique requires a reasonably high temperature scan rate (5–20 °C/min) to detect T_g; the higher scan rate reduces the resolution of the data.

3. The TBA technique gives the same value of T_g for an amorphous material by either scanning up or scanning down in temperature during measurements, whereas the DSC technique gives substantially different values.

4. In the TBA technique, the physical aging effect on T_g is essentially removed before T_g is identified by the maximum in the logarithmic decrement during a heating scan (6). In contrast, in the DSC technique, the effect of prior physical annealing and the onset of a reaction exotherm can directly affect the T_g measurement.

Results and Discussion

DSC Results. *Conversion vs. Cure Time.*   Figure 4 shows the extent of reaction, α, vs. cure time for three isothermal cure temperatures (136, 150, and 160 °C) under conditions far from diffusion control (no isothermal vitrification is encountered for these data). In this temperature range, the T_g of the material should eventually achieve $T_{g\infty}$ given sufficient

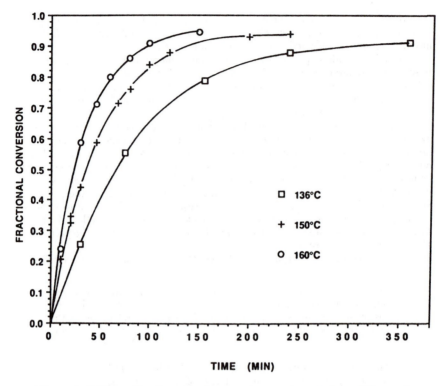

Figure 4. DSC results. Fractional conversion vs. time at three different iso-thermal temperatures.

time because T_{cure} is within 20 °C of $T_{g\infty}$. The results for high conversion ($\alpha > 95\%$) are not available because the residual exotherm becomes too small to be measured accurately by DSC; the DSC technique is not sensitive enough to monitor high conversions. However, curing at 150 and 160 °C for extended times raises T_g to its maximum value even though the change in conversion in the later stages is not measurable by DSC (e.g., for T_{cure} of 150 °C for 9 h, $T_g \sim 145$ °C $\equiv T_{g\infty,DSC}$). Moreover, topological constraints prevent complete chemical conversion. Therefore, ΔH_T as measured is underestimated.

The data in Figure 4 can be used to calculate the activation energy for the reaction. In the usual manner, assuming a single reaction with a single activation energy, the rate of a kinetically controlled reaction can be expressed by an Arrhenius rate expression:

$$\frac{d\alpha}{dt} = k\, f(\alpha) = A \times \exp\left(-\frac{E}{RT}\right) \times f(\alpha) \tag{1}$$

where α is the extent of reaction; k is the reaction rate constant, which is assumed to have Arrhenius temperature dependence; $f(\alpha)$ is a function of reactant concentration; A is the Arrhenius frequency factor in reciprocal seconds; E is the energy of activation in kilocalories per mole; R is the Boltzmann constant; and T is the cure temperature in kelvins. Rearranging equation 1 and integrating,

$$\int_0^\alpha \frac{d\alpha}{f(\alpha)} = A \times \exp\left(-\frac{E}{RT}\right) \times \int_{t=0}^{t=t_\alpha} dt = A \times \exp\left(-\frac{E}{RT}\right) \times t_\alpha \quad (2)$$

where t_α is the time at which the conversion reaches α. Taking the natural logarithm

$$\ln\left(\int_0^\alpha \frac{d\alpha}{f(\alpha)}\right) = \ln(A) - \frac{E}{RT} + \ln(t_\alpha) \quad (3a)$$

or

$$\ln t_\alpha = \frac{E}{RT} + \left[\ln\left(\int_0^\alpha \frac{d\alpha}{f(\alpha)}\right) - \ln(A)\right] \quad (3b)$$

For a fixed conversion, the terms in the square brackets are constants. Let $t_{\alpha 1}$ and $t_{\alpha 2}$ be the time to reach the same conversion at two different cure temperatures T_1 and T_2, respectively.

$$\ln t_{\alpha 1} - \ln t_{\alpha 2} = \frac{E}{R}\left[\frac{1}{T_1} - \frac{1}{T_2}\right] \quad (4)$$

Equation 4 shows that the difference between the ln(time) to reach the same conversion at two different cure temperatures is a constant for all conversions, α. In other words, for a kinetically controlled reaction, all conversion-vs.-ln(time) plots at different cure temperatures should be superposable by shifting all curves horizontally relative to a fixed reference temperature.

Figure 5 replots the data in Figure 4 as conversion vs. ln(time); the curves at 150 and 160 °C are then shifted to the right along the ln(time) axis in order to superpose with the curve at 136 °C as shown in Figure 6. All data appear to be superposable. The ln(time) shift factors used in superposing the three curves are plotted in an Arrhenius fashion against the reciprocal cure temperatures (kelvins) in Figure 7. The resulting plot can be correlated

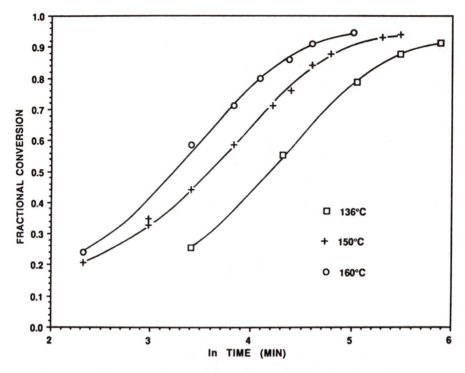

Figure 5. DSC results. Fractional conversion vs. ln(time) at three different temperatures (same data as in Figure 4).

with a straight line, the slope of which is equal to E/R (as in equation 4). The apparent activation energy thus determined is 15.4 kcal/mol.

The analysis of the chemical kinetics of the reaction from the available conversion data is not included, because the result is not conclusive due to the limited number of data points in a limited cure temperature range and the presence of residual solvent. A more extensive study (7) has been carried out on a similar epoxy–amine system without solvent, the chemical kinetics of which were determined to be second order autocatalyzed by the OH groups generated during the reaction.

T_g vs. Conversion. Figure 8 shows a plot of T_g vs. conversion for all data at the three cure temperatures from the DSC. All data appear to fall on a single curve. Thus, a one-to-one relationship exists between T_g and the chemical conversion, independent of the cure temperature (for the limited range of $T_{cure} = 136$–160 °C). This result is in agreement with the work of other investigators (8, 9). (All of the data in Figures 4–6 were obtained before vitrification. The T_g of the material after vitrification was reported (6) to depend on both conversion and physical aging. This dependence can affect the relationship between T_g and conversion.)

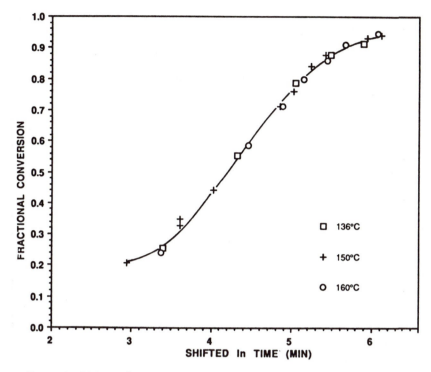

Figure 6. DSC results. Data in Figure 5 were shifted horizontally along the ln(time) axis to superpose all data at 136 °C. The single master curve represents the kinetically controlled reaction at 136 °C.

The significance of the relationship between T_g and conversion has been well recognized in the literature (3, 8–12). Fox and Loshaek (13) have shown that T_g (kelvins) can be directly related to the number-average molecular weight for linear homologous polymers, and also to the cross-link density for cross-linked materials. DiBenedetto (10) proposed a model relating T_g to chemical conversion:

$$\frac{T_g - T_{g0}}{T_{g0}} = \frac{(E_x/E_m - F_x/F_m)\alpha}{1 - (1 - F_x/F_m)\alpha} \tag{5}$$

where E_x/E_m is the ratio of lattice energies for cross-linked and uncross-linked polymers, and F_x/F_m is the corresponding ratio of segmental mobilities. The two ratios are generally treated as two constant empirical parameters for each system. Fitting equation 5 to the present T_g-vs.-conversion data, the values of E_x/E_m and F_x/F_m were determined to be 0.612 and 0.40, respectively. Equation 5 with these two values was used to calculate T_g as a function of conversion; the result of this calculation is shown in Figure 8 as a solid line. The calculated values provide a reasonable correlation with

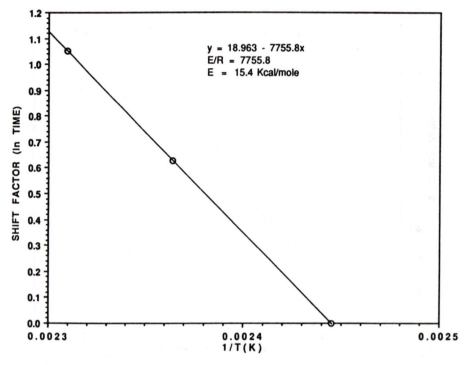

y = 18.963 - 7755.8x
E/R = 7755.8
E = 15.4 Kcal/mole

Figure 7. DSC results. Arrhenius plot of the shift factors (used in Figure 6 to superpose the data at different temperatures) vs. 1/T (K).

the experimental data. Previous (*14*) and current work (*7*) in this laboratory attempted to theoretically model the relationship between T_g and conversion to provide a clear fundamental basis for the variation of T_g with increasing conversion, and quantitatively take into account the effectiveness of various functional cross-linking units in raising the T_g of the material.

T_g increases nonlinearly with chemical conversion; the same change in the extent of conversion can raise the value of T_g more at high conversion than at low conversion (*3, 6, 15*). An explanation accounting for this nonlinear behavior is that T_g is more strongly dependent on the concentration of the highest-functional cross-linking units (i.e., for this system, tetrafunctional amine residues with all four of the amino hydrogens reacted) than on that of the lower-functional cross-linking units (i.e., amine residues with only three amino hydrogens reacted). In evaluating the effective cross-linking concentration, a trifunctional cross-linking unit is considered to be a half tetrafunctional unit because a tetrafunctional unit is mechanically equivalent to two trifunctional units (*7, 16*). It follows that T_g rises more sharply at high conversion with small changes in the overall conversion because the concentration of the higher-functional cross-linking units increases at high conversion at the expense of the lower-functional cross-linking units (*17*). As a

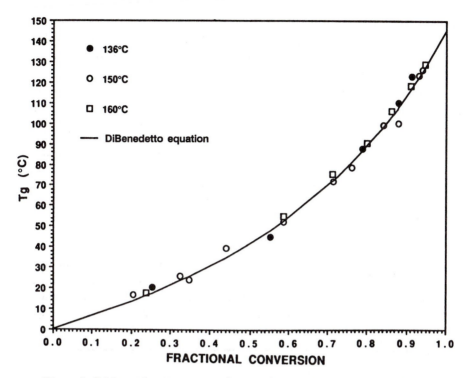

Figure 8. DSC results. Experimental T_g *vs. fractional conversion (symbols), and the computed fit using the DiBenedetto equation (solid line).*

consequence of this nonlinear behavior, T_g is sensitive to changes in conversion in the later stages of cure when the changes are small because of low reactant concentration at high conversion.

Moreover, the sensitivity of T_g to conversion and the advantage of the accuracy for measuring T_g make T_g a promising candidate for monitoring the slow rate of cure in the diffusion-controlled regime. Segmental diffusion mechanisms dominate the reaction, especially when T_g rises above T_{cure} (3, 18, 19), and can drastically slow down the reaction even at low conversion for low cure temperatures. Our current research (7) has successfully utilized the sensitivity of T_g to conversion and their unique one-to-one relationship to monitor the reaction, in both the chemical and diffusion-controlled regimes, of an epoxy–amine system similar to the system reported here. Quantitative expressions that satisfactorily described the kinetics of the material during isothermal cure both before and after vitrification have been derived.

For systems with unequally reactive functional groups, different time–temperature cure paths may result in materials with the same chemical conversion but different chemical structures. This outcome results when the relative reactivities of the functional groups change with the reaction temperature. In reacting epoxy with aromatic primary amine, if the relative reactivities of the primary amino hydrogen and the secondary amino hydrogen with epoxy change with temperature, then the one-to-one relationship between T_g and conversion would not be expected for the different time–temperature cure paths.

The DSC results reported here (with a curing temperature spread of 24 °C) do not suggest this effect. The validity of the unique relationship between T_g and conversion has been further tested (7) by expanding the range of T_{cure} investigated. Uniqueness of the one-to-one relationship between T_g and conversion for epoxy–amine systems would be a consequence of the ratio of the rate constant between epoxy and primary amine (k_1) to that between epoxy and secondary amine (k_2) being constant with respect to temperature. This result further implies the same activation energy for the two reactions. (Equal reactivity of all amino hydrogens is an extreme case of this criterion, in which $k_1/k_2 = 1$.)

TBA Results. T_g vs. *Cure Time*.   Figure 9 shows the progression of T_g as measured by TBA vs. cure time for five different isothermal cure temperatures (130, 140, 150, 160, and 200 °C). The data in Figure 9 are replotted as T_g vs. ln(time) in Figure 10. The resulting plots can be manip-

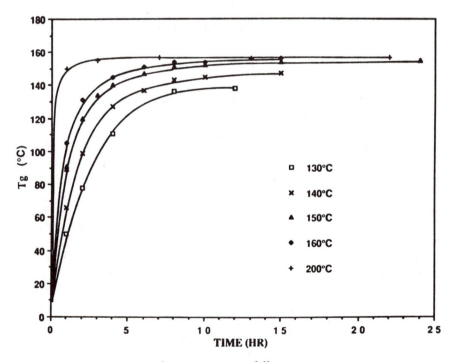

Figure 9. TBA results. T_g vs. time at different cure temperatures.

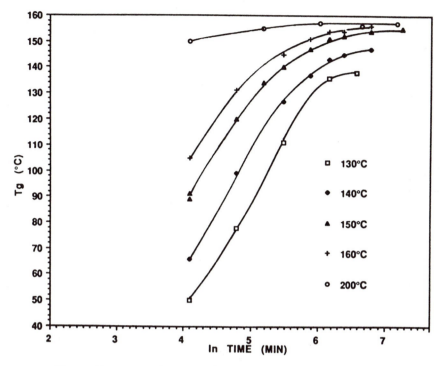

Figure 10. TBA results. T_g *vs. ln(time) (same data as in Figure 9).*

ulated in the same manner as the conversion-vs.-ln(time) plots, because T_g is equivalent to α. All curves are therefore shifted horizontally along the ln(time) axis relative to the curve at 130 °C in order to superpose all curves and thus form a master curve at 130 °C, as is shown in Figure 11. If the reaction is only kinetically controlled, all T_g-vs.-ln(time) curves should be superposable.

The superposition results in Figure 11 demonstrate that before the isothermal vitrification points (marked, when $T_g = T_{cure}$, by arrows in the figure), the data can be superposed to form a smooth master curve, which represents the reaction being only kinetically controlled, as implied by equations 1–3. For times for which T_g is greater than T_{cure} and for cure temperatures below $T_{g\infty,TBA}$ (156 °C), T_g increases at a much slower rate; the result is a deviation from the master curve, which is an indication that the reaction is becoming diffusion controlled.

Determination of Reaction Activation Energy. From the DSC results (for limited $\Delta T_{cure} = 24$ °C), a one-to-one relationship exists between T_g and the reaction conversion. Therefore, T_g can be considered as a unique function of conversion. Consequently, the time to reach a fixed T_g is equiv-

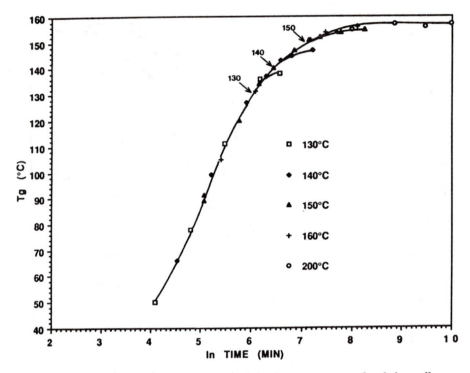

Figure 11. TBA results. Superposition of the data in Figure 10 by shifting all curves horizontally relative to the curve at 130 °C. Arrows represent the vitrification points ($T_g = T_{cure}$) *at the three lower cure temperatures. Data start to deviate from the kinetically controlled master curve after vitrification because of diffusion control.*

alent to the time to reach a fixed conversion. Therefore, the activation energy of the reaction can be determined from the TBA data in Figure 9 in the same way as the calculation using the conversion data.

The ln(time) shift factors used in constructing the TBA T_g-vs.-ln(time) master curve at 130 °C were plotted against the reciprocal cure temperatures (K), as shown in Figure 12. The results can be correlated with a straight line, the slope of which yields an apparent activation energy for the reaction equal to 15.2 kcal/mol. The value of the activation energy obtained by this method is close to that found by using the DSC conversion data at different cure temperatures ($E = 15.4$ kcal/mol from DSC). These results indicate that, for this system, the use of T_g as an index of changes in chemical conversion is valid.

Figure 13 is a plot of ln(t_{gel}), the time to reach the macroscopic gelation peak in the logarithmic decrement curve in an isothermal TBA scan, vs. the reciprocal cure temperature (K). The resulting plot can also be correlated with a straight line. However, the activation energy determined from the

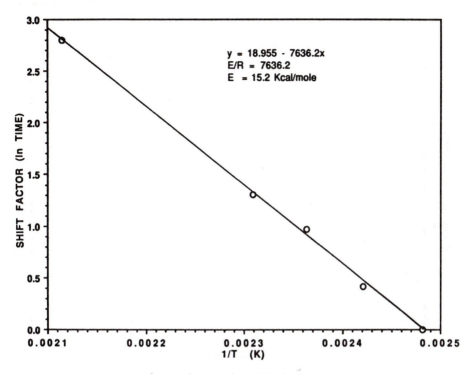

Figure 12. TBA results. Arrhenius plot of the ln(time) shift factors (used in Figure 11 to superpose data at different temperatures) vs. 1/T (K).

slope of this line is 13.7 kcal/mol, which is significantly different from the values determined from iso-T_g or isoconversion data. According to Flory's theory of gelation for a unireaction system (*20*), the onset of molecular gelation occurs at a fixed conversion for all cure temperatures, and therefore, the times to reach molecular gelation for such systems should be the same as the times to reach the fixed conversion corresponding to molecular gelation (and equivalently, to the times to reach the fixed T_g, $_{gel}T_g$).

The fact that the activation energy determined from the times to reach the macroscopic gelation peaks of the TBA isothermal scans is different from those determined from iso-T_g and isoconversion data suggests that the macroscopic gelation peak does not correspond to molecular gelation. The macroscopic gelation relaxation has been shown to be an isoviscous event (*1, 21*); in contrast, molecular gelation is an isoconversion event. An apparent viscous activation energy, determined from an Arrhenius plot of the log time to any observed isoviscous event vs. 1/*T* (K), approaches the value for the chemical activation energy leading to molecular gelation only if the level of the viscosity of the observed event is infinitely high (*1*), approaching the viscosity at the molecular gel point.

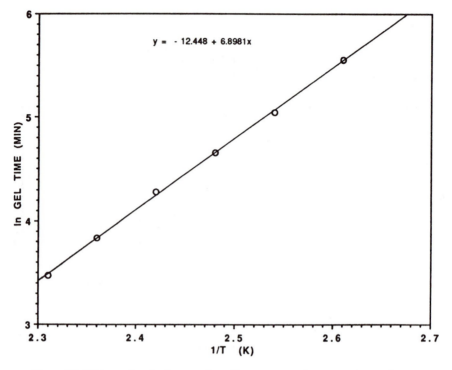

Figure 13. TBA results. Arrhenius plot of ln time to reach macroscopic gelation vs. 1/T (K).

Construction of the Vitrification and Iso-T$_g$ Contours in the TTT Diagram.   *Iso-T$_g$ Contours.*   The reaction activation energy determined in the previous section can be used for constructing an iso-T$_g$ contour if, in addition, a data point on the contour is known (4). From the integral form of the rate expression (equation 2),

$$\int_0^\alpha \frac{d\alpha}{f(\alpha)} = A \times \exp\left(-\frac{E}{RT}\right) \times t_\alpha \tag{2}$$

The left side of the equation is only a function of conversion, α, and thus, also a function of T_g only [$\equiv F(T_g)$]. The time to reach a fixed conversion, t_α, can be replaced by the time to reach a fixed T_g, t_{T_g}. Thus, equation 2 can be written as

$$F(T_g) = A \times \exp\left(-\frac{E}{RT}\right) \times t_{T_g} \tag{6}$$

This equation is valid for all cure temperatures as long as the reaction mechanism remains kinetically controlled, with the ratio of the rate constants for the reactions of epoxy with primary and secondary amines being independent of the cure temperature.

Let $t_{T_{g,1}}$ be the time needed to reach a given T_g at cure temperature T_1, and let $t_{T_{g,2}}$ be the time needed to reach the same T_g at cure temperature T_2. Then, from equation 6,

$$F(T_g) = A \times \exp\left(-\frac{E}{RT_1}\right) \times t_{T_{g,1}} \tag{7a}$$

and

$$F(T_g) = A \times \exp\left(-\frac{E}{RT_2}\right) \times t_{T_{g,2}} \tag{7b}$$

For a fixed T_g, equations 7a and 7b are equal.

$$A \times \exp\left(-\frac{E}{RT_1}\right) \times t_{T_{g,1}} = A \times \exp\left(-\frac{E}{RT_2}\right) \times t_{T_{g,2}} \tag{8}$$

Taking the natural logarithm gives

$$-\frac{E}{RT_1} + \ln\left(t_{T_{g,1}}\right) = -\frac{E}{RT_2} + \ln\left(t_{T_{g,2}}\right) \tag{9}$$

Equation 9 provides a relationship between the times to reach a fixed T_g and the corresponding cure temperatures. Thus if a time to reach a particular T_g at a cure temperature (i.e., $t_{T_{g,1}}$ at T_1) is known, then the times to reach the same T_g at different temperatures (i.e., $t_{T_{g,2}}$ at T_2) can be calculated from equation 9 provided that the reaction activation energy is available. (Similarly, if the time to reach a particular conversion at a cure temperature is known, then the times to reach the same conversion at different temperatures can be calculated provided that the reaction activation energy is available.)

The activation energy for the kinetically controlled reaction, which was determined in the previous section from the superposition of the TBA T_g-vs.-ln(time) data to be 15.2 kcal/mol, is used in the calculation of an iso-T_g contour. The calculated results are incorporated into the TTT diagram in the form of a series of iso-T_g contours by plotting the cure temperatures against the logarithm of the times to reach a particular T_g. The known data point ($t_{T_{g,1}}$ at T_1) for each iso-T_g contour is obtained from the interpolated data of the T_g-vs.-t_{cure} curve at 150 °C in Figure 9.

Figure 14 shows the calculated iso-T_g contours (from $T_g = 12\,°C$ to $T_g = 155\,°C$). The figure also shows selected experimental iso-T_g data obtained from interpolating the curves in Figure 9 for $T_{cure} = 130, 140, 160,$ and 200 °C. The calculation appears to correlate well with the experimental data (the range of T_{cure} extends beyond the ΔT_{cure} used for α vs. T_g). These results show that, for the cure temperature range in the experiment, the reactions prior to vitrification are mainly kinetically controlled. From a practical point of view, the degree of curing and the state of chemical conversion can be visualized through the levels of these iso-T_g contours.

Data from the macroscopic gelation peaks are also included in Figure 14. The macroscopic gelation curve appears to lie close (but not parallel) to the iso-$T_g = 70\,°C$ line, which corresponds to the chemical conversion of 65% (cf., the theoretical value for molecular gelation according to Flory's theory of gelation (20) is 58%). Extrapolation of the macroscopic gelation line to intersect the vitrification curve yields an apparent $_{gel}T_g{}'$ of approximately 70 °C (see Figure 14). However, caution in such extrapolation is to be noted. As mentioned earlier, the TBA macroscopic gelation curve is considered to be an isoviscous phenomenon and, therefore, cannot intersect the vitrification curve because the viscosity at vitrification is much higher than that at macroscopic gelation (1, 6, 22).

Vitrification Curve. The curing reaction of thermosets in general will be a combination of chemical kinetic and intersegmental diffusion controls (7, 18, 19). However, as has been shown in this and other work (3, 23, 24) for the curing of epoxy–amine systems, the reaction is primarily kinetically controlled with a single constant activation energy up to the vicinity of vitrification. Therefore, the iso-T_g relation (equation 9) is valid until the material vitrifies and thus can be used to estimate the isothermal times to vitrification. On the iso-T_g curve (equation 9), a point will correspond to a vitrification point when $T_{cure} = T_g$. Graphically, this point is equivalent to the point where the horizontal line at $T = T_{cure}$ intersects the iso-T_g line. The resulting points from all possible iso-T_g contours (from iso-$T_g = T_g$ to iso-$T_g = T_{g\infty}$) constitute the vitrification contour in the TTT diagram. The calculated vitrification curve is shown in Figure 14 together with the vitrification points that were determined experimentally from the second maximum in the logarithmic decrement of the TBA isothermal scans for different

---

Figure 14. TBA results. Summary TTT cure diagram for DER 337–TMAB,
showing the calculated iso-T$_g$ and vitrification contours (solid lines) (calculated from the activation energy and T$_g$ vs. t_{cure} *at 150 °C), and selected experimental iso-T$_g$ data (×) (from T$_g$ vs.* t_{cure} *at* $T_{cure} = 130, 140, 160,$ *and 200 °C), vitrification data (■), and macroscopic gelation (▲) data.* $_{gel}T_g{}' \sim 70\,°C$ *(determined from the intersection of the extrapolated macroscopic gelation line and the vitrification curve).*

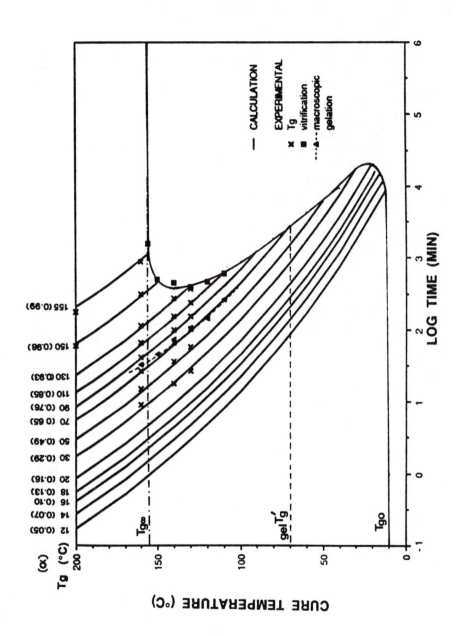

T_{cure}s (110, 120, 130, 140, and 160 °C). The calculated curve agrees with the experimental results. Thus, neglecting diffusion control prior to vitrification does not significantly affect calculation of the time to vitrify.

Previous approaches to calculation of the time to vitrification required analytical knowledge of both the kinetic rate law and the empirical relationship between T_g and conversion (3, 14). The kinetic relationship is quite complicated and difficult to obtain accurately. In those attempts, the conversion at vitrification (when $T_g = T_{cure}$) was obtained from the relationship between T_g and conversion. Once the conversion at T_g was known, then the time to vitrification was determined from an assumed kinetic rate law. All calculations also assumed a kinetically controlled reaction and one temperature-independent reaction mechanism. Thus, the approach described here using the iso-T_g contours to determine the time to vitrification is direct and uncomplicated. T_g is treated as a direct measure of conversion, and this approach eliminates the error associated with relating T_g to conversion at vitrification. It relies on the measurement of T_g, which can be made more accurately than the measurement of conversion, and on a one-to-one relationship between T_g and conversion.

Moreover, the procedure greatly simplifies the theoretical construction of the vitrification curve because it does not require knowledge of the kinetic rate law or the relationship between T_g and conversion (as long as it is one-to-one). The only required information is the reaction activation energy and some isolated data points, each of which relates a particular T_g, for example, T_g^*, to cure time at a cure temperature. Each data point together with the activation energy yields an iso-$T_g = T_g^*$ line, from which the vitrification point can be determined for $T_{cure} = T_g^*$. If the whole vitrification curve is desired, data points relating T_g–t_{cure}–T_{cure} over the whole range of T_g from T_g to $T_{g\infty}$ are required. These data points can be obtained by following the progress of the T_g of the material as a function of cure time at a cure temperature close to $T_{g\infty}$ (preferably $>T_{g\infty}$ to avoid vitrification) over the course of cure. Data points relating the whole range of T_g with cure time at one cure temperature are then available.

Conclusions

T_g is an appropriate parameter for monitoring the cure process for the following reasons:

1. T_g is easily measured by several techniques (TBA is especially convenient and sensitive).

2. T_g can be measured over the entire range of conversion from the unreacted state up to full conversion (cf., viscosity that can only be measured up to gelation).

3. The results show that T_g for this epoxy–amine system is a unique function of conversion for the cure temperature range investigated (ΔT_{cure} = 24 °C). The generality of this conclusion has been demonstrated elsewhere (7) for a wider range of cure temperatures (7).

4. The nonlinear dependence of T_g on conversion renders T_g more sensitive than other parameters (such as the heat of reaction and IR absorption bands of reactants and products) to the small changes in conversion with time when the reactant concentrations are low (i.e., kinetically limited at high conversion).

The same considerations also make T_g an appropriate parameter for monitoring the slow reaction rate in the diffusion-controlled region beyond vitrification [subject under current investigation (7)].

The times to reach a fixed T_g plotted against the reciprocal of the cure temperatures in an Arrhenius fashion can be used for determining the activation energy of the reaction, the value of which agrees well with that found from the times to reach a fixed conversion. This finding supports the one-to-one relationship between T_g and conversion.

The results show that neglecting diffusion control prior to vitrification does not significantly affect the calculation of the vitrification curve. Thus, the reaction kinetics up to vitrification is primarily kinetically controlled.

Acknowledgments

This project was supported in part by the Office of Naval Research.

References

1. Gillham, J. K. *Developments in Polymer Characterisation 3*; Dawkins, J. V., Ed.; Applied Science Publishers: Barking, Essex, England, 1982; pp 159–227.
2. Gillham, J. K. *Polym. Eng. Sci.* **1979**, *19*, 670; ibid. **1986**, *26(20)*, 1429.
3. Enns, J. B.; Gillham, J. K. *J. Appl. Polym. Sci.* **1983**, *28*, 2567.
4. Gan, S.; Gillham, J. K.; Prime, R. B. *J. Appl. Polym. Sci.* **1989**, *37*, 803.
5. Prime, R. B. *Thermal Characterization of Polymeric Materials*; Turi, E. A., Ed.; Academic: New York, 1981; p 435.
6. Pang, K. P.; Gillham, J. K. *J. Appl. Polym. Sci.* **1989**, *37*, 1969; ibid. **1990**, *39*, 909.
7. Wisanrakkit, G., Ph.D. Thesis, Princeton University, Princeton, NJ, 1990; Wisanrakkit, G.; Gillham, J. K. *J. Coat. Tech.* **1990**, *62(783)*, 35–50.
8. Lunak, S.; Vladyka, J.; Dusek, K. *Polymer* **1978**, *19*, 931.
9. Bair, H. E. *Polym. Prepr. Am. Chem. Soc. Div. Polym. Chem.* **1985**, *26*, 10.
10. DiBenedetto, A. T. In Nielsen, L. E. *J. Macromol. Sci. Rev. Macromol. Chem.* **1969**, *C3(1)*, 69.

11. Adabbo, H. E.; Williams, R. J. J. *J. Appl. Polym. Sci.* **1982**, *27*, 1327.
12. Bidstrup, S. A.; Sheppard, N. F., Jr.; Senturia, S. D. *Proc. Soc. Plast. Eng. 45th Ann. Tech. Conf.* **1987**, 987.
13. Fox, T. G.; Loshaek, S. *J. Polym. Sci.* **1955**, *15*, 371.
14. Aronhime, M. T.; Gillham, J. K. *J. Appl. Polym. Sci.* **1984**, *29*, 2017.
15. Chan, L. C.; Naé, H. N.; Gillham, J. K. *J. Appl. Polym. Sci.* **1984**, *29*, 3307.
16. Langley, N. R.; Polmanteer, K. E. *J. Polym. Sci. Polym. Phys. Ed.* **1974**, *12*, 1023.
17. Miller, D. R.; Macosko, C. W. *Macromolecules* **1976**, *9*, 206.
18. Kaelble, D. H. *Computer-Aided Design and Manufacture;* Marcel Dekker: New York, 1985; pp 113–148.
19. Mita, I.; Horie, K. *J. Macromol. Sci. Rev. Macromol. Chem. Phys.* **1987**, *C27(1)*, 91.
20. Flory, P. J. *Principles of Polymer Chemistry;* Cornell University Press: Ithaca, NY, 1953.
21. Stutz, H.; Mertes, J. *J. Appl. Polym. Sci.* **1989**, *38*, 781.
22. Enns, J. B.; Gillham, J. K. *Polymer Characterization: Spectroscopic, Chromatographic, and Physical Instrumental Methods;* Craver, C. D., Ed.; Advances in Chemistry 203, American Chemical Society: Washington, DC, 1983; pp 27–63.
23. Wisanrakkit, G.; Gillham, J. K.; Enns, J. B. *Prepr. Am. Chem. Soc. Div. Polym. Mat. Sci. Eng.* **1987**, *57*, 87.
24. Havlicek, I.; Dusek, K. In *Crosslinked Epoxies;* Sedlacek, B.; Kahovec, J., Eds.; Walter de Gruyter: New York, 1987; pp 417–424.

RECEIVED for review February 28, 1989. ACCEPTED revised manuscript September 27, 1989.

Characterization of Polymers by Thermally Stimulated Current Analysis and Relaxation Map Analysis Spectroscopy

J. P. Ibar[1], P. Denning[1], T. Thomas[1], A. Bernes[2], C. de Goys[2], J. R. Saffell[3], P. Jones[3], and C. Lacabanne[4]

[1]Solomat, Glenbrook Industrial Park, Stamford, CT 06906
[2]Solomat S.A., Ballainvilliers 91160, France
[3]Solomat Mfg. Ltd., Finnimore Industrial Estate, Ottery St. Mary, Devon, EX11 1AH England
[4]Paul Sabatier University, 31062 Toulouse Cédex, France

This chapter presents an overview of thermally stimulated current (TSC) analysis, which reveals the molecular mobility of a material's structure; and relaxation map analysis (RMA), which reveals structural transitions in polymers. Essentially, TSC and RMA are the same technique with two different focuses resulting in two types of analysis. The purpose of this chapter is to show that RMA is an exceptional technique that reveals more about the state of polymeric matter than previous methods. Equations are given for relaxation time, elementary retardation time, and temperature-dependent retardation time. The fully automated TSC/RMA spectrometer is described. Use of TSC/RMA spectroscopy for engineering applications is discussed. The influence of orientation, hydrostatic press, and processing conditions is described. A comparison of differential scanning calorimetric, thermal mechanical, and TSC/RMA spectroscopic characterization of latex copolymers is given.

THE MOLECULAR RESPONSE OF MATERIALS to physical or chemical influences can be analyzed by several techniques. Differential scanning calorim-

0065–2393/90/0227–0167$06.75/0

etry (DSC) and differential thermal analysis (DTA) are among the most popular choices in laboratories and on production sites. Other techniques include thermal mechanical analysis (TMA), stress relaxation or creep analysis, thermal expansion coefficient devices, dielectric constant analysis, and dynamic mechanical analysis (DMA).

Thermally Stimulated Current Analysis

Very special kinds of spectra can be obtained by recording the short-circuit current during warming-up after a material sample has been polarized at a constant direct-current (d.c.) field above a transition temperature and then quenched. Originally this technique was used to measure charge detrapping in low-molecular-weight organic and inorganic compounds. This method is called thermally stimulated current (TSC) analysis. It has also been referred to as thermostimulated current, thermocurrent, or electric depolarization current analysis.

Only since 1971 has the TSC technique been applied to the study of structural transitions in polymers (1–5). Thermocurrent studies have also been reported for crystals, polycrystals, semiconductors, and inorganic glasses; a thorough review of TSC communications published over the past 25 years is summarized in ref. 5. TSC is particularly suited for investigating the fine structure of polymers: semicrystalline polymers, copolymers and blends, polymer complexes, and resins. TSC also appears to be uniquely suited for determining the influence of additives, dopants, plasticizers, water content, and cross-linking.

In a typical TSC experiment, a high-voltage stabilized d.c. supply is used for polarizing the sample generally above its main transition temperature. The sample is heated at constant rate to the polarization temperature under an electric field of about 4×10^6 V per meter of thickness. The sample is held at this temperature for a specified time and then cooled down at a controlled rate to -150 °C. At that stage the external field is removed, and an electrometer is connected to the sample to record the short-circuit current while the sample is heated at a constant rate. A current is created when the material depolarizes.

This thermally stimulated current reveals the molecular mobility of the material's structure. The rate of depolarization is related to the relaxation times of the internal motions; this approach provides a new opportunity to study the physical and morphological structure of materials. The current peaks recorded this way correlate well with the transition temperatures measured by mechanical relaxation, DSC, or conventional [alternating-current (a.c.)] dielectric or mechanical spectroscopy.

Relaxation Map Analysis (RMA): A New and Powerful Analytical Concept

Although more powerful in its characterizations, TSC technology has not acquired the reputation of other analytical methods such as DSC, TMA, or dynamic mechanical analysis (DMA).

In 1974, Lacabanne (3) and Chatain (4) applied a new method of "windowing polarization" (also referred to as "thermal cleaning") to study relaxation phenomena by TSC analysis. Lacabanne's concept of windowing made possible the isolation of elementary Debye-type relaxations of the molecules over the entire relaxation spectrum. In previous work, the TSC output consisted of unresolved broad peaks that are the result of the interaction between several relaxation modes. Lacabanne's idea was to submit the polarized specimen to a windowing treatment (Figure 1).

First, the sample is polarized at temperature T_p for a time t_p selected to allow orientation only of a certain fragment of the dipoles. The sample is quenched to temperature T_d, generally 5–10 °C below the polarization temperature T_p. The polarizing voltage is then cut off and T_d is maintained for a time t_d. This step allows the depolarization of another fragment of the oriented dipoles. Finally, the sample is quenched to $T_0 << T_d$. The sample is then reheated at constant rate, and the current is measured. When t_p, t_d, and $(T_p - T_d)$ are conveniently chosen, the spectrum of depolarization is "simple". The spectrum is described by a single relaxation time that is a function of temperature only. By varying the value of T_p and repeating the process, the elementary modes can be isolated one by one (Figure 2) and the material's relaxation map can be constructed (Figure 3).

A TSC without windowing polarization produces results similar to DSC, TMA, or DMA working at very low frequencies ($\simeq 10^{-3}$ Hz). It does indeed provide interesting results at an accrued sensitivity, but perhaps no more interesting than results obtained from other analytical instruments operating at the same low frequency. The concept of windowing polarization gives TSC another dimension. The relaxation map obtained with the windowing polarization concept reveals a material's physical properties established from its elementary relaxations.

From low temperatures to the molten state, RMA seems ideally suited to the study of structural transitions in polymers. The objective of this chapter is to show that RMA is an exceptional technique that reveals more about the state of polymeric matter than previous methods. It recognizes the fine relaxation differences between a slowly cooled and a quenched plastic, as well as the resulting influence on the internal stress and on orientation. RMA has enough sensitivity to monitor the influence of external physical parameters such as the degree of cooling, pressure, orientation, processing conditions, annealing treatment, chemical composition, tacticity, and percent-

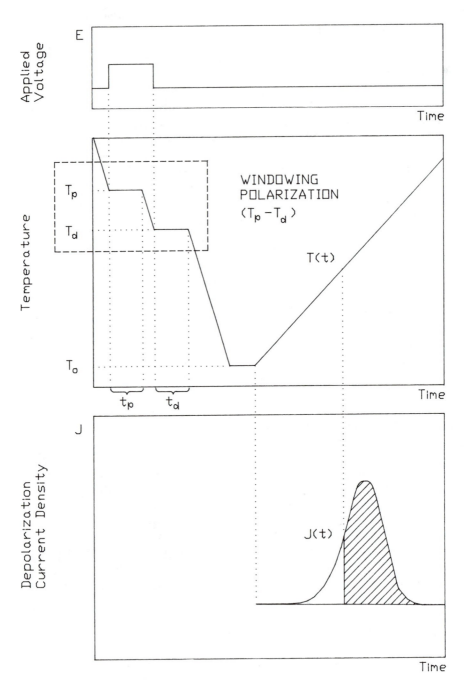

Figure 1. Principle of windowing polarization for relaxation map analysis.

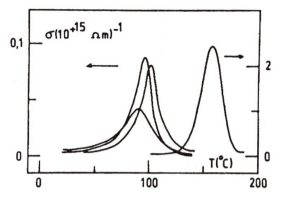

Figure 2. Deconvolution of global TSC peaks into their elementary Debye relaxation components. Relaxed polystyrene (see also Figure 5).

age of cross-linking. The influence of tacticity, molecular weight, and chemical structure are discussed elsewhere (5). The sensitivity of TSC is also compared to DSC and DMA in the investigation of the microstructure of latex block copolymers.

Principles of TSC and RMA

For both the TSC and RMA techniques, the current, I, and the temperature, T, are recorded versus time, t.

TSC.   To observe the various thermally stimulated current peaks, the mobile units of the sample are oriented by a constant electrostatic field, E,

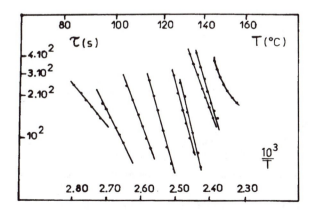

Figure 3. Relaxation map for PMMA. The spectral lines are relaxation curves obtained at various polarization temperatures T_p.

at a given polarization temperature, T_p. When the polarization, P, has reached its equilibrium value, the temperature is decreased to freeze this configuration. Then the field is cut off. The polarization recovery is induced by increasing the temperature in a controlled manner. The depolarization current, I, flowing through the external circuit is measured by an electrometer, and allows measurement of the dipolar conductivity, σ. If the isothermal polarization varies exponentially with time, then its relaxation time, τ, is deduced from the measure of σ:

$$\tau = \frac{P}{E \cdot \sigma} \tag{1}$$

RMA.   When the polarization is due to a distribution of relaxation times, the technique of windowing polarization is used for the experimental resolution of spectra (5) and production of the relaxation map (Figure 3). For a simple behavior described by a Kelvin–Voigt model, the elementary retardation time, τ_i, is given by:

$$\tau_i(T) = \dot{P}\,\frac{(T)}{J(T)} \tag{2}$$

where $J(T) = \dot{P}(T)$, the rate of depolarization. The analysis of each resolved spectrum gives a temperature-dependent retardation time, $\tau_i(T)$, that follows either an Arrhenius equation:

$$\tau_i(T) = \tau_{0i} \cdot \exp \frac{\Delta H}{kT} \tag{3}$$

(where τ_{0i} is the preexponential factor, ΔH is the activation enthalpy, and k is the Boltzmann constant) or a Vogel equation:

$$\tau_i(T) = \tau_{0i} \cdot \exp \left[\alpha_s \cdot (T - T_\infty)\right]^{-1} \tag{4}$$

where τ_{0i} is the preexponential factor, α_s is the average thermal expansion coefficient of the free volume, and T_∞ is the critical temperature at which the retardation time becomes infinite, that is, where there is no mobility. RMA determines the variation of the elementary enthalpies, preexponential factors (related to the entropy of activation), coefficients of free-volume expansion, and temperatures of zero mobility, with respect to a parameter under investigation, whether it be the dependence on molecular weight, chemical structure, orientation, or thermodynamic history.

The TSC/RMA Spectrometer: An Automated Instrument

In obtaining a material's relaxation map, a large number of experiments must be performed. Better resolution of the relaxation spectrum requires smaller experiment windows $(T_p - T_d)$ and more experiments. The solution is to automate the RMA process. Solomat has designed a fully automated TSC/RMA spectrometer (Figure 4).

The spectrometer's hardware features an IBM-type 80286 microprocessor with 1 Mbyte of RAM (random access memory) for high-speed real-time analysis, with a 40-Mbyte hard disk and a 1.2-Mbyte floppy disk. This computer system, connected to the cell head and to the electrometer, is driven by software that makes the instrument easy to use. Once the experiment has begun, the computer systems constantly monitor and control vacuum levels, cooling liquid, helium, and PID temperatures. Automatic analytical functions include digital data collection, data graphing, data transfer, data analysis, and slide preparation. The electrometer measures current with 10^{-16}-amp sensitivity. The cell head developed by the researchers at the Laboratory of Physics of Solids in Toulouse is reliable, precise, and simple to use.

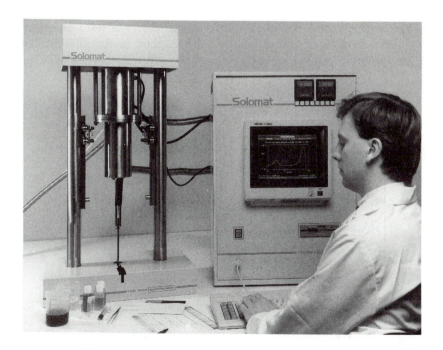

Figure 4. Solomat TSC/RMA spectrometer.

Liquid nitrogen, helium, and vacuum supplies are all that are needed for ready-to-run experiments, and a complete family of sample holders can accommodate most solids, liquids, or coatings.

TSC/RMA Spectroscopy for Engineering Applications

The application of windowing polarization technology (RMA) to the study of amorphous polymers has revealed properties of the glassy state never observed previously, with investigations on polystyrene (PS) (6–10), poly(methyl methacrylate) (PMMA) (11–15), poly(vinyl chloride) (PVC) (16), polycarbonate (17, 18), and poly(ethylene terephthalate) (PET) (19, 20). This new type of TSC analysis brings a new light to the following questions: Why is polycarbonate so tough? Why are polystyrene and Plexiglas so brittle? RMA shows that, in most amorphous polymers, the major relaxation modes responsible for internal flow decompose into a variety of elementary mechanisms well described by the activated-state theories (Figure 5). However, in a few instances the molecular processes obey a Williams–Landel–Ferry (WLF)-type equation, which reveals the dominance of a free-volume effect over an activated process for that relaxation mode (upper temperature relaxation curve in Figure 5). Such WLF activities observed for motions below the glass transition temperature seem to dominate in tough polymers (Figure 6).

For semicrystalline polymers, the resolution power of RMA ascertains the difference between the macromolecules trapped in the interlamellar regions and those that belong to the true amorphous region (Figure 7). The

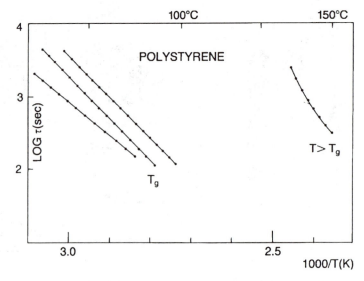

Figure 5. Relaxation map for annealed polystyrene. This map is the result of the analysis of the deconvoluted curves in Figure 2.

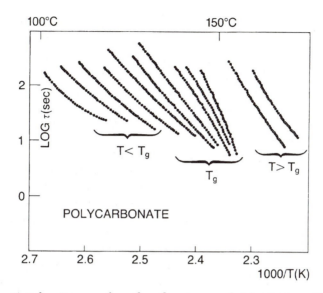

Figure 6. Relaxation map for polycarbonate annealed for 5 min at 235 °C.

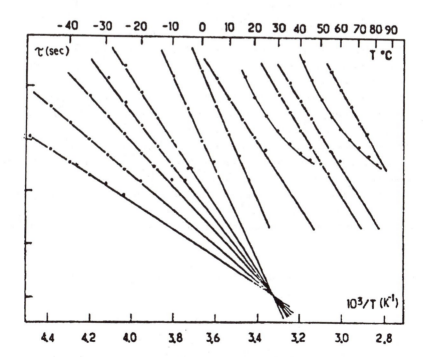

Figure 7. Relaxation map for polypropylene. The compensation point at $T_c =$ 20 °C corresponds to the T_g of the interspherulitic amorphous phase.

reason is simple. In TSC, amorphous polymers display a very strong relaxation mode at the glass transition temperature (T_g). This mode is attributed to micro-Brownian motions of the amorphous chains. Understanding the relative intensity of the interlamellar tissue versus the true amorphous component is crucial to determining the end-use properties.

In a semicrystalline material, window polarization analysis reveals two distinct relaxation modes at lower temperatures (Figure 7), a result clearly indicating the existence of a fine structure for the amorphous region. The relaxation component observed at lower temperatures is attributed to the regions free from constraint, that is, to the interspherulitic regions. This region of the polymer may be mechanically strained, as evidenced in polypropylene (Figure 7) by the presence of a law of compensation for the lower relaxation modes. As will be shown later, amorphous matter under internal stress, because of either orientation or thermal stresses, demonstrates such a compensation phenomenon.

TSC/RMA is therefore a unique technique to "measure" the extent of internal stress built up in a material, whether at the boundary between two phases or in the bulk. This technique is particularly suitable to study the quality of interfaces between the matrix and the fiber for a composite material, or to quantify the quality of the bonding phase for coatings, paints, and adhesives. For semicrystalline polymers, the component at higher temperatures corresponds to the amorphous chains under constraints from crystallites, that is, intercrystalline regions. No compensation effect exists for these relaxations, because the macroscopic stress will affect the interspherulitic material first.

The free-volume content in glassy thermoplastics and in the glassy phase of semicrystalline polymers depends upon the rate of cooling through the glass transition temperature and the processing conditions. It strongly affects certain macroscopic mechanical properties. RMA very easily identifies the relaxation modes that are free-volume controlled. These relaxations show up curved on an Arrhenius plot (Figure 7). The free-volume parameters can easily be found from analysis of the curved relaxation modes. The dependence of the processing variable on free-volume parameters for both the interspherulitic and the intraspherulitic amorphous phases can be identified and characterized.

Influence of Orientation, Hydrostatic Pressure, and Processing Condition

Polystyrene was used as the model for investigating the influence of a mechanical stress field on the transition maps. Figure 8 shows the relaxation map of oriented polystyrene. The orientation ($\simeq 30\%$) was induced by the extrusion process. The elementary processes isolated between 91 and 102 °C are characterized by relaxation times following a compensation law; they

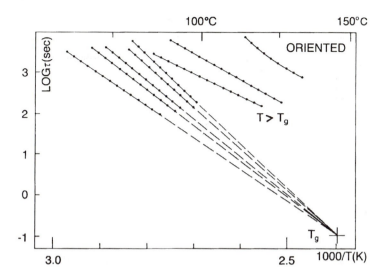

Figure 8. Effect of 30% orientation on the RMA plot of PS.

have the same relaxation time ($\tau_c = 0.11$ s) at the compensation temperature, $T_c = 145$ °C. This mode is the dielectric manifestation of the glass transition. The elementary peak isolated at $T_p = 135$ °C is well described by a Fulcher–Vogel equation with a critical temperature, $T_\infty = 50$ °C, a thermal expansion coefficient of free volume of 1.7×10^{-3} K^{-1}, and a preexponential factor of 0.74 s.

Comparison of the relaxation maps of annealed polystyrene (reference) and oriented polystyrene (Figures 5 and 8, respectively) shows that orientation induces a significant broadening of the distribution of relaxation times around the glass transition. Such compensation phenomena were found (5) to be accentuated by the presence of crystallinity, dopant, or plasticizer and are rather representative of an amorphous phase brought out of equilibrium by mechanical or thermal means. It is not surprising that such compensation phenomena appear in polystyrene under mechanical orientation. The samples designated as pressurized polystyrene were first annealed at 180 °C for 20 min to remove any sample molded in orientation, then heated to 122 °C, after which a hydrostatic pressure of 226 bar was applied. After the samples were quenched to room temperature, the pressure was released. For the samples designated as processed polystyrene, the thermal history was the same, but the hydrostatic pressure was modulated by a 55-bar signal at 30 Hz during cooling.

Pressurized Polystyrene.   Figure 9 shows the relaxation times isolated in pressurized polystyrene. The elementary processes isolated between 94 and 103 °C are characterized by relaxation times following a compensation

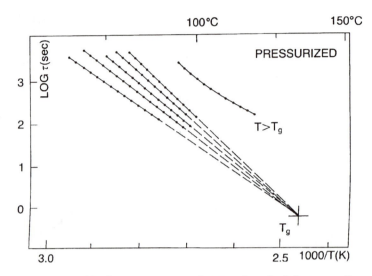

Figure 9. Effect of hydrostatic pressure (3300 psi) applied during cooling on the RMA plot of PS.

law at the compensation temperature, T_c = 135 °C. All the elementary processes have the same value (τ_c = 0.75 s). This mode corresponds to the dielectric manifestation of the glass transition. The elementary process isolated at 123 °C is well described by a relaxation time following a Fulcher–Vogel equation, with a critical temperature T_∞ = 16 °C and a thermal expansion coefficient of free volume α_s = 10^{-3} K^{-1}.

Processed Polystyrene. The relaxation map for the processed polystyrene represented in Figure 10 is qualitatively analogous with that of pressurized polystyrene. The relaxation times isolated in the glass transition mode follow a compensation law with T_c = 145 °C and τ_c = 0.19 s. This relaxation time is described by a Fulcher–Vogel equation. The thermal expansion coefficient of free volume (α_s = 4.8 × 10^{-3} K^{-1}) remains comparable with the WLF value; however, the critical temperature decreases in a spectacular manner (T_∞ = −22 °C).

In pressurized and processed polystyrene, the width of the distribution function of the relaxation time and also the order parameter are significantly increased. The compensation temperatures (145 and 135 °C) defined by such distributions correspond to a discontinuity in the recovery procedure.

In pressurized and processed polystyrene, the TSC peak associated with the liquid–liquid transition is shifted toward lower temperatures. Figure 11 illustrates this shift for processed polystyrene (solid line spectra) in comparison with reference polystyrene (dashed line spectra).

This evolution reflects a strong decrease in the critical temperature down from 50 °C for the reference polystyrene to 16 and −22 °C for the pressurized

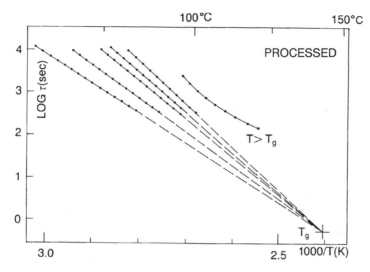

Figure 10. Effect of processing conditions during cooling on the RMA plot of PS.

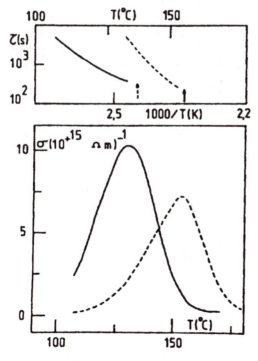

Figure 11. Comparison of the $T > T_g$ *elementary peak for annealed (– – –) and processing (—) conditions applied during the cooling of PS.*

and processed polystyrene, respectively. The spectacular decrease of T_∞ in processed polystyrene, illustrated in the Arrhenius diagram of Figure 11, might be responsible for the large improvement in the mechanical behavior observed for these treated materials (21).

Influence of Thermal Stresses on Cooling Conditions

It is difficult to measure the amount of thermal stresses induced by cooling. Yet, the cooling rate can determine the structure and the morphology of materials. Quickly cooled semicrystalline polymers, for instance, result in different morphologies, typically a larger number of smaller spherulites. The impact resistance of polycarbonate vanishes when it is cooled very slowly in the mold. The level of internal stress can be correlated with some success with birefringence measurements, but the one-to-one correspondence between stress and birefringence applies well only for pure elastic materials, which is not the case for the majority of transparent plastics.

DSC measurements are somewhat sensitive to the degree of cooling (Figure 12), but the basic problem remains to relate the heat capacity measurement to a parameter that could be used to measure the extent of thermal stress induced. Figure 13 shows RMA spectra obtained on polystyrene samples cooled from 180 °C under various cooling conditions. Figure 13a applies to glycol cooling (glycol at –22 °C), Figure 13b corresponds to forced-air cooling (1 bar of air pressure), and Figure 13c applies to water cooling. The RMA spectra are remarkably different. Figure 14 is a plot of the activation energy of the various Arrhenius lines drawn from the figures in Figure 13 versus the polarization temperature. The correlation between the activation

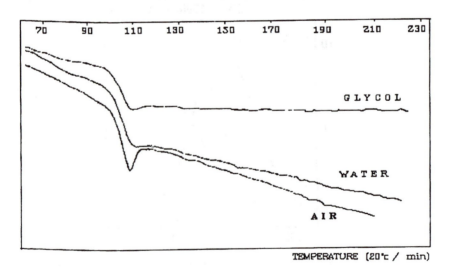

Figure 12. DSC curves for PS cooled at various rates.

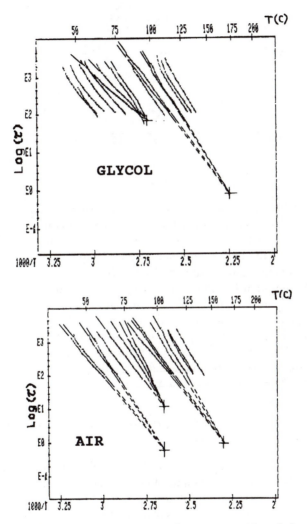

Figure 13. *Relaxation maps for the samples in Figure 12 cooled at various rates.* Continued on next page.

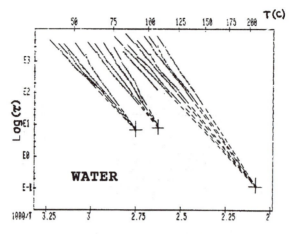

Figure 13. Continued.

enthalpy and T_p gives a better resolution of the various thermal transitions taking place in the material than DSC can provide (Figure 12).

Figure 13 reveals the presence of a maximum of three compensations: at slow cooling rate (air), the temperature of compensation for the $T < T_g$ compensation is the same as the one for the $T = T_g$ compensation, and the $T > T_g$ compensation has the same entropy (the mirror of the vertical coordinate) as the sub-T_g compensation. For water cooling, the entropy of the sub-T_g compensation is identical to the one at $T = T_g$. These observations are probably the result of competing mechanisms to establish the kinetics. One of the effects of cooling rate is to favor one mechanism or the other, and as a consequence, determine whether the properties below T_g are modulated by the $T > T_g$ history or the properties at T_g.

Characterization of Latex Copolymers by DSC, DMA, and TSC/RMA Spectroscopy

The microstructure of statistical and block copolymers is the key to their macroscopic behavior. The amount of alloying and segregation and the degree of interpenetration of the phases are crucial questions. Yet, in several instances, the traditional techniques of microstructure analysis such as DSC, TMA, or DMA have reached their limit of resolution, sometimes to the point that it is impossible to define even the fundamental question of whether there is segregation of the phases.

We will consider here three types of amorphous copolymers of styrene, S, and *n*-butyl acrylate, A (22). The conditions of polymerization are responsible for the change in the microstructures of these three latices. The statistical copolymer SC is prepared by a method of emulsion copolymeri-

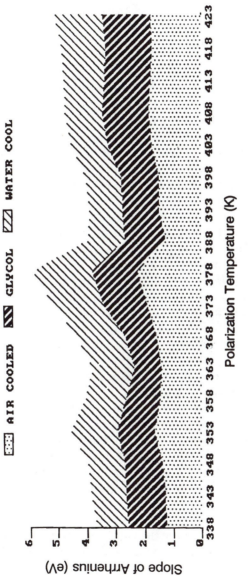

Figure 14. Variation of the slope of the Arrhenius lines of Figure 13 versus polarization temperature T_p.

zation in which both monomers are fed randomly in a semicontinuous manner
(23). Block copolymers are prepared by step emulsion copolymerization of
A first, then S, to form A.S.S. block copolymers, or S first, then A, to
synthesize the S.A.S. block copolymers. Our objective was to characterize
and compare the microstructures of these three latices. DSC, TMA, and
TSC/RMA spectroscopy all were used. It will be shown that for these latices,
the power of resolution and characterization of the TSC/RMA technique far
exceeds the other methods.

DSC.   Figure 15 displays the three traces of heat capacity C_p versus
temperature (the heating rate was 20 °C/min) recorded on a differential
scanning calorimeter (Dupont 990) for the SC, A.S.S., and S.A.S. samples.
The thermogram for the statistical copolymer (SC) is characteristic of a quasi-
homogeneous structure, with only one T_g drop-off observed at 20 °C, a
temperature located between the T_g values of the polyacrylate and polystyrene homopolymeric phases. This behavior indicates the presence of a single
homogeneous phase for this copolymer. The drop-off of the baseline for the
A.S.S. block copolymer spreads over 70 °C, and a kink is visible around
15 °C, which might be an indication of the presence of a heterogeneous
microstructure. However, it is difficult to assess where the T_gs of the two
phases are located, perhaps −21 and + 15 °C.

The alternate block copolymer S.A.S. is even more troublesome to
analyze. Again the drop-off at T_g is very broad, over 85 °C, but there is
apparently no indication of the presence of two T_gs. In such a case, we can
only guess that there are two segregated phases, but we have no indication
of the microstructural differences between the S.A.S. and A.S.S. samples.

DMA.   Figure 16 is a representation of the storage modulus E' and
loss modulus E'' for the three latices SC, A.S.S., and S.A.S. as function of
temperature. Figure 16 does not reveal much at all. The case of the statistical
random copolymer SC is the only conclusive one: the semicontinuous introduction of monomer gives a relatively homogeneous phase with a single
T_g. The other curves (Figure 16) do not indicate the influence of polymerization conditions on the microstructure of those block copolymers. As for
DSC, only very vague statements on the presence of at least two phases can
be made.

TSC.   The TSC spectra of the copolymers were recorded after polarization ($E = 3$ MV/m) for 2 min at the temperatures indicated by the arrows
in Figure 17.

The variation of the dynamic conductivity in the temperature range −150
to + 150 °C shows a single TSC peak for the SC copolymer and two resolved
peaks for both A.S.S. and S.A.S. The low-temperature peak observed in
both A.S.S. and S.A.S. is associated with the glass transition of butyl acrylate

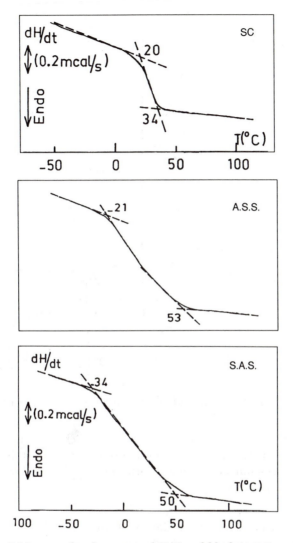

Figure 15. DSC traces for the statistical (SC) and block (A.S.S. and S.A.S.) copolymers.

sequences, and the high-temperature peak with the glass transition of styrene sequences. Comparison of Figures 15, 16, and 17 clearly reveals the superior sensitivity of the TSC method.

RMA.   The TSC analysis revealed the global microstructure of the materials under investigation: SC is a relatively homogeneous statistical random copolymer with a T_g located between the T_gs of the homopolymers. Both S.A.S. and A.S.S. are block copolymers with segregation of two phases,

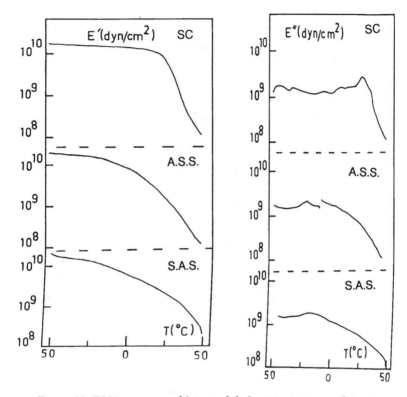

Figure 16. DMA storage and loss moduli for SC, A.S.S., and S.A.S.

one rich in the styrene component, and the other rich in acrylate sequences. The amount of styrene in acrylate (or acrylate in styrene) is, however, unknown. The technique of "window polarization" can be employed to answer this important question.

Compensation Point and Compensation Search.   When several Arrhenius lines converge into a single point, this point is called a compensation point. In such a case, the entropy (the negative or mirror image of the intercept of the Arrhenius line) and the enthalpy (the slope of the Arrhenius line) are linearly related to each other. Hence, a very simple and practical way to see whether a set of Arrhenius lines obtained at various T_p values converge is to plot intercept versus slope for these lines and to try to draw a straight line through the points (Figure 18). This drawn line is the compensation line. The coordinates of the compensation point are calculated from the slope and intercept of the compensation line. This type of analysis is called a compensation search. In general, for amorphous polymers, the behavior at T_g is characterized by a compensation phenomenon, as clearly demonstrated for oriented polystyrene under internal stress (Figure 8).

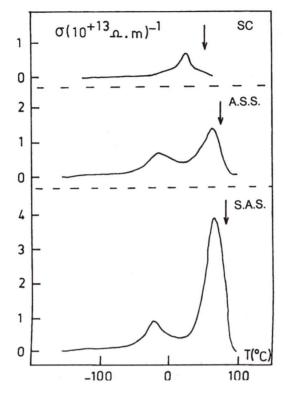

Figure 17. TSC traces for SC, A.S.S., and S.A.S.

To resolve the global TSC spectra of Figure 17, the technique of windowing polarization is applied in the temperature range –45 to +75 °C, with a temperature window of 5 °C. The electrical voltage applied has the same intensity as for the global TSC thermogram (Figure 17). Figure 19 shows, for A.S.S., the deconvolution of the global peaks into elementary Debye peaks for the two T_gs, and Figure 18 is a compensation search, a plot of intercept versus slope for all Arrhenius lines obtained for all copolymers studied. Table I presents the results of the analysis of the Arrhenius lines.

For the SC copolymer (the straight line in the middle of Figure 18), the single compensation line corresponds to one T_g, one single phase, but for the A.S.S. and S.A.S. copolymers, two compensation lines are observed. They correspond to the low A and high S temperature peaks observed with TSC (Figure 17). This behavior is characteristic of a biphasic structure.

The microstructure of the two phases in A.S.S. and S.A.S. is different, as shown in Table I for the value of the compensation parameters:

1. The relaxation modes associated with the A and S phases are significantly different in the A.S.S. and S.A.S. copolymers from those in the homopolymers.

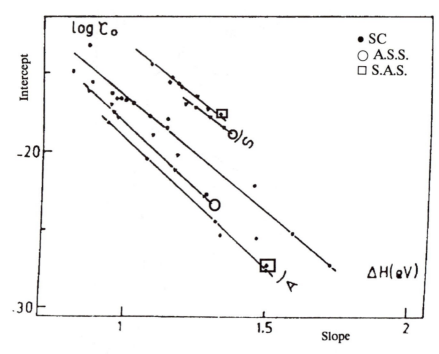

Figure 18. Compensation lines to determine the coordinates of the compensation points for the relaxation map analysis of SC, A.S.S., and S.A.S.

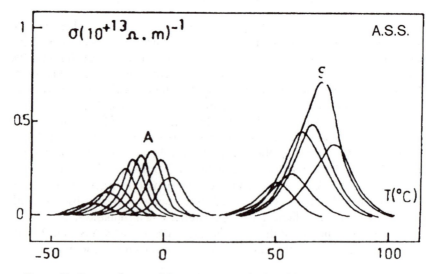

Figure 19. Deconvolution of the two main peaks found by TSC into elementary Debye components (S.A.S.).

Table I. Compensation Parameters for Copolymers

			A Line		S Line	
Sample	T_c (°C)	τ_c (s)	T_{cA} (°C)	τ_{cA} (s)	T_{cS} (°C)	τ_{cS} (s)
SC	64	0.035	—	—	—	—
A.S.S.	—	—	32	6.0×10^{-3}	105	0.045
S.A.S.	—	—	28	4.5×10^{-3}	104	0.55

NOTE: Subscript A refers to the A phase, S to the S phase. The A and S lines are those of Figure 18.

2. The phase segregation is not complete in the block copolymers.

3. The compensation diagram (Figure 18) quantifies the difference between the structure of the amorphous phases in A.S.S. and S.A.S. The preexponential factors are higher for A.S.S. (the compensation line is situated above), and, because the entropy of activation is the mirror of the log of the preexponential factor, the activation entropies are lower in A.S.S. than in S.A.S.

4. The phase segregation is lower in A.S.S. than in S.A.S.

These results indicate a core shell structure and suggest a morphology in which latex particles have a (butyl acrylate)-rich core and a styrene-rich shell for both A.S.S. and S.A.S. These results also suggest that the phase segregation is more pronounced for S.A.S. These conclusions are consistent with those drawn from other results published separately (*21, 22*).

References

1. Vanderschueren, J., Ph.D. Thesis, University of Liege, Belgium.
2. van Turnhout, J. *Thermally Stimulated Discharge of Polymer Electrets*; Elsevier: New York, 1975.
3. Lacabanne, C., Ph.D. Thesis, University of Toulouse, France, 1974.
4. Chatain, D., Ph.D. Thesis, University of Toulouse, France, 1974.
5. Bernes, A.; Boyer, R. F.; Chatain, D.; Lacabanne, C.; Ibar, J. P. In *Order in the Amorphous State of Polymers*; Keinath, S. E., Ed.; Plenum: New York, 1987; pp 305–326.
6. Diaconu, I.; Dumitrescu, S. V. *Eur. Polym. J.* **1978**, *14*, 971–975.
7. Goyaud, P., M.S. Thesis, University of Toulouse, France, 1979.
8. Lacabanne, C.; Goyaud, P.; Boyer, R. F. *J. Polym. Sci. Polym. Phys. Ed.* **1980**, *18*, 277–284.
9. Shrivastava, S. K.; Ranade, J. D.; Srivastava, A. P. *Thin Solid Films* **1980**, *67*, 201–206.
10. Jeszka, J. K.; Ulanski, J.; Glowacki, I.; Kryszewski, M. *J. Electrost.* **1984**, *16*, 89–98.
11. Kryszewski, M.; Zielinski, M.; Sapieha, S. *Polymer*, **1976**, *17*, 212–216.
12. Ohara, K.; Rehage, G. *Colloid Polym. Sci.* **1981**, *259*, 318–325.

13. Biros, J.; Larina, T.; Trekoval, J.; Pouchly, J. *Colloid Polym. Sci.* **1982**, *260*, 27–30.
14. Gourari, A., M.S. Thesis, University of Algeria, 1982.
15. Gourari, A.; Bendadaoud, M.; Lacabanne, C.; Boyer, R. F. *J. Polym. Sci. Polym. Phys. Ed.* **1985**, *23*, 889–916.
16. Barandiaran, J. M.; Del Val, J. J.; Colmenero, J.; Lacabanne, C.; Chatain, D.; Millan, J.; Martinez, G. *J. Macromol. Sci. Phys. Ed.* **1984**, *B22*, 645–663.
17. Aoki, Y.; Brittain, J. O. *J. Polym. Sci. Polym. Phys. Ed.* **1976**, *14*, 1297–1304.
18. Guerdoux, L.; Marchal, E. *Polymer*, **1981**, *22*, 1199–1204.
19. Sawa, G.; Nakamura, S.; Nishio, Y.; Ieda, M. *Jpn. J. Appl. Phys.* **1978**, *17*, 1507–1511.
20. Belana, J.; Colomer, P.; Pujal, M.; Montserrat, S. *J. Macromol Sci.* in press.
21. Ibar, J. P. *Polym. Plast. Technol. Eng.* **1981**, *17*, 1, 11.
22. Cebeillac, P., Ph.D. thesis, Paul Sabatier University, 31062 Toulouse, France, 1989.
23. Materials were synthesized by Rhone-Poulenc Recherches, Aubervilliers, France.

RECEIVED for review February 14, 1989. ACCEPTED revised manuscript July 26, 1989.

11

Thermally Stimulated Creep for the Study of Copolymers and Blends

Ph. Demont, L. Fourmaud, D. Chatain, and C. Lacabanne

Solid State Physics Laboratory, Paul Sabatier University, 31062 Toulouse Cédex, France

Amorphous phase segregation in polyamide-based copolymers and blends was investigated by thermally stimulated creep. Because of its resolving power, this technique allows complex retardation time spectra to be resolved. A series of poly(ether-amide) block copolymers with constant stoichiometry was studied as a function of mean block length. The thermally stimulated creep values of the polyamide–poly(vinylidene fluoride) dominant phase were compared. In both cases, the dominant phase was found to keep its own amorphous phase structure. Activation entropy–enthalpy compensation diagrams are well suited for determining amorphous phase separation in copolymers and blends.

Mᴜʟᴛɪᴘʜᴀꜱᴇ ᴍᴀᴛᴇʀɪᴀʟꜱ ꜱᴜᴄʜ ᴀꜱ ʙʟᴏᴄᴋ ᴄᴏᴘᴏʟʏᴍᴇʀꜱ or blends are widely applied in various fields as functional materials. In this chapter, we will consider, as examples of block copolymers, poly(ether-*block*-amide) (PEBA) copolymers, which constitute a new class of thermoplastic elastomer (*1*). PEBA copolymers consist of alternating linear soft polyether (PE) and hard polyamide (PA) blocks. The soft segments, which possess a relatively low glass transition temperature, are in their rubbery state at use temperature; they impart the elastomeric properties to the copolymers. The hard segments are semicrystalline; they can undergo some kind of intermolecular association with other such hard blocks and thereby form physical cross-links like hydrogen bonds. The physical cross-linking of the copolymers provides dimensional stability and minimizes cold flow.

0065–2393/90/0227–0191$06.75/0

The incompatibility of the two different chain segments causes micro-phase separation with the formation of hard- and soft-segment-rich domains. The extent of this microphase separation will be influenced by the block length, the crystallizability of the soft segment, and the overall hard-segment content. The degree to which the dissimilar blocks segregate into their respective domains will determine the thermal and mechanical properties of the block copolymers (2–4).

In polyamide–poly(vinylidene fluoride) blends, both parent homopol-ymers are semicrystalline. Despite this similarity, they also show strong differences, such as a glass transition temperature that is above room tem-perature for polyamide and below room temperature for poly(vinylidene fluoride). Moreover, the structure of the amorphous phase is stabilized by hydrogen bonds in polyamide but not in poly(vinylidene fluoride). Most likely, those blends also will demonstrate microphase segregation. The in-tegrity of the amorphous phase will be strongly related to the chemical composition.

Thermally stimulated creep (TSCr) analysis has been applied to the investigation of phase segregation in multiphasic polymers (5–9). Indeed, because of its very low equivalent frequency (10^{-3} Hz) and its high resolving power, it is very well suited for exploring broad retardation modes generally found in multiphasic polymers.

Experimental Details

Materials. Copolymer.   The poly(ether-amide) block copolymers used in this study were synthesized by ATOCHEM (France). A detailed procedure was published by Deleens et al. (1). This kind of block copolymer consists of sequences of soft oligoether and hard oligoamide segments:

$$\text{OH}-[[-\underset{\underset{O}{\|}}{C}-(CH_2)_{11}-NH]_m-\underset{\underset{O}{\|}}{C}-O-[-(CH_2)_4-O]_n]-H$$

Chemical structure of poly(ether – amide) block copolymers

The poly(tetramethylene oxide) (PTMO) soft segment has an average degree of po-lymerization specified by n. In this series of copolymers, n was 28, 14, and 9, giving average soft-segment molecular weights ($M_{n,PE}$) of 2000, 1000, and 650, respectively.

The polyamide-12 hard segment was prepared for values of m varying from 7 to 21, giving an average hard-segment molecular weight ($M_{n,PA}$) range of 600 to 4000. A summary of the structural parameters is given in Table I. The total molecular weight of the block copolymers was 20,000. The samples for TSCr experiments were prepared by compression molding under a pressure of 100 bar and at a temperature of 20 °C above the melting point of the respective copolymer, followed by quenching in ambient air. Films having dimensions of 60 × 5 × 0.5 mm were obtained and dried under vacuum for 2 h at 380 K.

Blends.   Polyamide-12 (PA12), poly(vinylidene fluoride) (PVDF), and their blends with or without compatibilizer were obtained from ATOCHEM. The char-

Table I. Structural Parameters of Poly(ether-*block*-amide) Copolymers

Sample	$\overline{M}_{n,PA}$	$\overline{M}_{n,PE}$	m	n	$W_H{}^a(\%)$	$\eta_{inh}{}^b$
4000–2000	4200	2032	21	28	67	1.33
2000–1000	2135	1000	10	14	67	1.37
1300–650	1360	600	7	9	67	1.30

[a] Weight fraction of the hard segment.
[b] Inherent viscosity in *m*-cresol solvent at 20 °C.

acteristics of homopolymers and blends are reported in Table II. The composition of blends is expressed in weight percent. Compression-molded sheets of homopolymers and blends were made in a press at a 200 °C melt temperature. The samples were then cut to a dimension of 60 × 6 × 0.5 mm for TSCr experiments. Before measurements, samples were dried under vacuum (10^{-4} torr, 13.3 mPa) at 400 K for 1 h.

Table II. Weight Fraction of Polyamide (PA),
Poly(vinylidene fluoride) (PVDF), and Compatibilizer (A)
in Blends

Sample	W_{PA}	W_{PVDF}	W_A
PA–PVDF	85	15	0
PVDF–PA	36	64	0
PA–PVDF–A	84.5	14.5	1
PVDF–PA–A	35.5	63.5	1

NOTE: All values are given in percents.

Methods.   *Differential Scanning Calorimetry (DSC).* DSC thermograms over the temperature range from –140 to about 200 °C were recorded on a Perkin Elmer DSC II. Calibration was performed with indium and mercury as standards. The experiments were carried out at a heating rate of 20 °C/min under a helium purge on 10-mg samples.

Thermally Stimulated Creep (TSCr) Recovery. The technique has been described elsewhere (5, 6, 10). The samples of copolymers were heated to a temperature T_σ, and a shear stress was applied for 2 min. The samples were then quenched from T_σ to $T_0 \ll T_\sigma$, thus freezing-in mechanical strain. At T_0, the stress was removed. The samples were then heated in a slow and controlled manner up to T_σ at a rate of 7 K/min. As the specimens were heated, the decay of the frozen-in strain $\gamma(t)$ was recorded: differentiation of $\gamma(t)$ yielded the rate of release of strain $\dot\gamma(t)$. All measurements were made in the linear region at strains lower than 10^{-3}. The temperature dependence of TSCr recovery rate $\dot\gamma(T)$ was normalized to the stress to obtain thermally stimulated creep complex spectra.

The technique of fractional loading (FL) (5–12) was used to study the kinetics of mechanical relaxations in copolymers. In FL experiments, a narrow packet of the distribution of retardation times is stimulated by an imposed mechanical stress history. The recovery kinetics of this narrow packet of relaxation times is then recorded experimentally.

FL experiments were performed as follows:

1. A stress σ was applied at stressing temperature T_σ during time t_σ = 3 min.

2. The sample was cooled to the temperature $T_d = T_\sigma - \Delta T$ (5 K) under constant stress. $T_\sigma - T_d$ is called the stress window.

3. At T_d the stress was removed and the sample was permitted to recover partially for a time $t_d = 1$ min.

4. The sample was then quenched to 50 K below T_d, and viscoelastic strain consequently was frozen in.

5. The heating run at a rate of 7 K/min then commenced, during which the frozen-in strain $\gamma(t)$ was slowly released. The heating run was complete at 50 K above T_σ.

By varying T_σ in the temperature range with a "loading step" of 5 K, the whole TSCr spectrum can be resolved into "elementary spectra".

The FL data show a series of sharp elementary peaks. Every thermally stimulated process can be characterized by a single retardation time $\tau(t)$, which is then determined by

$$\tau(t) = \frac{\gamma(t)}{\dot{\gamma}(t)} \tag{1}$$

Because the recovery of the strain of the sample is thermally stimulated, time and temperature are related by a linear relationship. Therefore, the retardation time $\tau(t)$ is temperature dependent and may be written as $\tau(T)$. In oligomers and copolymers, an Arrhenius type dependence on temperature is assumed for the retardation time:

$$\tau(T) = \tau_0 \exp \frac{\Delta H}{kT} \tag{2}$$

where k is the Boltzmann constant, τ_0 is the preexponential factor, and ΔH is the activation enthalpy and the preexponential factor.

These activation parameters are characteristic of the relaxing unit and have been determined by using a least-squares fitting procedure taking retardation times between 150 and 5000 s. This time range corresponds to the linear portion of the ln τ − T^{-1} plot. In the barrier theories for relaxation phenomena in solid-state polymers, the temperature dependence of the retardation time is also expressed by the Eyring equation:

$$\tau(T) = \frac{h}{kT} \exp \frac{-\Delta S}{k} \exp \frac{\Delta H}{kT} \tag{3}$$

where ΔS is the activation entropy and h is the Planck constant. It follows from equations 2 and 3 that

$$\tau_0 = \frac{h}{kT} \exp \frac{-\Delta S}{k} \tag{4}$$

The preexponential factor can be considered to be related to the number of available orientations for the relaxing element in the chain rotation (7, 13–15).

As already reported, the τ_0 value determined in thermally stimulated current (TSC) and TSCr experiments contains information about the short-range organization of the amorphous phase in polymers (5, 7, 13, 15).

Compensation Law. The most striking conclusion of thermal stimulation experiments on dielectric and mechanical relaxations in polymers is the fact that, within a distribution of relaxation retardation times, the various units have different activation enthalpies (*5, 16*).

When an activation enthalpy distribution is observed in polymer relaxations, a compensation law exists, and it has been found to be a general phenomenon in polymer-distributed relaxations (*5–10, 13, 16*). When several Arrhenius lines converge into a single point, this point is called a compensation point. In general, for amorphous polymers, or for the amorphous region in semicrystalline polymers, the behavior at T_g is characterized by a compensation phenomenon (*5–9, 14, 16*).

The compensation effect is a linear relationship between the logarithm of τ_0 and ΔH in a temperature-activated process. For retardation times, the compensation law is expressed as

$$\ln \tau_0 = -\frac{\Delta H}{kT_c} + \ln \tau_c \tag{5}$$

where T_c is the compensation temperature and τ_c is the compensation time. It follows from equations 4 and 5 that

$$\Delta H = T_c \Delta S + \Delta H_0 \tag{6}$$

where ΔH_0 is a constant. Then, the ith retardation time τ_i of the distribution of retardation times depends on temperature as

$$\tau_i = \tau_c \exp \frac{\Delta H_i}{k} \left[\frac{1}{T} - \frac{1}{T_c} \right] \tag{7}$$

At the compensation temperature T_c, all retardation times take the same value τ_c. If a compensation phenomenon occurs, all Arrhenius straight lines converge to a point located at $T = T_c$, on a $\ln \tau$ vs. $1/T$ diagram.

The compensation temperature T_c is obtained from the slope of the straight line on a $\ln \tau_0$ vs. ΔH diagram by least-squares fitting. τ_c is then calculated from equation 5.

In the polymers studied, the fine structure of the distributed TSCr process was systematically characterized by such $\ln \tau_0$–ΔH compensation diagrams.

PEBA Copolymers—Results

Differential Scanning Calorimetry (DSC). The results of DSC studies on poly(ether-*block*-amide) copolymers are shown in Figure 1 and are summarized in Table III. Three thermodynamic events may be observed: the glass transition of soft segments (T_g), the melting of soft segments ($T_{m,\text{PE}}$), and the melting of hard segments ($T_{m,\text{PA}}$).

Glass Transition of Soft Segments (T_g). Decreasing the block length of soft segments increases the T_g and the heat capacity change $\Delta C_p(T_g)$ through the transition zone. The fact that T_g shows a shift to a higher tem-

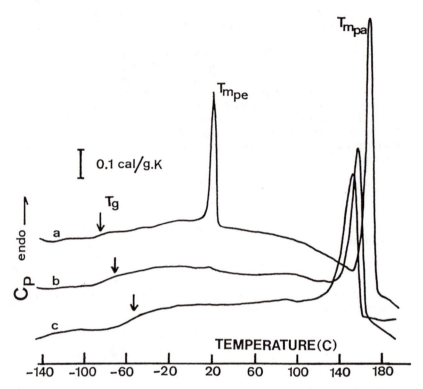

Figure 1. DSC thermograms of poly(ether-block-amide) copolymers (a) 4000–2000, (b) 2000–1000, and (c) 1300–650. The heating rate was 20 °C/min.

Table III. DSC Results for PEBA Copolymers

Sample	T_g (°C)	$\Delta C_p (T_g)$ (cal/g deg)	$T_{m,PE}$ (°C)	$T_{m,PA}$ (°C)	$X_{PE}{}^a$ (%)	$X_{PA}{}^b$ (%)
4000–2000	−83 (−83)	1.7×10^{-2}	24 (25)	172 (173)	16.8 (57)	21.5(26)
2000–1000	−70 (−88)	5.8×10^{-2}	—^{c} (19)	158 (163)	— (55)	23 (25)
1300–650	−53 (−90)	8.0×10^{-2}	— (14)	153 (155)	— (60)	23 (25)

NOTE: The values in parentheses are the DSC results for the PTMO and PA12 oligomers used for the preparation of the PEBA copolymers, reported here for reference.

^{a} $X_{PE} (PEBA) = \left[\dfrac{\Delta H_m}{\Delta H^0_{m,PE}} \right] \times \dfrac{1}{W_S}$; $W_S = 1 - W_H$, where W_S is the weight fraction of the soft segment and W_H is the weight function of the hard segment; $\Delta H_{m,PA} = 53.8$ cal/g (17).

^{b} $X_{PA} (PEBA) = \left[\dfrac{\Delta H_m}{\Delta H^0_{m,PA}} \right] \times \dfrac{1}{W_H}$; $\Delta H_{m,PE} = 41.15$ cal/g (18).

^{c} — indicates no value available.

perature with respect to the molecular weight of soft PTMO sequences suggests that solubilized hard segments are included within the amorphous soft phase. Most likely, short PTMO and PA12 sequences act as a random copolymer. A comparison of the T_g values of PTMO oligomers and copolymers indicates the increase of $T_g = T_{g,\text{PEBA}} - T_{g,\text{PTMO}}$ from 0 to 37 °C when soft-segment molecular weight is reduced by a factor of 3.

For relatively short sequence lengths, the poly(ether-*block*-amide)s reveal extensive phase mixing in the amorphous phase. The progressive influence of the hard segments on the mobility of the amorphous soft-segment units is accompanied by the gradual broadening of the glass transition domain.

Melting of Soft Segments $(T_{m,\text{PE}})$.   A clearly defined low-temperature glass transition occurs for the soft segment, but no melting endotherm is observed in the shorter block copolymers. These data indicate the absence of any soft-segment crystallinity in these samples. As in segmented polyurethanes, the soft PTMO blocks crystallize only when the molecular weight is greater than about 1000 (*4, 19–22*).

This result supports the hypothesis of increasing strength of the glass transition with decreasing block length. The PTMO constituents are present in a partially crystalline form only for the 4000–2000 sample. The results of DSC study listed in Table II show a large decrease of the soft-segment crystallinity in this sample compared to the corresponding pure PTMO 2000 oligomer, which has a very high crystallinity of 57%.

As shown in a previous study on poly(ether-*block*-amide) copolymers (*15*), the constraint on soft-segment mobility caused by the covalent joints between the hard and soft segments reduces the crystallization of soft segments.

Melting of Hard Segments $(T_{m,\text{PA}})$.   The high-temperature endotherm is identified as the melting of the crystalline polyamide phase. Values of the melting point $T_{m,\text{PA}}$ compared to the corresponding oligoamides are summarized in Table III. A decrease of the melting temperature $T_{m,\text{PA}}$ is observed with the decreasing polyamide block length. This lowering of the $T_{m,\text{PA}}$ of partially crystalline block copolymers is explained by the oligomer melting behavior. Indeed, the hard structural phase melts and crystallizes independently of the presence of soft segments.

This assumption is verified by the dependence of the reciprocal melting temperature on the reciprocal average degree of polymerization of oligoamide structural units, as can be expected according to Flory (*23*). Hence, the depression of the melting point of hard sequences is only the consequence of reduced chain length.

Thermally Stimulated Creep Complex Spectra.   Figures 2 and 3 show the TSCr spectra of series of poly(ether-*block*-amide) samples. For the 4000–2000 sample, four peaks are designated by γ, β$_{PE}$, β$_{PA}$, and α for increasing temperature. Values of their maximum temperatures are listed in Table IV. With decreasing soft- and hard-block length, the following features can be noted:

γ Peak.   The γ peak observed at about 120 K is almost independent of sequence length. As shown by results in Table IV, the TSCr peak lies between those of the pure oligomers. The γ relaxation may be associated with local motions of the $(CH_2)_4$ sequences of the soft PTMO segments and the $(CH_2)_{11}$ sequences of the hard PA12 segments, which occur in the amorphous phase (*24, 25*).

β Peaks.   The β$_{PE}$ peak is shifted to higher temperatures because of the restrictions that the hard segments impose on the amorphous soft phase. At the same time the intensity of the β$_{PE}$ peak significantly increases.

Throughout the series of 1300–650 samples studied, the β$_{PA}$ peak is rich in polyamide hard segments. The observation of this broad retardation process is coherent with the broad transition zone shown by the DSC thermogram.

The absence of any soft-segment crystallinity can explain the increase

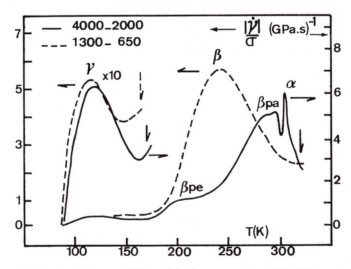

Figure 2. Low-temperature (LT) and high-temperature (HT) TSCr complex spectra of PEBA copolymers. Key: —, 4000–2000, LT T$_\sigma$ = 170 K, σ = 0.8 MPa; HT T$_\sigma$ = 320 K, σ = 0.4 MPa; and ---, 1300–650, LT T$_\sigma$ = 170 K, σ = 0.9 MPa; HT T$_\sigma$ = 320 K, σ = 0.4 MPa.

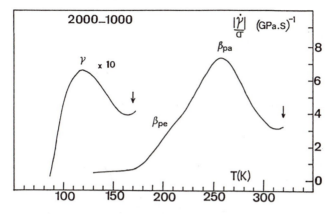

Figure 3. TSCr complex spectra of 2000–1000 PEBA copolymer. LT T$_\sigma$ = 170 K, σ = 0.6 MPa; HT T$_\sigma$ = 320 K, σ = 0.3 MPa.

of intensity of the β$_{PE}$ peak for short soft-segment lengths and so the absence of any β$_{PA}$ peak.

The progressive shift of the β$_{PE}$ peak to higher temperature with decreasing segment length can be attributed to a greater fraction of hard segments dissolved in the soft phase. Table IV gives a comparison of the peak position of the β$_{PA}$ processes in copolymers with the corresponding temperature observed in pure PA12 oligomers. A significant decrease of β$_{PA}$ peak temperature is observed, and this result confirms the improved amorphous phase mixing in short-sequence materials. On the other hand, the 4000–2000 block copolymer has hard and soft segments with lengths long enough to form two crystalline phases and exhibit a better phase separation. This behavior may be explained in terms of higher hard-domain interconnectivity and soft-segment crystallinity.

α *Peak.* The α peak occurs only in copolymers containing a relatively high soft-block length: $M_{n,PE} > 2000$ (*4, 15*). It is associated with the melting

Table IV. TSCr Peak Temperatures for the γ, β$_{PE}$, and β$_{PA}$ Relaxation Regions of the PEBA Block Copolymers and the Pure PE and PA Oligomers

	T$_\gamma$			T$_{\beta.PE}$		T$_{\beta.PA}$	
Sample	PEBA	PE	PA	PEBA	PE	PEBA	PA
4000–2000	119	126	114	200	215	289	307
2000–1000	119	121	110	215	209	257	302
1300–650	117	119	107	—[a]	208	242	296

NOTE: All values are given in kelvins.
[a] — indicates no value available.

of PTMO crystallites. At lower molecular weight, no process is observed, a result confirming the DSC results (no melting endotherm).

If the crystalline phase separation is clearly defined in these copolymers, the amorphous phase separation is more difficult to ascertain. Accordingly, the complex spectral study of poly(ether-*block*-amide) copolymers was completed by an analysis of the fine structure of the β_{PE} and β_{PA} processes associated with the glass transition of the amorphous polyether and polyamide phases, respectively.

Fine Structure Analysis of TSCr Complex Spectra.   Table V gives the values of the activation enthalpy ΔH and of the preexponential factor τ_0 determined in a scanning of β_{PE} and β_{PA} processes of the poly(ether-amide) block copolymers by a convenient fractional loading program. ΔH and τ_0 are reported as a function of the loading temperature T_σ. Figures 4–6 show compensation diagrams, reporting the results of Table V for the 4200–2000, 2000–1000, and 1300–650 PEBA samples, respectively. The activation parameters determined in the β_{PE} and β_{PA} processes of pure oligomers are also reported as references.

The fine structure analysis of the various relaxations clearly shows the strong discrepancy in the behavior of the activation parameters of β_{PE} and β_{PA} relaxations in poly(ether-*block*-amide) copolymers:

- The β_{PE} retardation mode is characterized by both an activation enthalpy and preexponential factor distributions. In fact, all copolymers investigated in this study show a compensation phenomenon in their β_{PE} relaxation region as in pure PTMO oligomers. Each retardation time of the distribution of retardation times isolated in β_{PE} mode obeys equation 7. Table VI gives the compensation temperature T_c and compensation time τ_c determined from equation 5 in β_{PE} retardation mode of PTMO oligomers.

- The β_{PA} retardation mode is characterized by a sharp activation enthalpy distribution and a τ_0 distribution. In the three samples, no compensation phenomenon was revealed in the β_{PA} process of the copolymers, as opposed to the oligoamide samples. Only an increase of τ_0 with a slight decrease of ΔH was observed when T_σ increased.

Influence of Sequence Length on β_{PE} and β_{PA} Processes.   With decreasing soft- and hard-sequence length at constant composition, the following features can be noted for β_{PE} processes:

- No difference in activation–enthalpy distribution can be observed in the three samples, although the length of PTMO segments is different.

Table V. Activation Parameters of β_{PE} and β_{PA} Retardation Modes of PEBA Copolymers

T_σ (K)	4000–2000		2000–1000		1300–650	
	ΔH (eV)	τ_0 (s)	ΔH (eV)	τ_0 (s)	ΔH (eV)	τ_0 (s)
			β_{PE} Retardation Mode			
170	0.615	7.7×10^{-17}	0.61	1.4×10^{-16}	0.61	1.0×10^{-16}
175	0.72	3.5×10^{-19}	0.665	5.2×10^{-18}	0.67	4.2×10^{-18}
180	0.83	7.8×10^{-22}	0.78	2.1×10^{-20}	0.74	2.1×10^{-19}
185	0.995	1.9×10^{-25}	0.88	2.9×10^{-22}	0.82	5.2×10^{-21}
190	1.185	4.4×10^{-30}	1.03	1.9×10^{-25}	0.90	1.2×10^{-22}
195	1.495	4.9×10^{-37}	1.215	6.5×10^{-30}	1.025	4.3×10^{-25}
200	1.44	4.9×10^{-35}	1.515	1.2×10^{-36}	1.19	2.4×10^{-28}
205	1.455	1.3×10^{-33}	1.455	1.8×10^{-34}	1.36	7.3×10^{-32}
210	1.38	1.7×10^{-31}	1.39	7.3×10^{-32}	1.54	7.3×10^{-36}
215	1.41	1.5×10^{-31}	1.43	1.2×10^{-31}	1.52	3.6×10^{-34}
220	1.30	3.2×10^{-28}	1.265	1.4×10^{-27}	1.57	1.6×10^{-34}
225	1.35	1.2×10^{-28}	1.37	3.6×10^{-29}	1.60	1.8×10^{-34}
			β_{PA} Retardation Mode			
230	1.27	2.2×10^{-26}	1.385	2.9×10^{-29}	1.57	4.4×10^{-33}
235	1.215	1.9×10^{-24}	1.365	2.4×10^{-28}	1.525	5.3×10^{-30}
240	1.17	4.7×10^{-23}	1.36	5.9×10^{-27}	1.50	4.4×10^{-30}
245	1.40	7.2×10^{-27}	1.335	1.9×10^{-26}	1.46	1.1×10^{-28}
250	1.27	4.8×10^{-24}	1.365	1.8×10^{-26}	1.50	7.2×10^{-29}
255	1.51	3.9×10^{-28}	1.30	1.1×10^{-24}	1.47	1.3×10^{-27}
260	1.43	3.9×10^{-26}	1.335	7.1×10^{-25}	1.475	3.6×10^{-27}
265	1.33	1.2×10^{-23}	1.30	9.6×10^{-24}	1.49	7.2×10^{-27}
270	1.35	1.3×10^{-23}	1.42	1.8×10^{-25}	1.42	5.3×10^{-25}
275	1.40	4.8×10^{-24}	1.43	7.9×10^{-29}	1.665	5.9×10^{-25}
280	1.495	2.1×10^{-25}	1.44	5.8×10^{-25}	—[a]	—

[a] — indicates no value available.

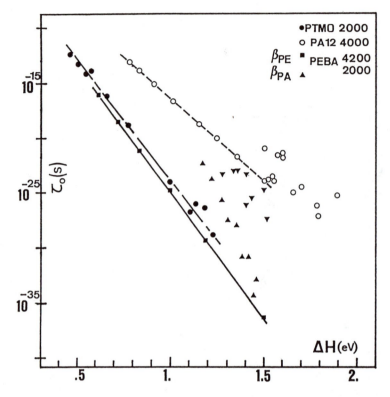

Figure 4. Compensation diagrams for the activation parameters of the β_{PE} *processes of 4000–2000 PEBA (■) and the 2000 PTMO oligomer (●), and the* β_{PA} *processes of 4000–2000 PEBA (▲) and the 4000 PA12 oligomer (○).*

- An increase of the compensation temperature T_c and a decrease of the compensation time τ_c can be deduced from Table VI.

- For a given higher value of activation enthalpy spectra, the τ_0 values increase; that is, the activation entropy decreases according to equation 4.

The following features can be noted for β_{PA} processes:

- The activation–enthalpy distribution slightly shifts toward higher values. On the other hand, the τ_0 distribution is shifted toward lower values.

- The existence of two distinct distributions of activation parameters is observed only for the 4000–2000 sample.

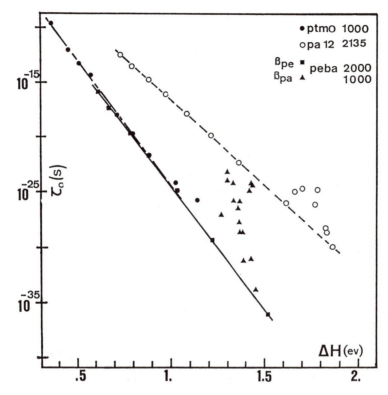

Figure 5. Compensation diagrams for the activation parameters of the β_{PE} processes of 2000–1000 PEBA (■) and the 1000 PTMO oligomer (●), and the β_{PA} processes of 2000–1000 PEBA (▲) and the 2135 PA12 oligomer (○).

Comparison with Homopolymers.   A comparison of the fine structure of copolymers with those of their parent oligomers also was performed. The main feature for β_{PE} PTMO oligomer relaxation is that, for a given specific ΔH value, the τ_0 is similar for 1300–650 and 2000–1000 copolymers (600 and 1000 PTMO oligomers), but lower for 4000–2000 copolymers (2000 PTMO oligomers).

The unexpected high values of τ_0 for the PTMO oligomers can be explained by the very high crystallinity of these samples with regard to the inhibition of crystallization of soft blocks in copolymers (*see* Table II and Figure 1). Concerning the β_{PA} relaxation of oligoamides, the values of τ_0 are always higher than in copolymers when ΔH is maintained constant.

PEBA Copolymers—Discussion

The fine structure analysis shows that retardation preexponential factors contain information about the environment of chain segments involved in

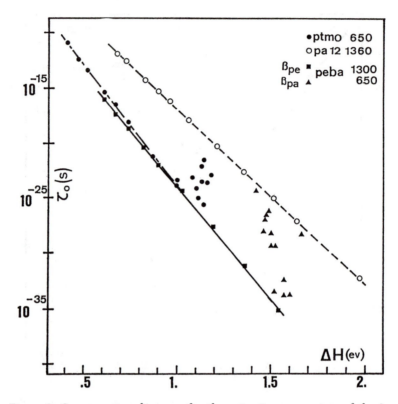

Figure 6. Compensation diagrams for the activation parameters of the β_{PE} processes of 1300–650 PEBA (■) and the 650 PTMO oligomer (●), and the β_{PA} processes of 1300–650 PEBA (▲) and the 1360 PA12 oligomer (○).

relaxation. The modification of the fine structure of β_{PE} and β_{PA} retardation modes as a function of the molecular weight of PTMO and polyamide segments can be interpreted in terms of degree of phase separation. The decrease of the activation entropy in β_{PE} processes with shortening soft sequences indicates that more order exists in the PTMO amorphous phase of the 1300–650 copolymer than the 4000–2000 copolymer. This conclusion

Table VI. Compensation Temperature and Time in the
β_{PE} Relaxation Region of the PEBA Copolymers and the
β_{PE} Retardation Mode of the PTMO Oligomers

Sample	T_c (K)		τ_c (s)	
	PEBA	PTMO	PEBA	PTMO
4000–2000	220	216	9.1×10^{-3}	2.1×10^{-1}
2000–1000	228	225	5.4×10^{-3}	1.7×10^{-2}
1300–650	247	230	3.0×10^{-4}	9.6×10^{-3}

is consistent with the shift of the β_{PE} peak toward higher temperatures observed in TSCr complex spectra.

We conclude that the mobility of the soft segment is restricted by the presence of polyamide segments in the PTMO amorphous phase. The high values of $T_{\beta,PE}$ and $T_{c,PE}$ are indicative of a relatively poor phase separation. Because the 1300–650 sample has the shortest PTMO and PA12 sequences of the series studied, it has the highest interblock junction point density. For phase separation to proceed to any given extent, a high number of these junction points must be immobilized at interphase boundaries, accompanied by a large decrease in entropy.

For the same composition by weight, as the chain length of the soft segments decreases, the shifting of $T_{\beta,PE}$ (or T_c) to higher values corresponds to an increased content of polyamide segments in amorphous PTMO phase.

The progressive influence of hard-segment structure on the mobility of the amorphous soft-segment unit is noted by the gradual broadening of the glass transition responses.

The large decrease of peak temperature $T_{\beta,PA}$ and the disappearance of the compensation phenomenon in β_{PA} processes prove clearly that soft segments have largely penetrated the amorphous polyamide phase. This finding implies that the polyamide phase in the 1300–650 sample contains a relatively high concentration of polyether units. As in polyether–ester copolymers (26–29), the rigidity of the polyamide segment domain seems to be controlled by the soft PTMO segment domain, because a strong interfacial mixing between the PTMO and polyamide PA12 segment domains occurs.

In conclusion, the fine structure analysis of the β_{PE} and β_{PA} retardation modes of block copolymers confirms the conclusions about the evolution of amorphous phase segregation in the poly(ether-amide) block copolymers deduced from the complex spectral study.

The short block poly(ether-amide) copolymers at high polyamide hard segments contain a crystalline polyamide phase, an amorphous PTMO-rich phase, and an amorphous polyamide-rich phase. The composition of these two amorphous phases depend on the molecular weight of the PTMO and PA12 parent oligomers.

PA–PVDF Blends—Results

Thermally Stimulated Creep Complex Spectra.   TSCr complex spectra of PA–PVDF, PA–PVDF–A, PVDF–PA, and PVDF–PA–A blends are shown in Figures 7–10, respectively. For each sample, four retardation processes were observed. The temperatures of the corresponding peaks are reported in Table VII. The peaks observed at the lowest temperatures (γ_1 and γ_2 in order of decreasing temperatures) are indicative of a complex mode due to localized movements of methylenic sequences of the polyamide chain

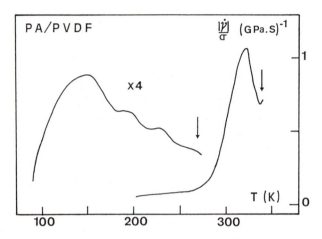

Figure 7. TSCr complex spectra of PA–PVDF blends. LT T$_\sigma$ = 270 K; HT T$_\sigma$ = 340 K; in both cases, σ = 1 MPa.

(24, 25). The β$_{PVDF}$ retardation mode was observed in the vicinity of the glass transition temperature of the PVDF homopolymer (235 K), as shown by Table VII. This β$_{PVDF}$ mode has been attributed to the anelastic manifestation of the glass transition of PVDF (30–34).

A last retardation mode is found in the vicinity of the glass transition temperature of the PA12 homopolymer (320 K) (cf. Table VII). This β$_{PA}$ mode has been associated with the anelastic effect of the glass transition of polyamides (35–39). Comparing PA–PVDF and PVDF–PA complex TSCr

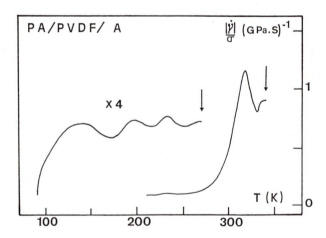

Figure 8. TSCr complex spectra of PA–PVDF–A blends. LT T$_\sigma$ = 270 K; HT T$_\sigma$ = 340 K; in both cases, σ = 0.8 MPa.

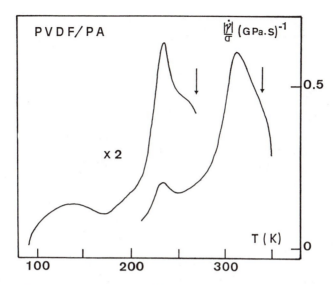

Figure 9. TSCr complex spectra of PVDF–PA blends. LT T_σ = 270 *K; HT* T_σ = 340 *K; in both cases,* σ = 1.1 *MPa.*

spectra shows that the intensity of β modes varies in the same manner that the weight fraction does.

In blends that are predominantly PA, although the intensity of the β_{PA} mode remains constant, that of the β_{PVDF} mode increases when compatibilizer is added. These data show that molecular mobility of the PVDF amorphous chains is improved by the presence of compatibilizer.

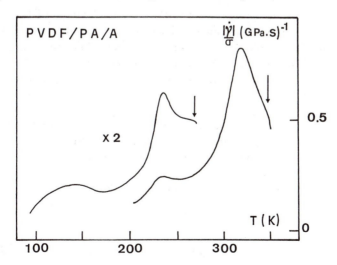

Figure 10. TSCr complex spectra of PVDF–PA–A blends. LT T_σ = 270 *K; HT* T_σ = 340 *K; in both cases,* σ = 8 *MPa.*

Table VII. TSCr Peak Temperature in the γ_{PA}, β_{PVDF}, and β_{PA} Relaxation Regions of Homopolymers and Blends

Sample	$T_{\gamma,PA}$ 2	1	$T_{\beta,PVDF}$	$T_{\beta,PA}$
PA	125	180	—[a]	320
PVDF	—	—	235	—
PVDF–PA	155	200	230	325
PVDF–PA	145	200	235	310
PA–PVDF–A	145	200	230	320
PVDF–PA–A	145	200	235	315

NOTE: All values are in kelvins.
[a] — indicates no value available.

In blends that are predominantly PVDF, although the intensity of the β_{PVDF} mode remains constant, that of the β_{PA} mode increases by 60% when compatibilizer is added. The presence of compatibilizer improves the molecular mobility of amorphous polyamide chains responsible for the β_{PA} retardation mode.

Fine Structure Analysis of TSCr Complex Spectra. The β modes were experimentally resolved into elementary TSCr peaks by using fractional loading programs. Each spectrum can be characterized by a retardation time following an Arrhenius equation (eq 2); the corresponding activation parameters are listed in Tables VIII and IX, and the compensation diagrams are represented in Figures 11–14.

In blends that are predominantly PA, the PA–PVDF and PA–PVDF–A compensation diagrams show the existence of two compensation phenomena for both samples. In other words, for each straight line, the experimental points are well described by equation 5, with compensation temperatures and times indicated in Table X. The two compensation diagrams are not so different: the compatibilizer does not greatly influence the amorphous phases of the blends.

Because of the relative values of the compensation temperatures, the compensation line corresponding to the highest activation enthalpies is associated with the polyamide amorphous phase, and the line corresponding to the lowest activation enthalpies is associated with the poly(vinylidene fluoride) amorphous phase. This assignment is coherent with the fact that the former compensation line corresponds to the PVDF mode, and the latter to the β_{PA} mode.

In Table X, the compensation parameters of the parent homopolymers are given for comparison. For both samples, the compensation line corresponding to the β_{PVDF} mode is characterized by parameters that are quite different from those of the homopolymer. In particular, for a given activation

Table VIII. Activation Parameters of β_{PVDF} and β_{PA} Retardation Modes of PA–PVDF and PA–PVDF–A Blends

	PA–PVDF		PA–PVDF–A	
T_σ (K)	ΔH (eV)	τ_0 (s)	ΔH (eV)	τ_0 (s)
	β_{PVDF} Retardation Mode			
215	0.77	1.1×10^{-16}	—[a]	—
220	0.825	1.5×10^{-17}	0.805	4.5×10^{-17}
225	0.89	7.4×10^{-19}	0.91	4.5×10^{-19}
230	1.04	1.6×10^{-21}	1.10	4.1×10^{-23}
235	1.08	4.1×10^{-22}	1.16	1.2×10^{-23}
240	1.01	5.6×10^{-20}	1.01	5.3×10^{-20}
245	0.95	2.5×10^{-18}	1.00	2.5×10^{-19}
250	1.00	6.5×10^{-19}	0.99	6.4×10^{-19}
255	0.94	2.0×10^{-17}	0.91	1.0×10^{-16}
260	0.935	8.4×10^{-17}	0.91	1.9×10^{-16}
265	0.95	7.5×10^{-17}	0.90	8.2×10^{-16}
270	0.97	9.0×10^{-17}	0.93	4.5×10^{-16}
	β_{PA} Retardation Mode			
275	0.98	1.4×10^{16}	1.02	2.5×10^{-17}
280	1.04	2.6×10^{-17}	1.03	4.4×10^{-17}
285	1.09	1.3×10^{-17}	1.06	2.3×10^{-17}
290	1.15	1.1×10^{-18}	1.21	1.0×10^{-19}
295	1.36	7.7×10^{-22}	1.34	2.2×10^{-21}
300	1.60	1.9×10^{-25}	1.52	3.1×10^{-24}
305	1.81	1.4×10^{-28}	1.83	6.0×10^{-29}
310	2.02	1.2×10^{-31}	2.025	1.1×10^{-31}
315	2.21	3.7×10^{-34}	2.23	2.0×10^{-34}
320	2.14	1.2×10^{-32}	2.18	3.7×10^{-33}
325	2.15	2.9×10^{-32}	2.225	1.8×10^{-33}

[a] — indicates no value available.

enthalpy, the preexponential factor, τ_0, is higher in blends than in the PVDF homopolymer; in other words (cf. eq 4), the activation entropy is lower in the blend than in pure PVDF. So, the amorphous PVDF phase of the blends is more "structured" than the amorphous phase of the PVDF homopolymer: this phase is more rigid in blends than in the homopolymer.

Blending is also accompanied by a significant decrease of the maximum activation enthalpy: 1.9 eV in the PVDF homopolymer and 1.1 eV in blends. So, the mobile units are shorter in blends than in the homopolymer. This result has been attributed to the effect of stresses exerted by the amorphous polyamide matrix onto the PVDF nodules.

In blends that are predominantly PVDF, in contrast to blends that are predominantly PA major differences were observed between the compensation diagrams of PVDF–PA and PVDF–PA–A samples. Indeed, although

Table IX. Activation Parameters of
β_{PVDF} and β_{PA} Retardation Modes of PA–PVDF and PA–PVDF–A Blends

	PA–PVDF		PA–PVDF–A	
T_σ (K)	ΔH (eV)	τ_0 (s)	ΔH (eV)	τ_0 (s)
		β_{PVDF} Retardation Mode		
170	0.43	4.7×10^{-12}	—[a]	—
175	0.49	2.6×10^{-13}	—	—
180	0.54	7.3×10^{-14}	—	—
185	0.58	1.4×10^{-14}	0.44	1.2×10^{-11}
190	0.63	2.2×10^{-15}	0.48	3.8×10^{-12}
195	0.68	3.6×10^{-16}	0.53	3.7×10^{-13}
200	0.73	5.2×10^{-17}	0.58	5.4×10^{-14}
205	0.78	6.2×10^{-18}	0.65	2.2×10^{-15}
210	0.86	2.4×10^{-19}	0.78	7.5×10^{-18}
215	0.97	1.5×10^{-21}	0.83	1.8×10^{-18}
220	1.09	1.0×10^{-23}	0.98	2.3×10^{-21}
225	1.295	8.3×10^{-28}	1.21	2.5×10^{-26}
230	1.475	4.3×10^{-32}	1.25	9.3×10^{-27}
235	1.66	1.7×10^{-34}	1.435	3.1×10^{-30}
240	1.64	2.5×10^{-35}	1.68	5.7×10^{-35}
245	1.44	1.3×10^{-28}	1.55	1.5×10^{-31}
250	1.295	1.2×10^{-28}	1.23	3.1×10^{-24}
255	1.11	1.0×10^{-20}	1.16	4.3×10^{-22}
260	1.14	5.5×10^{-21}	1.11	5.9×10^{-21}
265	1.09	1.4×10^{-19}	1.05	4.2×10^{-19}
270	1.09	4.0×10^{-19}	1.05	7.7×10^{-19}
275	1.11	4.3×10^{-19}	0.99	3.4×10^{-17}
		β_{PA} Retardation Mode		
280	1.045	1.7×10^{-17}	0.90	2.3×10^{-15}
285	1.08	6.9×10^{-18}	1.05	4.1×10^{-17}
290	1.1	1.1×10^{-17}	1.06	4.1×10^{-17}
295	1.13	6.6×10^{-18}	1.18	8.0×10^{-19}
300	1.18	1.6×10^{-18}	1.36	1.5×10^{-21}
305	1.19	2.1×10^{-18}	1.55	2.6×10^{-24}
310	1.63	3.0×10^{-25}	1.65	1.4×10^{-25}
315	1.755	6.4×10^{-27}	1.77	3.8×10^{-27}
320	1.73	3.9×10^{-26}	1.68	2.9×10^{-25}
325	1.71	2.2×10^{-25}	1.66	1.2×10^{-24}

[a] — indicates no value available.

the compensation line associated with the PVDF amorphous phase is observed in both PVDF–PA and PVDF–PA–A blends, the line associated with the PA amorphous phase is found only in the PVDF–PA–A blend. Therefore, in blends without compatibilizer, PVDF inhibits any organization of the amorphous polyamide phase. Compatibilizer allows polyamide chains to recover a structure like that of the parent homopolymer. Nevertheless, the compensation parameters are not the same as shown in Table X. In contrast,

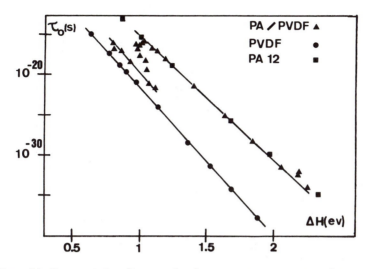

Figure 11. *Compensation diagrams for the activation parameters of* β$_{PVDF}$ *and* β$_{PA}$ *processes of homopolymers and blend.*

for blends that are predominantly PVDF, the compensation phenomenon of both blends has the same characteristics as the PVDF homopolymer. In fact, the predominant phase is segregated independently from the presence of compatibilizer.

This study of polyamide–poly(vinylidene fluoride) blends shows that the effect of compatibilizer depends on the predominant polymer: in blends that

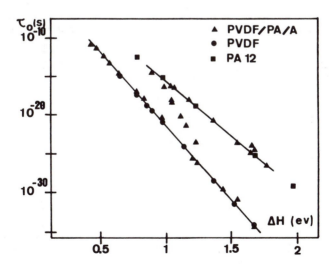

Figure 12. *Compensation diagrams for the activation parameters of* β$_{PVDF}$ *and* β$_{PA}$ *processes of homopolymers and blend.*

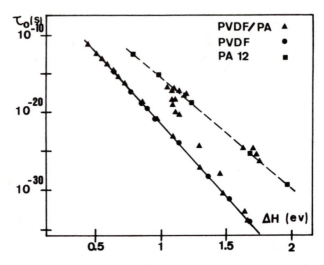

Figure 13. Compensation diagrams for the activation parameters of β_{PVDF} *and* β_{PA} *processes of homopolymers and blend.*

are predominantly polyamide, compatibilizer does not introduce a significant difference in the structure of the amorphous phases, but in blends that are predominantly poly(vinylidene fluoride), compatibilizer allows the amorphous polyamide phase to recover its structure.

The differences in the efficiency of compatibilizer may be explained by studies of morphology of the blends. In fact, in blends that are predominantly

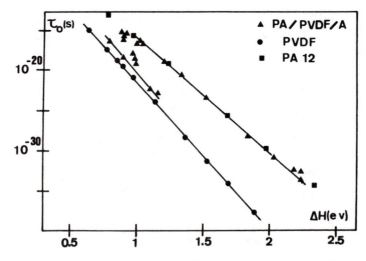

Figure 14. Compensation diagrams for the activation parameters of β_{PVDF} *and* β_{PA} *processes of homopolymers and blend.*

Table X. Compensation Temperature and Time in the β_{PVDF} and β_{PA} Relaxation Region of Homopolymers and Blends

Sample	PVDF		PA	
	$T_c\ (K)$	$\tau_c\ (s)$	$T_c\ (K)$	$\tau_c\ (s)$
PA	—[a]	—	348	0.3
PVDF	271	1.4×10^{-3}	—	—
PVDF–PA	275	6.6×10^{-4}	338	0.14
PA–PVDF–A	248	1.19	343	7.0×10^{-2}
PVDF–PA–A	265	5.1×10^{-3}	354	4.6×10^{-2}

[a] — indicates no value available.

PA, the compatibilizer acts as a nodule in the polyamide matrix, so it does not perturb the "structures" of the amorphous phases. In contrast, in blends that are predominantly PVDF, the compatibilizer is disposed at the interface of the PA and PVDF phases. At the same time, the size of the PVDF nodules decreases and thus allows a short-range organization in the amorphous polyamide phase to occur as in the PA12 homopolymer.

Conclusion

By TSCr analysis, the amorphous phase of polymers can be characterized by the corresponding compensation diagram in which the parameters of the retardation times isolated around the glass transition temperature are reported. Any modification of the integrity of this phase is accompanied by a remodeling of the compensation diagram. Analyzing the compensation diagrams of copolymers and blends can provide a quantitative estimation of phase segregation.

Acknowledgments

This work was partly supported by Centre d'Etudes et de Recherches d'ATOCHEM–Groupe ELF–AQUITAINE, 27460 Serquigny, France.

References

1. Deleens, G.; Foy, P.; Marechal, E. *Eur. Polym. J.* 1977, *13*, 343.
2. Yui, N.; Tanaka, J.; Sanui, K.; Ogota, N. *Macromol. Chem.* 1984, *185*, 2259.
3. Bornschlegl, E.; Goldbach, G.; Meyer, K. *Progr. Coll. Polym. Sci.* 1885, *79*, 119.
4. Xie, M.; Camberlin, Y. *Makromol. Chem.* 1986, *187*, 383.
5. Demont, P.; Chatain, D.; Lacabanne, C.; Ronarch, D.; Moura, J. L. *Polym. Eng. Sci.* 1984, *24*, 127.
6. Stefenel, M.; Thesis, Paul Sabatier University, Toulouse, France, 1984.
7. Ronarch, D.; Audren, P.; Moura, J. L. *J. Appl. Phys.* 1985, *58*, 474.

8. Faruque, H. S.; Lacabanne, C. *Polymer* **1986**, *27*, 527.
9. Faruque, H. S.; Lacabanne, C. *J. Phys. D: Appl. Phys.* **1987**, *20*, 939.
10. Lacabanne, C.; Chatain, D.; Monpagens, J. C.; Hiltner, A.; Baer, E.; *Solid State Commun.* **1978**, *27*, 1055.
11. Hino, T. *J. Appl. Phys.* **1975**, *46*, 1956.
12. Zielinski, M.; Swiderski, T. *Phys. Stat. Solidi* **1977**, *A42*, 305.
13. Zielinski, M.; Swiderski, T.; Kryszewski, M. *Polymer* **1978**, *19*, 883.
14. Ronarch, D.; Audren, P.; Haridoss, S.; Herrou, J. *J. Appl. Phys.* **1983**, *54*, 4439.
15. Demont, P.; Chatain, D.; Lacabanne, C.; Glotin, M. *Makromol. Chem. Makromol. Symp.* **1989**, *25*, 167.
16. McCrum, N. *Polymer* **1984**, *25*, 299.
17. Trick, G. S.; Ryan, J. M. *J. Polym. Sci. Part C* **1967**, *18*, 93.
18. Van Krevelen, D. W. In *Properties of Polymers—Correlations with Chemical Structure;* Elsevier: Amsterdam, 1972; pp 47–49.
19. Huh, D. H.; Cooper, S. L. *Polym. Eng. Sci.* **1971**, *11*, 369.
20. Hesketh, T. R.; Van Bogart, J. W. C.; Cooper, S. L. *Polym. Eng. Sci.* **1980**, *20*, 190.
21. Vallance, M. A.; Yeung, A. S.; Cooper, S. L. *Coll. Polym. Sci.* **1983**, *261*, 564.
22. Bandara, U.; Droscher, M. *Coll. Polym. Sci.* **1983**, *261*, 26.
23. Flory, P. J. *J. Chem. Phys.* **1949**, *17*, 223.
24. Willbourn, A. H. *Trans. Faraday Soc.* **1958**, *54*, 717.
25. McCrum, N. G.; Read, B. E.; Williams, G. In *Anelastic and Dielectric Effects in Polymeric Solids;* Wiley: London, 1967; pp 496–497 and 561–565.
26. Shen, M.; Mehra, U.; Niinomi, N.; Koberstein, J. R.; Cooper, S. L. *J. Appl. Phys.* **1974**, *45*, 4182.
27. Seymour, R. W.; Oerton, J. R.; Corley, L. S. *Macromolecules* **1975**, *8*, 331.
28. Lilaonitkul, A.; Cooper, S. L. *J. Am. Chem. Soc.* **1976**, *37*, 30.
29. Wegner, G.; Fujii, T.; Meyer, W.; Lieser, G. *Angew Makromol. Chem.* **1978**, *74*, 295.
30. Mandelkern, L.; Martin, G. M.; Quin, F. A., Jr. *J. Res. Nat. Bur. Std. U.S.* **1957**, *58*, 137.
31. Pfister, G.; Abkowitz, M. A. *J. Appl. Phys.* **1974**, *3*,1001.
32. Callens, A.; Debatist, R.; Eersels, L. *Nuovo Cimento* **1976**, *33B*, 434.
33. Callens, A.; Eersels, L.; Debatist, R. *J. Mat. Sci.* **1977**, *12*, 1361.
34. El Sayed, T., Thesis, Paul Sabatier University, Toulouse, France, 1987.
35. Illers, K. H. *Polymer* **1977**, *18*, 551.
36. Goldbach, G. *Angew Makromol. Chem.* **1973**, *32*, 37.
37. Greco, R.; Nicolais, L. *Polymer* **1976**, *17*, 1049.
38. De Rong, S.; Leverne Williams, H. *J. Appl. Polym. Sci.* **1985**, *30*, 2575.
39. Cazzitti, A., Thesis, Paul Sabatier University, Toulouse, France, 1988.

RECEIVED for review February 14, 1989. ACCEPTED revised manuscript September 19, 1989.

Analysis of the Glass Transition Temperature, Conversion, and Viscosity during Epoxy Resin Curing

B. Fuller[1,3], J. T. Gotro[1]*, and G. C. Martin[2]

[1]Systems Technology Division, IBM Corporation, Endicott, NY 13760
[2]Department of Chemical Engineering and Materials Science, Syracuse University, Syracuse, NY 13244

Differential scanning calorimetry (DSC) and oscillatory parallel-plate rheometry were used to investigate the curing of an epoxy resin (Dow Quatrex 5010). The viscosity was correlated with the glass transition temperature (T_g) and conversion by aborting the rheometer runs at specified intervals during the cure. The quenched samples were analyzed by DSC to determine the T_g and the conversion. The isothermal T_g–conversion relationship was modeled with the DiBenedetto equation. The T_g and conversion during nonisothermal (dynamic) curing could be modeled by using the DiBenedetto equation and a second-order kinetic equation. The model constants determined from isothermal experiments were used in the nonisothermal calculations.

$\mathbf{M}$ULTILAYERED PRINTED CIRCUIT BOARDS are complex structures consisting of layers of prepreg and copper. The prepreg is fabricated by impregnating woven glass cloth with a catalyzed epoxy resin solution. The solvent is removed, and the epoxy partially advanced in a treater tower, yielding a tack-free, stable prepreg. Typically, several sheets of prepreg are placed between two thin copper foils and laminated under heat and pressure with a hydraulic press. As the prepreg is heated during lamination, the

[3]Current address: Hercules Aerospace, Magna, UT 84044
*Corresponding author.

0065–2393/90/0227–0215$06.00/0

partially cured (B-staged) epoxy softens and flows. At elevated temperatures the cross-linking reaction causes the viscosity to increase. The rheological properties during the lamination process are a complex function of the resin chemistry, the cure kinetics, and the macroscopic flow of the resin.

Because of the complexity of the lamination process, the ability to predict material properties during processing is a great benefit to material designers and process engineers. To facilitate the development of such a process model, the objectives of this work were

- to examine the curing of a commercial epoxy resin system with differential scanning calorimetry (DSC) and oscillatory parallel-plate rheometry;

- to measure the viscosity, the glass transition temperature (T_g), and the conversion under isothermal and nonisothermal (dynamic) conditions;

- to model the conversion and the glass transition temperature during isothermal curing and to use the models to predict these properties for nonisothermal curing.

Experimental Details

A solution of epoxy resin Quatrex 5010 in methyl ethyl ketone (MEK) was obtained from the Dow Chemical Company (1). The system is based on tris(hydroxyphenyl)methane epoxy and a brominated bisphenol A hardener. The resin was catalyzed with 0.3 phr (parts per hundred parts of resin) of 2-methylimidazole (2-MI). The glass transition temperature of the fully cured resin was found to be 180 °C by DSC measurements.

Prepreg was fabricated by using standard solvent impregnation techniques. The woven E-glass fabric was coated with the catalyzed resin solution in a large dip pan. Metering rolls were used to control the thickness of the web as it emerged from the dip tank. The web traveled through a four-zone treater tower. The temperatures in the first two zones were adjusted to remove the solvent (MEK). The resin was partially reacted (advanced to the B-stage) in the second two zones. The prepreg emerged as a tack-free, stable material that was cut into the appropriate sizes.

Epoxy resin in powder form was obtained by shaking the partially cured, epoxy resin from woven glass cloth and removing extraneous pieces of broken glass fiber with a fine mesh sieve. The powder was then molded for 3 min at 95 °C to form disks with a thickness of 1.5 mm and a diameter of 25.4 mm. At these molding conditions, the resin softened and flowed, forming a void-free disk without advancing the epoxy. The T_g of the powder was 57 °C.

Rheological experiments were conducted in a Rheometrics System Four rheometer with a forced-convection oven. The oven was continuously purged with nitrogen. The rheometer was operated in the oscillatory parallel-plate geometry. Measurements were made every 30 s with a 2% strain amplitude at a frequency of 6.3 rad/s (1 Hz). Disposable aluminum plates were used to ensure quick separation and sample removal during aborted experiments. The bottom plates had machined slots to accommodate a thermocouple allowing direct measurement of the sample temperature.

Isothermal experiments were conducted at 123, 135, 145, 157, and 162 °C. At specified time intervals the experiment was stopped, and the sample was removed from the oven and quenched in liquid nitrogen. Thermal analysis was performed on the quenched samples with a DuPont 910 differential scanning calorimeter (DSC) at a heating rate of 20 °C/min from 0 to 300 °C. Nitrogen was used to purge the sample chamber of the DSC. The T_g was determined as the onset of the endothermic base-line deflection. The conversion (α) was calculated from

$$\alpha = \frac{\Delta H_0 - \Delta H_r}{\Delta H_0} \tag{1}$$

where ΔH_0 is the total heat of reaction and ΔH_r is the residual heat of reaction.

The total heat of reaction ΔH_0 was determined from DSC experiments on the unreacted catalyzed resin solution. The MEK was removed from the catalyzed varnish in a vacuum oven at room temperature. DSC samples were prepared by using the dried varnish. The total heat of reaction was 144 J/g.

Nonisothermal experiments were conducted by equilibrating the sample at 30 °C and heating the sample at rates of 4.9, 9.8, and 13.3 °C/min to a final temperature of 170 °C. At specific time intervals, the experiment was stopped, and the sample was removed from the fixtures and quenched in liquid nitrogen to stop the reaction.

Results and Discussion

Conversion Results. The complex viscosity (η) is plotted as a function of curing time (t_c) for various isothermal cure temperatures in Figure 1. As expected, the higher cure temperatures cause a faster rise in the viscosity with time. Each data point in Figure 1 represents an aborted sample; therefore, T_g and conversion data are available for each data point. The corresponding T_gs are plotted as a function of time in Figure 2. At cure temperatures below the ultimate T_g, the resin will not fully cure. In Figure 2, the final T_g value is a strong function of the cure temperature. The fully cured T_g of this resin was 180 °C. Because the isothermal experiments were performed at temperatures below the ultimate T_g, as the T_g approaches the cure temperature, vitrification may take place, an effect that would cause a marked decrease in the reaction rate. The T_g values become nearly constant with time because of the diffusion-controlled reaction in the glassy state. Longer times or higher temperatures are then required to drive the cure reaction to completion.

For thermosetting polymers, curing can be divided into two general categories, nth order and autocatalytic. Generally, most commercial epoxies have quite complex curing mechanisms, and often only the kinetics of the overall reaction are determined when chemical reactions occur simultaneously. The focus of our kinetic modeling was to determine whether the overall reaction kinetics could be fitted to a general kinetic model. In a subsequent section, the conversion will be related to the T_g of the growing network.

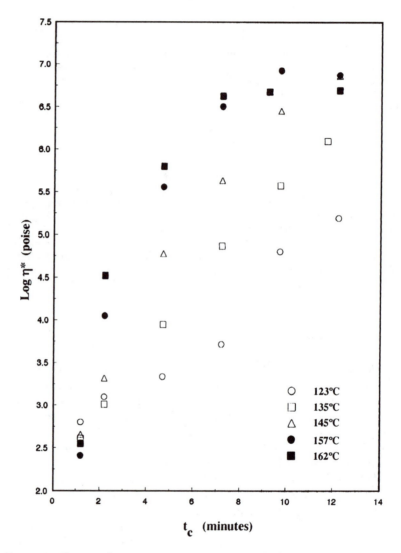

Figure 1. The complex viscosity versus curing time for various isothermal temperatures.

The general kinetic equation for the case of nth-order kinetics is given by (2)

$$\frac{d\alpha}{dt} = k(1 - \alpha)^n \qquad (2)$$

where α is the epoxy conversion, K is the reaction rate constant, n is the reaction order, and t is the reaction time.

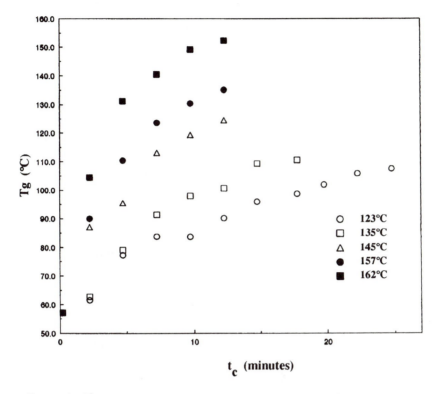

Figure 2. *Glass transition temperature versus curing time for various iso-thermal temperatures.*

The general expression for an autocatalytic reaction is

$$\frac{\mathrm{d}\,\alpha}{\mathrm{d}\,t} = k\alpha^m(1 - \alpha)^n \tag{3}$$

where n and m are reaction orders. Isothermal conversion data were analyzed for first-order, nth-order, and autocatalytic behavior. The conversion data could adequately be modeled with an nth-order model. To determine the value of n, the isothermal conversion data were plotted by using

$$\ln \frac{\mathrm{d}\,\alpha}{\mathrm{d}\,t} = \ln k + n \ln (1 - \alpha) \tag{4}$$

where n was found (by least-squares fitting) to be approximately 2.0, a result indicating that the reaction was second order. For a second-order reaction

($n = 2$), the relationship between the rate constant and the conversion is given by

$$kt = \frac{1}{1 - \alpha} - \frac{1}{1 - \alpha_0} \tag{5}$$

where α_0 is the B-stage conversion. The total time is the combination of the B-staging time and the curing time, so for a B-staged resin, if the curing time is defined as t_c, then

$$\frac{\alpha}{1 - \alpha} = kt_c + C \tag{6}$$

where C is a constant.

To determine the rate constant for each isothermal temperature, $\alpha/(1 - \alpha)$ is plotted versus t_c, as shown in Figure 3. The slopes (k) were determined

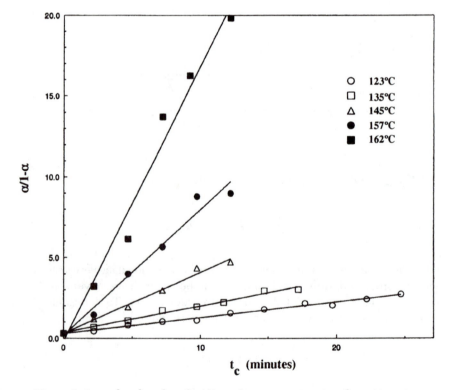

Figure 3. Second-order plot of $\alpha/(1 - \alpha)$ versus curing time for various iso-thermal cure temperatures. The solid lines are the least-squares fit to the experimental data.

by least-squares fitting. The rate constant is given by

$$k = k_0 \exp\left[-\frac{E_a}{RT}\right] \qquad (7)$$

where k_0 is the pre-exponential factor, E_a is the activation energy, R is the gas constant, and T is the temperature. From an Arrhenius plot, $k_0 = 3.96 \times 10^{10}$ s^{-1} and $E_a = 24.3$ kcal/g mol were determined.

For second-order kinetics, equation 2 may be integrated from $\alpha = \alpha_0$ at $t = 0$ to $\alpha = \alpha$ at $t = t$ and solved for α to yield

$$\alpha = 1 - \left(Kt + \frac{1}{1 - \alpha_0}\right)^{-1} \qquad (8)$$

where α_0 is the prepreg conversion at the start of the isothermal experiments. The initial conversion, $\alpha_0 = 0.235$, was determined from the heats of reaction. In Figure 4, the experimental conversion data are compared with

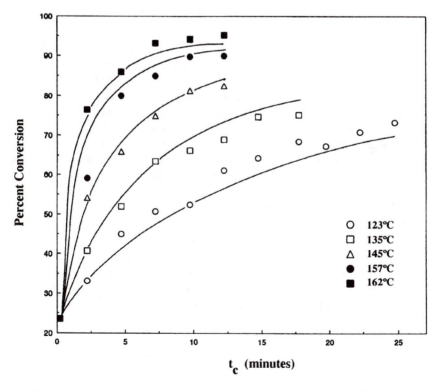

Figure 4. *Conversion versus curing time during isothermal curing. The symbols are the experimental data, and the lines were calculated with equation 8.*

the conversion calculated with equation 8. The predicted values show good correlation with the experimental data. Good agreement is seen between the calculated and experimental data over the regime where the reaction is not diffusion controlled. For isothermal experiments conducted below the ultimate T_g, the kinetic models presented here will not predict the experimental conversions when the reaction becomes diffusion controlled near vitrification. With this condition in mind, the data at high conversions were excluded from the kinetic modeling.

The accuracy of the isothermal models establishes a reliable starting point for modeling nonisothermal behavior. One of the objectives of this work was to test the validity of using the isothermal model parameters to predict the conversion during nonisothermal curing.

For a second-order reaction, equation 8 may be modified to include a nonisothermal temperature profile, $f(t)$. Substitution of equation 7 into equation 8 yields a relationship for the conversion during nonisothermal conditions

$$\alpha = 1 - \left[K_0 \int_0^t \exp\left(\frac{-E_a}{Rf(t)}\right) \, dt + \frac{1}{1 - \alpha_0} \right]^{-1} \tag{9}$$

The conversion during nonisothermal curing was determined from aborted rheometer runs at sample heating rates of 1.6, 4.9, 9.8, and 13.3 °C/min. The temperature profile in the rheometer oven was linear. In Figure 5, the nonisothermal conversion data from the aborted rheometer runs are plotted versus temperature. The symbols are the experimental data, and the solid lines represent the conversions calculated with equation 9. The second-order equation accurately predicts the conversion as a function of temperature for the heating rates of 4.9 and 9.8 °C/min. However, at the fastest heating rate, the calculated conversions do not predict the experimental data.

Glass Transition Temperature Results.   The glass transition temperature and the epoxy conversion during isothermal curing were modeled with the DiBenedetto equation (3–5). The segmental mobility of the polymer chains, which is determined in part by the cross-link structure, decreases as the cross-linking reactions proceed. This decrease is reflected by the increase in T_g as the network builds. The model relates the polymer structure to the cross-link network formation and the chain segmental mobility by the relation

$$\frac{T_g - T_{g0}}{T_{g0}} = \frac{\left(\dfrac{E_x}{E_m} - \dfrac{F_x}{F_m}\right)\alpha}{1 - \left(\dfrac{F_x}{F_m}\right)\alpha} \tag{10}$$

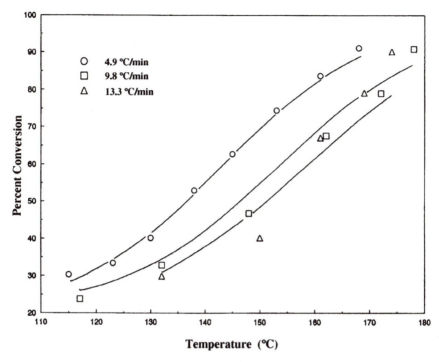

Figure 5. Conversion versus temperature for nonisothermal curing. The symbols represent the experimental data. The curves were calculated with the nonisothermal second-order kinetic equation and the model parameters obtained during isothermal curing.

where E_x/E_m is the ratio of lattice energies for the cross-linked and uncross-linked polymers, and F_x/F_m is the ratio of the segmental mobilities for the same two polymers. T_{g0} is the glass transition temperature of a polymer of the same chemical composition as the cross-linked polymer, except that the cross-links themselves are absent. The three model constants were determined from experimental T_g and conversion data with the Powell conjugate direction search algorithm (6).

Figure 6 is a plot of T_g versus conversion during isothermal curing. The symbols are the experimental data, and the solid line is the DiBenedetto model prediction. The following values were obtained for the model constants: $C_1 = 1.18$, $C_2 = 0.56$, and $T_{g0} = 43.39$ °C. From these, the values $E_x/E_m = 1.61$ and $F_x/F_m = 0.44$ were calculated. The T_g of the unreacted dried varnish was found to be 9 °C from DSC measurements. Enns and Gillham (7) reported DiBenedetto parameters for various types of epoxies. The values of E_x/E_m varied from 0.34 to 1.21, and F_x/F_m ranged from 0.19 to 0.74. The values for F_x/F_m in this work are within the range reported by

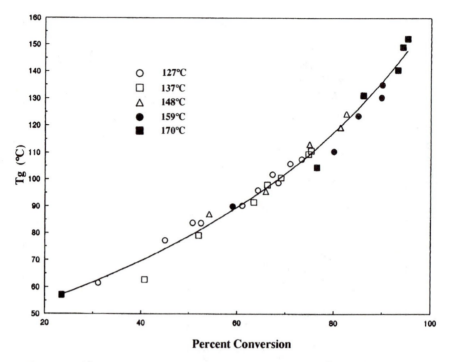

Figure 6. Glass transition temperature versus conversion data. The experimental data are represented by the symbols, and the solid line is the fit of the DiBenedetto equation to the experimental data.

Enns and Gillham, but the value of E_x / E_m is slightly higher than the literature values.

The relationships between T_g, conversion, and heating rates were established by using aborted nonisothermal rheometer runs. In Figure 7, T_g is plotted as a function of temperature for three heating rates. The symbols are the experimental results, and the solid lines were calculated by using a combination of the kinetic expression (equation 9) and the T_g–conversion expression (equation 10). The model parameters determined from isothermal experiments were used in the calculations. At incremental times during the nonisothermal cure, equation 9 was used to calculate the conversion, the conversion was substituted into equation 10, and the T_g was calculated. The T_g values determined by using the combined models were plotted as a function of temperature in Figure 7. The calculated T_gs for the two slower heating rates were in good agreement with the experimental data. At the fastest heating rate, the calculated T_gs underestimate the T_g–temperature relationship over the entire temperature range.

In Figure 8, the experimental T_g–conversion data are plotted for both the isothermal and nonisothermal aborted rheometer runs. The DiBenedetto

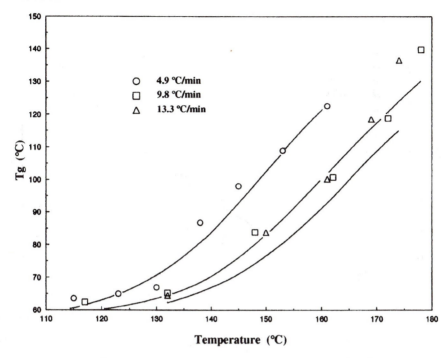

Figure 7. *Glass transition temperature versus temperature under noniso-thermal conditions. The symbols represent the experimental data. The curves were calculated with a combination of the second-order kinetic equation and the DiBenedetto equation.*

equation was fitted to the combined experimental data. The following values were obtained for the model constants, $T_{g0} = 49.1$ °C, $E_x/E_m = 1.18$, and $F_x/F_m = 0.34$. The model parameters fall in the range reported by Enns and Gillham (7). Good correlation is seen between the T_g–conversion data for both isothermal and nonisothermal curing. The solid line in Figure 8 is the calculated T_g–conversion relationship from equation 10 and the model parameters determined from both the isothermal and nonisothermal data. Figure 8 can be viewed as a master curve for determining the T_g–conversion relationship for a given cure cycle. The agreement between the isothermal and nonisothermal data indicates that the T_g–conversion relationship is cure-path independent for this resin system.

Viscosity Results.   The complex viscosity–curing time relationship for various heating rates is shown in Figure 9. The minimum viscosity varies by approximately 2 orders of magnitude from the slowest to the fastest heating rate. This behavior is typical of epoxy-based laminating resins (8–10). Because the flow is governed largely by the magnitude of the minimum viscosity, the heating rate is an important process variable. The "flow win-

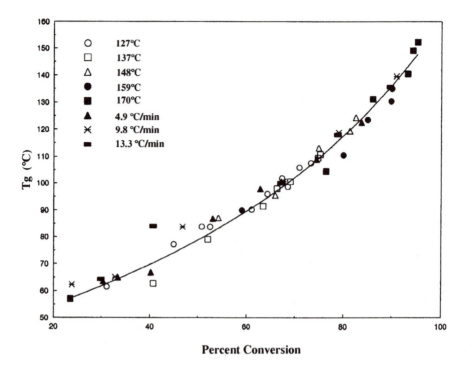

Percent Conversion

Figure 8. Glass transition temperature versus epoxy conversion for isothermal and nonisothermal conditions. The symbols are the experimental data. The solid line represents the T_g–conversion relation calculated with equation 10.

dow", or time when the resin can flow, is longer for the slower heating rates, but the minimum viscosity is relatively high, a situation that causes less flow. At high heating rates, the resin is fluid for a shorter time, but the minimum viscosity is much lower, and more flow is observed. Thus, the resin flow and the gel time are directly affected by the heating rate.

To correlate the conversion, T_g, and viscosity during nonisothermal curing, the rheometer runs were aborted at specified time intervals, and the samples were quenched and analyzed by DSC. In Figure 10, the T_g and complex viscosity are plotted as a function of curing time for a sample heating rate of 9.8 °C/min. During the initial softening, the viscosity decreased by approximately 2 orders of magnitude as a result of its strong temperature dependence. Although there was a large change in the viscosity, the T_g remained constant during the early portion of the curing. As the temperature continued to rise, the onset of the rapid cross-linking reaction caused a sharp increase in T_g with a subsequent increase in the viscosity.

In Figure 11, the complex viscosity and conversion are plotted as a function of curing time for a sample heating rate of 9.8 °C/min. The initial

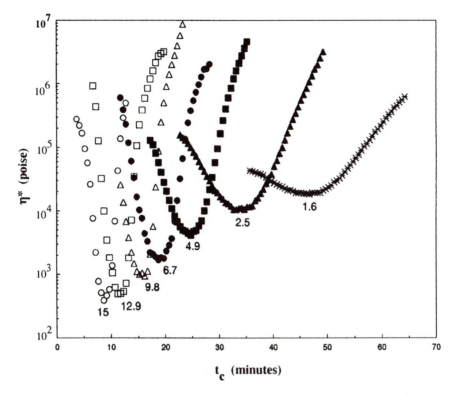

Figure 9. Complex viscosity versus curing time. Heating rates are noted in degrees Celsius per minute to a cure temperature of 175 °C.

conversion was 0.23. The conversion does not change during the initial softening of the resin. When the viscosity approaches the minimum value, the conversion begins to increase. Although the conversion is increasing, the viscosity of the growing network is more strongly governed by the temperature dependence of the viscosity. For a short time, the viscosity continues to decrease while both the T$_g$ and the conversion are increasing. After the minimum viscosity, network formation causes a rapid increase in the viscosity.

The glass transition temperature is plotted versus curing time for heating rates of 1.6, 4.9, 9.8, and 13.3 °C/min in Figure 12. The T$_g$, at the minimum viscosity, and the point where $G' = G''$ (the dynamic loss and storage moduli are equal) are noted in the plot. All of the heating rates display an initial plateau where the T$_g$ does not increase. This plateau corresponds to the softening of the resin, as was shown in Figure 10. The T$_g$, at the minimum viscosity, was higher for the fastest heating rate and decreased slowly at progressively slower heating rates.

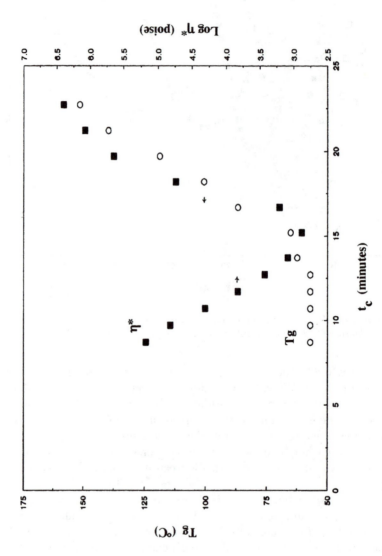

Figure 10. Complex viscosity and glass transition temperature as a function of curing time for a sample heating rate of 9.8 °C/min.

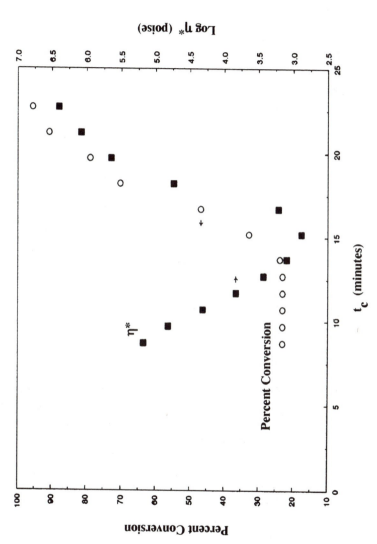

Figure 11. Complex viscosity and epoxy conversion as a function of curing time for a sample heating rate of 9.8 °C/min.

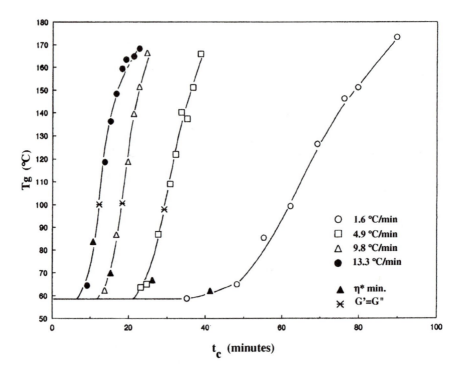

Figure 12. Glass transition temperature versus curing time for heating rates of 1.6, 4.9, 9.8, and 13.3 °C / min.

Two competing factors control the viscosity: the decrease in the viscosity caused by the increasing temperature profile and the increase in the viscosity due to cross-linking. The higher heating rates give rise to a higher T_g at the minimum viscosity because the temperature dependence of the viscosity plays a more dominant role. The decrease in viscosity due to the temperature increase overshadows the incremental increase in the viscosity due to the cross-linking. At the slower heating rates, the temperature dependence of the viscosity is less pronounced, and the cross-linking reaction controls the viscosity increase. For a series of heating rates, a steady progression of decreasing T_gs at the minimum viscosity was observed.

To estimate the T_g and conversion at the gel point, the intersection of the dynamic loss and storage moduli ($G' = G''$) was used. Some evidence (11–14) supports and disproves whether the point where tan delta (the ratio of loss to storage modulus) is unity is the gel point. For the three fastest heating rates, the dynamic moduli exhibited a crossover ($G' = G''$) during the nonisothermal curing. In Figure 13, the dynamic storage and loss moduli are plotted as a function of curing time for a 9.8 °C / min heating rate. The T_g at $G' = G''$ was found to be approximately 100 °C and independent of

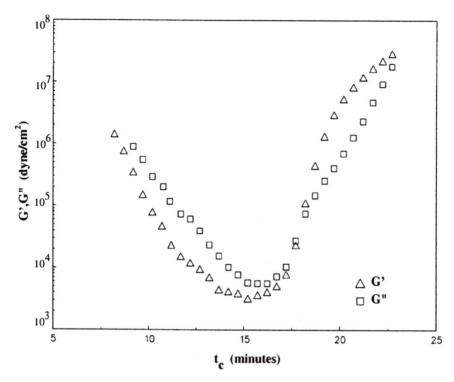

Figure 13. Dynamic loss and storage moduli versus curing time for a sample heating rate of 9.8 °C / min.

the heating rate. The T_g data for all of the heating rates are given in Table I. No crossover point was observed for the 1.6 °C / min heating rate, as shown in Figure 14. In this case, the T_g increased at a rate similar to that of the sample temperature, and the final T_g for this sample was identical to the final T_gs measured on samples cured at higher heating rates.

If the material is in the glassy state or the sample T_g is close to the test temperature, the elastic (storage) modulus will be larger than the corre-

Table I. Thermal and Rheological Data for Nonisothermal Curing

| Sample Heating Rate (°C/min) | Minimum Viscosity | | G' = G'' | |
	T_g (°C)	Conversion (%)	T_g (°C)	Conversion (%)
1.6	62.2	27.0	—[a]	—
4.9	65.0	33.4	98.0	62.8
9.8	65.2	32.8	100.7	67.6
13.3	83.8	40.3	100.2	67.6

[a] — indicates that dynamic moduli did not exhibit a crossover.

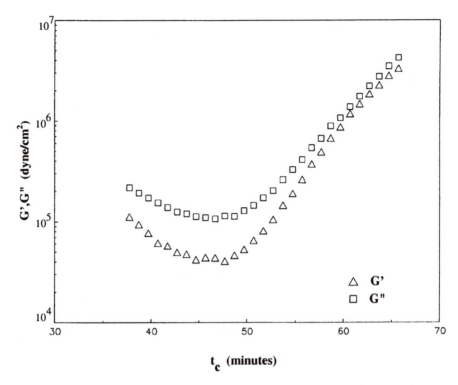

Figure 14. Dynamic loss and storage moduli versus curing time for a sample
heating rate of 1.6 °C / min.

sponding viscous (loss) modulus. Gillham (15) postulated that for curing
systems, the sample T_g will equal or be close to the cure temperature. If
the sample T_g approaches the cure temperature, vitrification with a subsequent slowing of the cross-linking reaction will occur. The sample T_g can
surpass the cure temperature, but the reaction rate in the glassy state will
be much slower. At the faster heating rates, the sample temperature is
increasing faster than the sample T_g, and thus, the sample remains in the
rubbery region. During nonisothermal curing, the dynamic moduli crossover
is caused by the rapidly increasing elastic modulus and the decreasing contribution from the viscous component. Past the gel point (when rubbery),
the elastic component of the complex modulus governs the dynamic mechanical response.

The dynamic moduli data presented here indicate that using $G' = G''$
as a definition of the gel point may be not appropriate. Recent results of
Heise, Martin, and Gotro (16) on the curing of epoxies with imidazoles
demonstrated that the point where $G' = G''$ was sensitive to the resin
stoichiometry. Winter (14) also found that the resin stoichiometry had an
effect on the point where $G' = G''$. The heating rate dependence and the

effects of resin stoichiometry indicate that using the point where $G' = G''$ may not be an appropriate way to measure the gel point.

Summary and Conclusions

The viscosity, T_g, and conversion were correlated by using aborted rheometer runs during isothermal and nonisothermal curing. The isothermal conversion data was modeled with a second-order kinetic expression. A kinetic expression for the nonisothermal conversion was developed for a linear temperature rise to the cure temperature. The kinetic parameters determined from isothermal experiments were used to predict the conversion during nonisothermal curing.

The conversion and T_g during isothermal curing were modeled with the DiBenedetto equation. The isothermal model accurately predicted the experimental T_g–conversion data. To predict the T_g during nonisothermal curing, the conversion at a given temperature was calculated, and the T_g was determined with the DiBenedetto equation. There was agreement between the calculated and experimental data.

The correlation of the viscosity, T_g, and conversion during nonisothermal curing indicated that the initial softening is controlled by the temperature dependence of the viscosity. After the initial softening, the cross-linking reaction causes the T_g and conversion to rapidly increase.

For a slow heating rate (1.6 °C / min) the dynamic loss and storage moduli did not exhibit a crossover. At faster heating rates, however, a crossover was observed. The heating rate dependence of the crossover point indicates that this method may not be appropriate to determine the gel point during nonisothermal curing.

Acknowledgments

The authors thank Aroon Tungare (Syracuse University) and Gerard Kohut (IBM) for their assistance with numerical modeling and data analysis. This work was performed as part of the Syracuse University–IBM Graduate Work Study Program.

References

1. Schrader, P., U.S. Patent 4 393 396, 1972.
2. Prime, R. B. In *Thermal Characterization of Polymeric Materials*; Turi, E., Ed., Academic: Orlando, FL, 1981; p 435.
3. Nielson, J. *J. Macromol. Sci. Rev. Macromol. Chem.* 1969, *C3*, 69.
4. Adabbo, H.; Williams, R. *J. Appl. Polym. Sci.* 1982, *27*, 1327.
5. DiBenedetto, A. J. *J. Polym. Sci. Polym. Phys. Ed.* 1987, *25*, 1949.
6. Powell, M. *Comp. J.* 1964, *7*, 155.
7. Enns, J.; Gillham, J. K. *J. Appl. Polym. Sci.* 1983, *28*, 2567.

8. Gotro, J.; Appelt, B.; Yandrasits, M.; Ellis, T. *Polym. Comp.* **1987**, *8*, 222.
9. Tungare, A.; Martin, G.; Gotro, J. *Tech. Pap. Soc. Plast. Eng.* (ANTEC 88) **1988**, *34*, 1075.
10. Appelt, B.; Gotro, J.; Schmitt, G.; Ellis, T.; Wiley, J. *Polym. Comp.* **1987**, *7*, 91.
11. Tung, C.; Dynes, P. J. *Appl. Polym. Sci.* **1982**, *27*, 569.
12. Winter, H.; Chambon, F. *J. Rheol.* **1986**, *30*, 367.
13. Chambon, F.; Winter, H. H. *J. Rheol.* **1987**, *31*, 683.
14. Winter, H. H. *Polym. Eng. Sci.* **1987**, *27*, 1698.
15. Gillham, J. K. In *Developments in Polymer Characterisation 3;* Dawkins, J. V., Ed; Applied Science: London, 1982; p 159.
16. Heise, M. S.; Martin, G. C.; Gotro, J. T. *Polym. Eng. Sci.*, in press.

RECEIVED for review February 14, 1989. ACCEPTED revised manuscript December 14, 1989.

13

Modeling Rheological and Dielectric Properties during Thermoset Cure

G. C. Martin[1]*, A. V. Tungare[1], and J. T. Gotro[2]

[1]Department of Chemical Engineering and Materials Science, Syracuse University, Syracuse, NY 13244
[2]Systems Technology Division, IBM Corporation, Endicott, NY 13760

For optimal processing of fiber-reinforced resin composites used in multilayer circuit boards, it is necessary to understand the dependence of the viscosity and the dielectric properties on the curing conditions. Three commercial epoxy resins (Dow Quatrex 5010 and two Ciba-Geigy FR-4 resins) were used in this study. Simultaneous conversion, glass transition temperature, viscosity, and ionic conductivity data were obtained under isothermal and dynamic (non-isothermal) curing conditions. These data were analyzed by using the dual Arrhenius viscosity model and the Williams–Landel–Ferry (WLF) models for the rheological and dielectric behavior. The model parameters were evaluated with numerical optimization techniques. The dual Arrhenius parameters were used to demonstrate the effects of B-staging (partial curing) on the flow behavior of FR-4 resins.

THERMOSETTING RESINS ARE WIDELY USED as matrix materials in structural composites and in the packaging of electrical circuits into multilayered circuit boards. During the manufacture of multilayered circuit boards, the resin must flow around the circuit information and cure to form a highly cross-linked network. The flow of the resin is governed by its viscosity history and fluidity. Fluidity is the time integral of the reciprocal viscosity and is a measure of the amount of resin flow. For cross-linking systems, the resin viscosity and fluidity depend on the degree of B-staging (partial curing), the

*Corresponding author

0065–2393/90/0227–0235$06.00/0

curing conditions, and the progress of the curing reaction. To determine the resin flow, it is essential to understand the viscosity and the curing behavior of the resin.

In recent years, rheological and dielectric measurements have been used to study the curing reaction (1–3). In the study reported here, simultaneous viscosity and ionic conductivity data obtained under isothermal and dynamic (nonisothermal) temperature conditions were analyzed with models for rheological and dielectric behavior. The model parameters were evaluated from the experimental data by using numerical optimization and linear regression techniques. Epoxy resins with different degrees of B-staging were used to study the effects of B-staging on the model parameters and the resin fluidity.

Experimental Details

Three commercial resins, Dow Quatrex 5010 and two Ciba-Geigy FR-4 resins, Resins A and B, which were B-staged to 20% and 25% conversions, were studied. The FR-4 resins consisted of a mixture of brominated diglycidyl ether of bisphenol A and an epoxidized cresol novolac with dicyandiamide as the hardener. Dow Quatrex 5010 has a high glass transition temperature (T_g) and contains a tris(hydroxyphenyl)methane-based epoxy with brominated bisphenol A as the hardener. The resin is catalyzed with 0.3 phr (parts per hundred parts of resin) of 2-methylimidazole (4).

The viscosity tests were conducted in the parallel-plate dynamic oscillatory mode with a Rheometrics System Four rheometer. B-staged resin disks of 2.54-cm diameter and 0.15-cm thickness were used. Simultaneous viscosity and dielectric data were obtained by embedding a small interdigitated comb-electrode dielectric sensor in the bottom plate of the rheometer so that it was flush with the resin disk. The sensor was connected to a Micromet Instruments Eumetric System II microdielectrometer. The dielectric measurements were obtained at a frequency of 100 Hz under isothermal and increasing temperature curing conditions. This simultaneous viscosity and dielectric measuring technique was discussed in detail by Gotro and Yandrasits (3).

To determine the resin conversions and the glass transition temperatures, the isothermal and dynamic viscosity tests were aborted at fixed time intervals, and the resin disks were quenched in liquid nitrogen and then scanned at a heating rate of 20 °C/min in a Dupont 910 differential scanning calorimeter (DSC) to determine the glass transition temperatures and the residual heats of reactions. The resin conversions were determined from the residual heat data; the glass transition temperatures were obtained from the onset of the endothermic deflection in the dynamic DSC scans. The procedures for determining the glass transition temperatures and the conversions were discussed in detail by Fuller et al. (4).

Theory

For thermosetting resins, the viscosity (η) can be characterized by the dual Arrhenius viscosity model (5, 6) given by

$$\ln \eta(t,T) = \ln \eta_0 + \int_0^t k \, dt \tag{1}$$

where

$$\eta_0 = \eta_\infty \exp\left(\frac{\Delta E_\eta}{RT}\right) \tag{2}$$

and

$$k = k_x \exp\left(\frac{\Delta E_k}{RT}\right) \tag{3}$$

In equations 2 and 3, η_∞ and k_x are the preexponential factors; ΔE_η and ΔE_k are the activation energies for flow and the cross-linking reaction, respectively; η_0 is the zero-time viscosity; k is the apparent reaction rate constant; R is the gas constant; T is the temperature; and t is time.

For a temperature profile given by $T = f(t)$, the dual Arrhenius model can be written as

$$\ln \eta(t,T) = \ln \eta_\infty + \frac{\Delta E_\eta}{Rf(t)} + k_x \int_0^t \exp\left(\frac{\Delta E_k}{Rf(t)}\right) dt \tag{4}$$

The model parameters can be determined from isothermal or increasing temperature viscosity data with numerical optimization techniques such as the Powell conjugate direction search algorithm (7, 8). Seed values of the parameters that are provided are then adjusted by the algorithm so as to minimize the standard deviation of the difference between the experimental and the predicted viscosity profiles.

With linear regression analysis (9), the model parameters can also be evaluated directly from viscosity–temperature–time data at the minimum viscosity at different heating rates. For a linear temperature rise, where B_1 is the starting temperature and B_2 is the heating rate, the condition for the minimum viscosity, which can be obtained by differentiating equation 4 with respect to time, is given by

$$\ln\left(\frac{B_2}{T_{min}^2}\right) = \ln\left(\frac{k_x R}{\Delta E_\eta}\right) - \frac{\Delta E_k}{RT_{min}} \tag{5}$$

The expression for the minimum viscosity is

$$\ln \eta_{min} = \ln \eta_\infty + \frac{\Delta E_\eta}{R}\left[\frac{1}{T_{min}} + \frac{k_x R}{\Delta E_\eta B_2}\left\{T_{min} \exp\left(\frac{-\Delta E_k}{RT_{min}}\right)\right.\right.$$
$$\left.\left. - B_1 \exp\left(\frac{-\Delta E_k}{RB_1}\right) + \frac{\Delta E_k}{R}\left\{E_i\left(\frac{-\Delta E_k}{RT_{min}}\right) - E_i\left(\frac{-\Delta E_k}{RB_1}\right)\right\}\right\}\right] \tag{6}$$

where $E_i(-\Delta E_k/RT_{min})$ and $E_i(-\Delta E_k/RB_1)$ are exponential integrals that can be approximated by

$$E_i(x) = \left(\frac{1}{xe^x}\right) \left[\frac{x^4 + a_1x^3 + a_2x^2 + a_3x + a_4}{x^4 + b_1x^3 + b_2x^2 + b_3x + b_4}\right] \tag{7}$$

where x is a dummy variable. In equation 7, a_1, a_2, a_3, a_4, b_1, b_2, b_3, and b_4 are tabulated constants (10). With linear regression analysis, equations 5 and 6 can be used to evaluate the dual Arrhenius model parameters from viscosity–temperature–time data at the minimum viscosity.

The Williams–Landel–Ferry (WLF) equation (11) relates the temperature dependence of polymer segmental mobility to mechanical and electrical relaxation processes. The WLF equation can be used to model both the viscous and the dielectric response of the resin. Tajima and Crozier (12) modeled the viscosity of an epoxy resin exhibiting second-order kinetic behavior with the WLF equation given by

$$\log\left(\frac{\eta(T)}{\eta(T_s)}\right) = \frac{C_1(T - T_s)}{C_2 + (T - T_s)} \tag{8}$$

In equation 8, C_1 and C_2 are the model constants, T_s is the reference temperature, and $\eta(T_s)$ is the viscosity at the reference temperature. Tajima and Crozier chose a value of T_s such that the viscosity data could be described by a single curve, and they then related T_s to the resin conversion. Lee and Han (13) used a similar approach to model the viscosity of an unsaturated polyester resin over a narrow range of conversions. Subsequently, they repeated the analysis by replacing the reference temperature, T_s in the WLF equation, with the resin glass transition temperature, T_g.

For curing systems with the glass transition temperature as the reference, the WLF models for the viscosity and the ionic conductivity can be written as

$$\log\left(\frac{\eta(T)}{\eta(T_g)}\right) = \frac{C_1(T - T_g)}{C_2 + (T - T_g)} \tag{9}$$

and

$$\log\left(\frac{\sigma(T)}{\sigma(T_g)}\right) = \frac{C_1(T - T_g)}{C_2 + (T - T_g)} \tag{10}$$

In equations 9 and 10, $\eta(T)$ and $\sigma(T)$ are the viscosity and ionic conductivity, respectively, at temperature T; and $\eta(T_g)$ and $\sigma(T_g)$ are the viscosity and ionic conductivity, respectively, at the glass transition temperature T_g. Sheppard (2), Bidstrup et al. (14), and Bidstrup et al. (15) observed that log $\sigma(T_g)$ and C_2 exhibited a linear dependence on T_g. Equation 10 can be then written as

$$\log \sigma(T) = C_5 + C_6 T_g + \frac{C_1(T - T_g)}{C_3 + C_4 T_g + (T - T_g)} \tag{11}$$

where C_3, C_4, C_5, and C_6 are constants. The model parameters in equations 9 and 11 can be determined from isothermal viscosity and ionic conductivity data by using the Powell conjugate direction search algorithm (7, 8). The parameters can also be calculated from the conditions at maximum ionic conductivity for different heating rates by using a procedure similar to that outlined in reference 9 for the dual Arrhenius parameters.

Results and Discussion

The optimum dual Arrhenius model parameters for the two FR-4 resins are reported in Table I. The parameters were determined from viscosity data obtained during dynamic cure at heating rates of 4.5, 6.8, 9.8, 12.3, and 13.2 °C/min. The predicted and the experimental viscosity profiles for the two FR-4 resins are compared in Figures 1 and 2. The curves represent the predicted profiles, and the discrete points represent the experimental data at 6.8 °C/min. The differences between the predicted and the experimental viscosity profiles in Figures 1 and 2 arise because the dual Arrhenius parameters used are average values, determined from experimental data varying over a wide range of temperatures and heating rates.

The compositions of Resins A and B are similar except that Resin A is B-staged to 20% conversion and Resin B is B-staged to 25% conversion. A comparison of the viscosity profiles in Figures 1 and 2 indicates that Resin A attains a lower minimum viscosity than Resin B. This difference in the

Table I. Dual Arrhenius Model Parameters Obtained with Numerical Optimization

Parameter	Quatrex 5010 (Dynamic Viscosity Data)	(Isothermal Viscosity Data)	Resin A (Dynamic Viscosity Data)	Resin B (Dynamic Viscosity Data)
ln η_0	−39.2	−33.1	−70.2	−67.8
ΔE_η (kcal/g mol)	34.9	30.8	58.8	57.7
k_x (s^{-1})	3.54×10^7	1.73×10^7	1.71×10^5	1.45×10^5
ΔE_k (kcal/g mol)	17.0	16.8	12.4	12.2

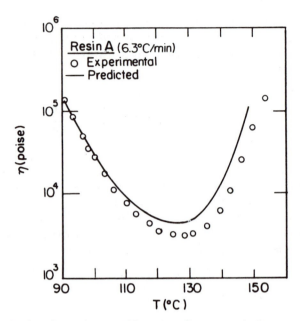

Figure 1. Predicted viscosity profile using equation 1 for Resin A at 6.8
°C/min.

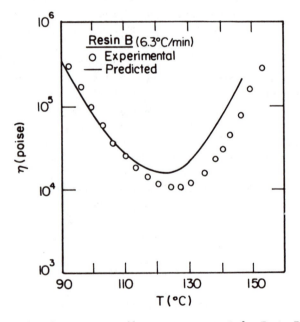

Figure 2. Predicted viscosity profile using equation 1 for Resin B at 6.8
°C/min.

minimum viscosities of Resins A and B is due to the difference in their degree of B-staging. The degree of B-staging also has a marked effect on the resin fluidity. The fluidity integrals for Resins A and B are compared in Figure 3. Resin A, which has a lower initial conversion, exhibits a much higher fluidity than Resin B.

The differences in the fluidity of Resins A and B can be explained by using their dual Arrhenius parameters. For isothermal cure at 125 °C, the value of the reaction rate constant, k, for Resin A is $2.43 \times 10^{-2} \text{ s}^{-1}$, and for Resin B it is $2.59 \times 10^{-2} \text{ s}^{-1}$. Because the rate constants are approximately equal, there is little difference in the curing reactions of Resins A and B. This result is expected because the resins have the same composition. The effects of B-staging are contained in the parameter η_0, which is related to resin softening and reflects the prior thermal history of the resin. For isothermal cure at 125 °C, the values of η_0 are 83 P (poise) for Resin A and 231 P for Resin B. Resin A exhibits more softening with temperature than Resin B and hence attains a lower viscosity and higher fluidity. Hence, the dual Arrhenius viscosity model can be used to characterize the resin chemorheology and to understand the effects of B-staging on the flow behavior of resins.

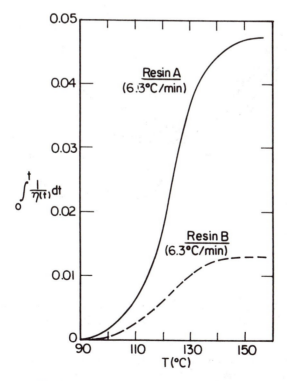

Figure 3. Fluidity integrals for Resins A and B at 6.8 °C/min.

The viscosity profiles of the Quatrex 5010 resin were characterized with the dual Arrhenius viscosity model. A numerical optimization technique was used to obtain the model parameters from the dynamic and the isothermal viscosity data. The dynamic viscosity data were obtained at heating rates of 2.5, 4.9, 6.7, 9.8, 12.9, and 13.3 °C/min, and the isothermal viscosity data were obtained at 123, 135, 145, and 157 °C. The average values of the parameters obtained from the two sets of data are listed in Table I. The viscosity profiles at heating rates of 4.9 and 9.8 °C/min, predicted by using the two sets of dual Arrhenius parameters, are compared with the experimental viscosity data in Figure 4. The solid lines were obtained with the dual Arrhenius parameters from dynamic viscosity data, and the dashed lines are the predictions using the dual Arrhenius parameters from isothermal viscosity data. The dynamic viscosity profiles predicted by using the dual

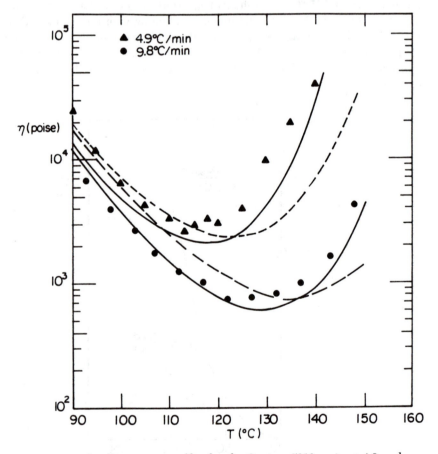

Figure 4. Predicted viscosity profiles for the Quatrex 5010 resin at 4.9 and 9.8 °C/min using dynamic (—) and isothermal (- -) dual Arrhenius parameters. The discrete points represent the experimental data.

Arrhenius parameters from isothermal viscosity data are not in good agreement with the experimental viscosity data.

The dual Arrhenius parameters for the Quatrex 5010 resin were also determined by using the linear regression analysis (equations 5 and 6). ΔE_k determined from the linear regression analysis was 17.8 kcal/g mol, and the preexponential factor, k_x, was 1.51×10^8 s^{-1}. These values are similar to those reported in Table I using the numerical optimization technique.

The experimental and predicted viscosity profiles are compared in Figure 5 at a heating rate of 12.9 °C/min. In Figure 5, the solid line is the viscosity profile predicted by using the dual Arrhenius parameters determined with the numerical optimization procedure; the dashed line is the viscosity profile predicted by using the dual Arrhenius parameters obtained from linear regression analysis; and the discrete points are the experimental data. The differences in the predicted profiles occur because, in the linear regression technique, only the conditions at the minimum viscosity are used to determine the dual Arrhenius parameters, whereas, in the numerical optimization procedure, the entire experimentally determined viscosity profile is used to determine the model parameters.

The isothermal viscosity–glass transition temperature data for the Quatrex 5010 resin were analyzed with the WLF equation. The log $\eta(T_g)$ term in equation 9 was observed to have a linear dependence on T_g. Equation 9

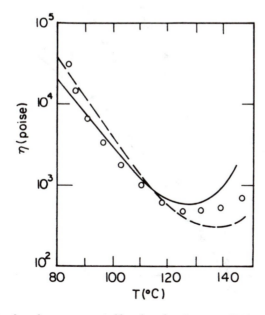

Figure 5. Predicted viscosity profiles for the Quatrex 5010 resin at 12.9 °C/min using dual Arrhenius parameters from optimization (—) and linear regression (– –) techniques. The discrete points represent experimental data.

can be then rewritten as

$$\log \eta(T) = C_5 + C_6 T_g + \frac{C_1(T - T_g)}{C_2 + (T - T_g)} \qquad (12)$$

Figure 6 shows the isothermal viscosity-versus-time data for the Quatrex 5010 resin at 123, 135, and 145 °C. The lines indicate the predictions using equation 12 and the optimum parameters, which are reported in Table II. The predictions and the experimental data are in good agreement. The WLF model described by equation 12 can, therefore, be used to model the viscosity of the Quatrex 5010 resin during isothermal cure.

In another analysis of the isothermal viscosity data, the Tajima and

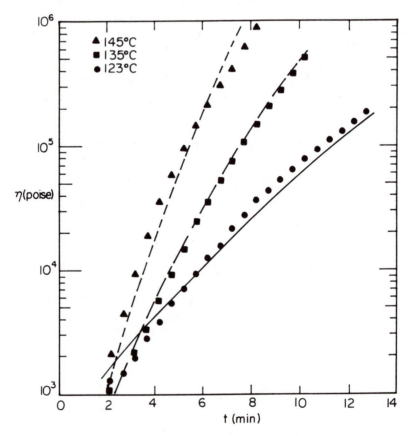

Figure 6. Viscosity predictions using equation 12 for the Quatrex 5010 resin during isothermal cures at 123, 135, and 145 °C.

**Table II. WLF Model Parameters for the
Quatrex 5010 Resin**

Parameter	Ionic Conductivity Model (eq 11)	Viscosity Model (eq 12)	Viscosity Model (eq 13)
C_1	13.15	−38.07	19.39
C_2	—	345.39	114.41
C_3	−153.87	—	24.13
C_4	0.77	—	166.71
C_5	−23.44	5.11	13.86
C_6	0.027	0.010	−2.757

Crozier formulation given in equation 8 was used. For low conversions, log $\eta(T_s)$ and T_s can be approximated as linear functions of the conversion, α. Equation 8 can then be rewritten as

$$\log \eta(T) = C_5 + C_6\alpha + \frac{C_1[T - (C_3 + C_4\alpha)]}{C_2 + [T - (C_3 + C_4\alpha)]} \tag{13}$$

The optimum values of the parameters in equation 13 are reported in Table II. The model predictions are compared with the isothermal viscosity data in Figure 7. The model predictions and the experimental data are in agreement. However, the model in equation 13 has six adjustable parameters, whereas the model in equation 12 has only four adjustable parameters.

Fuller et al. (4) observed that the curing behavior of the Quatrex 5010 resin can be modeled with a second-order kinetic model, and the relationship between the conversion and the glass transition temperature can be described by the DiBenedetto equation (16). They also observed that the thermal properties of the resin were cure-path independent; that is, irrespective of the curing conditions, for a given conversion, the resin had a constant glass transition temperature. Hence, the glass transition temperature can be predicted at any time and temperature during different curing conditions. The kinetic and the glass transition temperature models of Fuller et al. can be incorporated into the WLF viscosity models given by equations 12 and 13 to relate the viscosity to the curing chemistry and the thermal properties of the resin.

The variation of ionic conductivity with T_g at three isothermal temperatures is shown in Figure 8. The ionic conductivity decreases monotonically with the progress of the curing reaction. When cured to a fixed T_g, higher isothermal cure temperatures result in higher ionic conductivities. This effect is due to the increased ionic mobility resulting from the lower resin viscosity at higher temperatures.

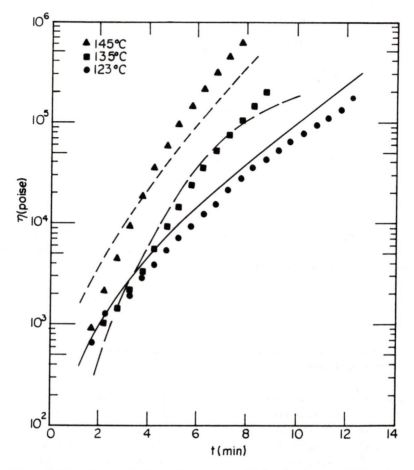

Figure 7. Viscosity predictions using equation 13 for the Quatrex 5010 resin during isothermal cures at 123, 135, and 145 °C.

The isothermal ionic conductivity data were modeled with equation 11. The model parameters were optimized with the Powell conjugate direction search algorithm. In the analysis, the standard deviation of the predicted and experimental ionic conductivities was minimized. The optimum WLF parameters for the Quatrex 5010 resin are reported in Table II. The experimental data and the model predictions at 123, 135, and 157 °C isothermal temperatures are compared in Figure 8. The discrete points represent the experimental data, and the curves represent the model predictions. The agreement between the model predictions and the ionic conductivity data in Figure 6 indicates that the WLF equation can be used to model the ionic conductivity changes during isothermal cure of the Quatrex 5010 resin.

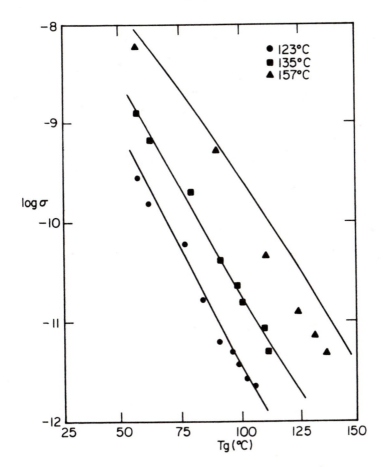

Figure 8. Ionic conductivity predictions using equation 11 for the Quatrex 5010 resin during isothermal cures at 123, 135, and 157 °C.

Acknowledgment

The support of this research by the IBM Corporation is gratefully acknowledged.

References

1. Senturia S. D.; Sheppard, N. F., Jr. *Adv. Polym. Sci.* **1986**, *80*, 1.
2. Sheppard, N. F., Jr., Ph. D. Thesis, Massachusetts Institute of Technology, 1986.
3. Gotro, J. T.; Yandrasits, M. *Tech. Pap. Soc. Plast. Eng.* **1987**, *33*, 1039.
4. Fuller, B. W.; Gotro, J. T.; Martin, G. C. *Proc. ACS Div. Polym. Mat. Sci. Eng.* **1988**, *59*, 975.
5. Roller, M. B. *Polym. Eng. Sci.* **1975**, *15*, 406.

6. Roller, M. B. *Polym. Eng. Sci.* **1986**, *26*, 432.
7. Tungare, A. V.; Martin, G. C.; Gotro, J. T. *Polym. Eng. Sci.* **1988**, *28*, 1071.
8. Powell, M. J. D. *Comp. J.* **1964**, *7*, 155.
9. Martin, G. C.; Tungare, A. V.; Gotro, J. T. *Tech. Pap. Soc. Plast. Eng.* **1988**, *34*, 1075.
10. Gautschi, W.; Cahill, W. F. In *Handbook of Mathematical Functions with Formulas, Graphs, and Mathematical Tables;* Abramowitz, M.; Stegun, L. A., Eds.; AMS Series 55; National Bureau of Standards: Washington, DC, 1970; p 231.
11. Williams, M. L.; Landel, R. F.; Ferry, J. D. *J. Am. Chem. Soc.* **1955**, *77*, 3701.
12. Tajima, Y. A.; Crozier, D. G. *Polym. Eng. Sci.* **1988**, *28*, 491.
13. Lee, D. S.; Han, C. D. *Polym. Eng. Sci.* **1987**, *27*, 955.
14. Bidstrup, S. A.; Sheppard, N. F., Jr.; Senturia, S. D. *Tech. Pap. Soc. Plast. Eng.* **1987**, *33*, 987.
15. Bidstrup, W. W.; Bidstrup, S. A.; Senturia, S. D. *Tech. Pap. Soc. Plast. Eng.* **1988**, *34*, 960.
16. DiBenedetto, A. T. *J. Polym. Sci.* **1987**, *25*, 1949.

RECEIVED for review February 14, 1989. ACCEPTED revised manuscript August 8, 1989.

14

Chemical and Rheological Changes during Cure in Resin-Transfer Molding

In Situ Monitoring

D. E. Kranbuehl[1], M. S. Hoff[1], T. C. Hamilton[1], W. T. Clark[1], and W. T. Freeman[2]

[1]Department of Chemistry, College of William and Mary, Williamsburg, VA 23185
[2]Langley Research Center, National Aeronautics and Space Administration, Hampton, VA 23665

Frequency-dependent electromagnetic sensing (FDEMS) is a convenient and sensitive technique for monitoring in situ infiltration and cure in the tool during the resin-transfer molding (RTM) process. The magnitude of the fluidity and viscosity as a function of the time and temperature, the time to infiltration at various ply depths, the effects of aging and elapsed time before infiltration of the RTM process, and monitoring of the cure cycle are four important areas where FDEMS is shown to significantly help in determining the RTM process procedure.

A NONDESTRUCTIVE IN SITU FREQUENCY-DEPENDENT impedance sensing technique was reported previously (*1–10*) for measuring cure-processing properties of both thermoset and thermoplastic resins. The technique uses the frequency dependence of the resin's impedance to measure molecular ionic and dipolar diffusion rates. These molecular parameters can be used continuously throughout the process cycle to monitor the time and temperature dependence of events such as reaction onset, maximum flow, viscosity, gel, the buildup in modulus, evolution of volatiles, and reaction completion (*1–10*). The measurements can be made in a research environment to evaluate resin processing properties and in the manufacturing tool

0065–2393/90/0227–0249$06.00/0
© 1990 American Chemical Society

on the plant floor (1, 9–11). In this chapter we discuss preliminary results on the use of frequency-dependent electromagnetic sensing (FDEMS) to monitor and determine the fabrication process in resin-transfer molding (RTM).

Resin-transfer molding of advanced fiber-architecture materials promises to be a cost-effective process for obtaining composite parts with exceptional strength. However, a large number of material processing parameters must be observed, known, or controlled during the resin-transfer molding process. These parameters include the viscosity during both impregnation and cure. In situ sensors that can observe these processing properties within the RTM tool during the fabrication process are essential. This chapter will discuss recent work on the use of FDEMS techniques to monitor these properties in the RTM tool. Our objective is to use these sensing techniques to address problems of RTM scale-up for large complex parts and to develop a closed-loop, intelligent, sensor-controlled RTM fabrication process.

Experimental Details

Dynamic dielectric measurements were made with a Hewlett-Packard 4192A LF impedance analyzer controlled by a 9836 Hewlett-Packard computer. Measurements at frequencies from 50 to 5×10^6 Hz were taken at regular intervals during the cure cycle and converted to the complex permittivity, $\epsilon^* = \epsilon' - i\epsilon''$. Measurements were made with a geometry-independent Dek Dyne frequency-dependent electromagnetic sensor (Polymer Laboratories, Amherst, MA) (12) that was embedded in the resin.

Dynamic mechanical measurements were made with a Rheometrics RDA-700 rheometer at 1.6 Hz and were used to compute the magnitude of the complex viscosity.

All measurements were made on a Shell Epon diglycidyl ether of bisphenol A (DGEBA) resin-transfer molding resin, RSL 1282, with the aromatic amine curing agent, 9470.

Theory

Measurements of capacitance, C, and conductance, G, were used to calculate the complex permittivity $\epsilon^* = \epsilon' - i\epsilon''$ where

$$\epsilon' = \frac{C_{\text{material}}}{C_0} \tag{1}$$

$$\epsilon'' = \frac{G_{\text{material}}}{C_0 2\pi f} \tag{2}$$

C_0 is the geometry-independent capacitance, and f is frequency, at nine frequencies between 125 Hz and 1 MHz.

This calculation is possible when using the Dek Dyne FDEMS probe whose geometry-independent capacitance, C_0, is invariant over all measurement conditions. Both the real and the imaginary parts of ϵ^* have an ionic and a dipolar component. The dipolar component arises from diffusion of bound charge or molecular dipole moments. The dipolar term is generally the major component of the dielectric signal at high frequencies and in highly viscous media. The ionic component often dominates ϵ^* at low frequencies, low viscosities, and/or higher temperatures.

Analysis of the frequency dependence of ϵ^* in the hertz to megahertz range is, in general, optimum for determining both the ionic mobility–conductivity, σ, and a mean dipolar relaxation time, τ. These two parameters are directly related on a molecular level to the rate of ionic translational diffusion and dipolar rotational mobility and thereby to changes in the molecular structure of the resin that reflect the reaction rate, changes in viscosity, and the degree of cure.

Results and Discussion

Figure 1 is a plot of the log of the dielectric loss factor ($\epsilon'' \times \omega$) of the RTM resin scaled by the frequency during a multiple ramp-hold cure cycle. Measurements were made over the frequency range of 125 Hz to 1 MHz. Plots of $\epsilon'' \times \omega$ times frequency are convenient, as previously discussed (*1, 2*),

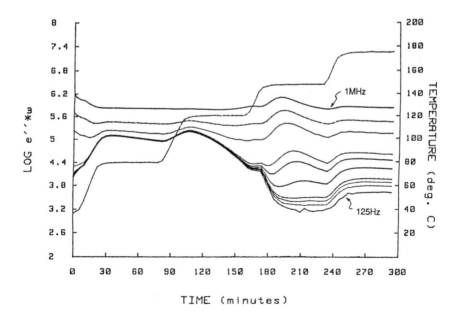

Figure 1. Log ($\epsilon'' \times \omega$) vs. time during multiple ramp-hold cure cycle 1 for the DGEBA epoxy resin. The frequencies were 125, 250, 500, 5×10^3, 25×10^3, 50×10^3, 125×10^3, 250×10^3, 500×10^3, and 1×10^6 Hz.

because overlapping lines indicate the frequencies and time and temperature periods during cure where $\epsilon'' \times \omega$ is dominated by ionic diffusion. Non-overlapping lines that exhibit a systematic series of peaks with frequency, time, and temperature can be used to determine a characteristic dipolar relaxation time. Figure 1 shows that the value of $\epsilon'' \times \omega$ is dominated by ionic contributions throughout the first hold and at the low frequencies in the second hold. The low-frequency values of ϵ'' monitor the ionic mobility and thus reciprocally monitor changes in viscosity. Figure 1 that shows the resin goes through an ionic mobility maximum, that is, a viscosity minimum, at the beginning of the first hold and at the beginning of the second hold.

Figure 2 displays the magnitude of the logarithmic complex viscosity η (1.6 Hz) versus time and temperature for cure cycle 1. The times of occurrence of viscosity minima are in good agreement with the maxima in ϵ'' (ionic mobility). The relative change in log ϵ'' and log viscosity during the initial hold indicates that for a given temperature the ionic mobility is directly proportional to the reciprocal of the viscosity in this highly fluid region of cure.

Dipolar relaxation times in the megahertz to kilohertz region are observed during the third hold. The occurrence of a peak in $\epsilon'' \times \omega$ for a particular frequency (f) indicates the time during cure when the relaxation time τ is equal to $\frac{1}{2}f$. The occurrence of these relaxation times indicates the onset of the relaxation region that is associated with glass formation. Glass formation occurs as τ becomes infinite.

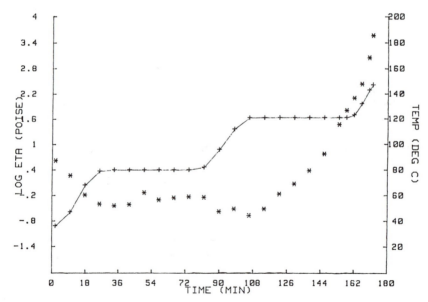

Figure 2. Log complex viscosity, η, vs. time during multiple ramp-hold cure cycle 1.

The approach of the values of ε″ to a constant value in the third and fourth holds indicates the time at which the reaction is approaching completion. This point is supported by the flatness of ε″ with time during the fourth hold.

The ability of the sensors to monitor the effects of a change in the cure cycle is shown in Figures 3 and 4. Figure 3 displays the loss factor ε″ for a cure cycle in which the initial hold temperature is increased to 100 °C and the second hold to 135 °C. Comparison of Figures 1 and 3 shows that higher temperature holds significantly increase the magnitude of the first ε″ maximum, a result indicating a decrease in the viscosity. The position of the second viscosity minimum, (ε″ maximum) occurs 10 min earlier. In addition, the reaction advances more quickly as the dipolar α relaxation peaks begin to occur during the second hold, an event indicating that the approach to glass transition temperature (T_g) occurs 1 h earlier. The reaction is more advanced and is setting up much more quickly. The time of occurrence of the first and second viscosity minima as shown in Figure 4 are in good agreement with the time of occurrence in the maxima in ε″ (ionic mobility). This finding is supported by the rapid buildup in viscosity occurring at 100 min as opposed to 160 min for cure cycle 1.

The ε″ values show that the cure is essentially complete in the third hold. This result suggests that if a high T_g is desired, the third hold could be eliminated and one could proceed directly to the 175 °C hold.

Next we examine the feasibility of using FDEMS techniques to measure

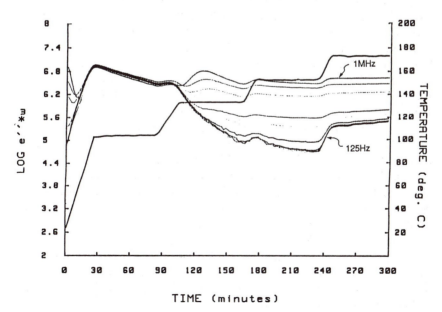

Figure 3. Log (ε″ × ω) vs. time during multiple ramp-hold cure cycle 2.

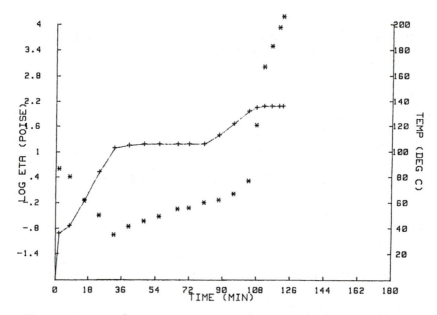

Figure 4. Log complex viscosity, η, vs. time during a multiple ramp-hold cure cycle 2.

and monitor the effects of individual component age on resin processing properties, the effect of layup time on resin processing properties, and on the ability of the sensor to monitor resin infiltration into the fiber architecture. Figure 5 is a plot of ϵ'' versus time and temperature during cure. The resin was mixed and held at room temperature for 1, 8, and 24 h before being cured. The sensor measurement output suggests that after 8 h the mixed resin begins to show signs of advancement. The value of ϵ'', which is reflecting ionic mobility and resin fluidity, is lower throughout the cure cycle following the onset of the first hold at 30 min. After 24 h, the mixed resin shows significantly less fluidity, that is, a higher viscosity. On the basis of the FDEMS and viscosity measurements, the viscosity has increased over a factor of 3 in the high-flow 20- to 120-min portion of the cure cycle.

Finally in the last and most important experiment, we demonstrate the ability of FDEMS to monitor both the impregnation and cure process including these processing properties continuously and in situ at various positions in RTM mold during fabrication. The RTM process involved placement of the RTM resin in the bottom of a mold (*see* Figure 6). An eight-ply 8-Harness satin graphite cloth layup was placed over the resin. Sensors were placed at the second, fourth, sixth, and top plies. After 4 min the mold was closed in a press, and cure cycle 1 was used to impregnate and cure the resin–cloth layup.

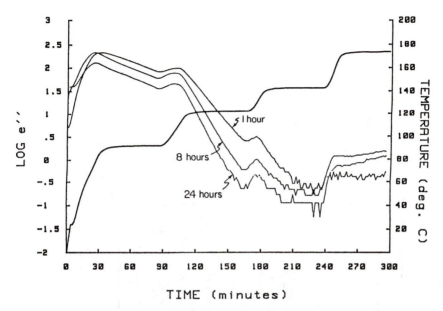

Figure 5. *Comparison of log ε″ (125 Hz) vs. time for a freshly mixed DGEBA resin and DGEBA resins mixed 8 and 24 h earlier.*

Figures 7 and 8 (Figure 8 is an enlargement of a portion of Figure 7) show that the weight of the top plate and the application of a vacuum to the closed mold were sufficient to partially consolidate the cloth layers and cause infiltration of the resin through the second ply. No further infiltration was observed until the temperature cycle was begun. It took approximately 4 min for the resin to flow from the second ply to the top ply. The differences in the viscosity between the second and top plies, as indicated by the magnitude of ε″ (125 Hz), occur only in the initial hold and are believed to reflect small variations in temperature due to the impregnation process. A pressure of 100 psi was applied at 57 min. For this resin–cloth layup, the FDEMS output shows that the cure was uniform at all ply positions throughout the remainder of the cure cycle.

In summary, FDEMS sensing is a convenient and sensitive technique for monitoring in situ infiltration and cure in the tool during the resin-transfer molding process. The magnitude of the fluidity–viscosity as a function of time–temperature, the time to infiltration at various ply positions, the effects of varying the time–temperature cure cycle, the effects of aging and elapsed time before infiltration on the RTM process, as well as monitoring the uniformity of the impregnation cure process are important areas where the FDEMS sensing technique can make a significant contribution to the development of an RTM process.

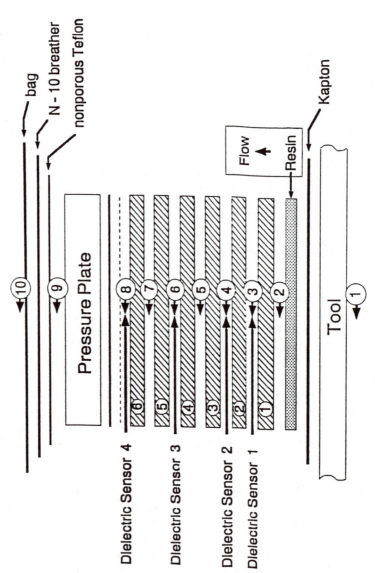

Figure 6. A diagram of the RTM tool.

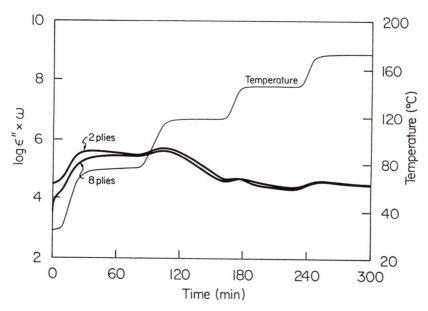

Figure 7. Comparison of log ($\epsilon'' \times \omega$) (125 Hz) for sensors placed two and eight plies from the bottom of an RTM tool.

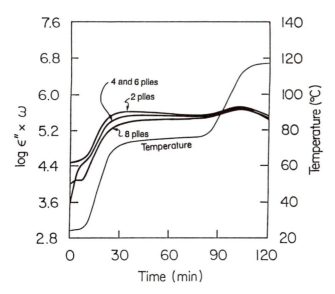

Figure 8. An expansion of Figure 7 including sensors placed four and six plies from the bottom of the RTM tool.

Acknowledgments

This work was made possible through support from a NASA Langley Research Center grant.

References

1. Kranbuehl, D. *Developments in Reinforced Plastics V;* Elsevier Applied Science: New York, 1986; pp 181–204.
2. Kranbuehl, D.; Delos, S.; Hoff, M.; Weller, L.; Haverty, P.; Seeley, J. In *Cross-Linked Polymers: Chemistry, Properties, and Applications;* ACS Symposium Series 367; American Chemical Society: Washington, DC, 1988, pp 100–119. (ACS Div. Polym. Mater. Sci. Eng. **1987,** 56, 163–168.)
3. Kranbuehl, D.; Delos, S.; Hoff, M.; Weller, L.; Haverty, P.; Seeley, J.; Whitham, B. *Natl. SAMPLE Sym. Ser.* **1987,** 32, 338–348.
4. Loos, A.; Kranbuehl, D.; Freeman, W. In *Intelligent Processing of Materials and Advanced Sensors Metallurgical Society;* Metallurgical Society: Warrendale, PA, 1987, pp 197–211.
5. Bidstrup, W. D.; Senturia, S. *SPE 45th Annu. Tech. Conf. Proc.* **1989,** 45, 1035.
6. Groto, J.; Yandrasits, M. *SPE 45th Annu. Tech. Conf. Proc.* **1989,** 45, 1039.
7. Nass, K.; Seferis, J.; Bachman, M. *SPE 45th Annu. Tech. Conf. Proc.* **1989,** 45, 1047.
8. Day, D. *SPE 45th Annu. Tech. Conf. Proc.* **1989,** 45, 1045.
9. Kranbuehl, D.; Delos, S.; Hoff, M.; Haverty, P.; Freeman, W.; Hoffman, R.; Godfrey, J. *Polym. Eng. Sci.* **1989,** 29(5), 285–289.
10. Kranbuehl, D.; Haverty, P.; Hoff, M.; Hoffman, R. *Polym. Eng. Sci.* **1989,** 29(5), 285.
11. Kranbuehl, D.; Hoff, M.; Haverty, P.; Loss, A.; Freeman, T. *Natl. SAMPE SYM. SER.* **1988,** 33, 1276.
12. Inquiries regarding the FDEMS and instrumentation can be directed to D. Kranbuehl.

RECEIVED for review February 14, 1989. ACCEPTED revised manuscript August 8, 1989.

Epoxy–Dinorbornene Spiro Orthocarbonate System

Fourier Transform Infrared Spectroscopy and Dynamic Mechanical Testing

Hatsuo Ishida and John Nigro

Department of Macromolecular Science, Case Western Reserve University, Cleveland, OH 44106–1712

Polymerization of an epoxy resin in the presence of dinorbornene spiro orthocarbonate was studied with both Fourier transform infrared spectroscopy and dynamic mechanical spectrometry. The kinetic study indicates that the epoxy and the spiro orthocarbonate compound polymerize at similar rates; however, most of the spiro orthocarbonate compound reacts before the gel point of the epoxy resin, and thereby negates the volume expansion effect of the spiro orthocarbonate compound. The activation energy of the reaction of the spiro orthocarbonate compound was found to be 18.1 kcal/mol based on first-order kinetics.

DINORBORNENE SPIRO ORTHOCARBONATE* (*see* Scheme I) counteracts the detrimental stress formation associated with cure-induced shrinkage of epoxy resins. During polymerization, the double-ring diether breaks open (*1–5*). This ring scission provided by the cross-linking reaction results in the breakage of two covalent bonds for every linkage (intermolecular bond) formed. Bailey and co-workers (*1, 2, 4, 5*) contend that this molecular "unpacking" manifests itself in an expansion that counteracts cure shrinkage.

---

*The current *Chemical Abstracts* index name of this compound is orthocarbonic acid, cyclic bis(5-norbornen-2-ylidenedimethylene) ester.

0065–2393/90/0227–0259$06.00/0

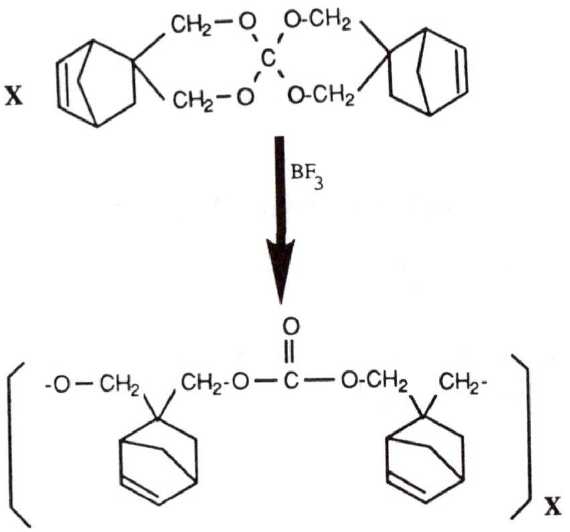

Scheme I. Molecular structure of dinorbornene spiro orthocarbonate before and after polymerization.

Thus, the spiro orthocarbonate acts primarily as a shrinkage-inhibiting agent. Reduced shrinkage leads to a reduction in internal resin stress.

Although resin shrinkage occurs in all phases of cure, only that occurring while the resin is relatively solid is responsible for stress formation. For Bailey's theory to accurately account for the observed stress reduction, a major portion of the spiro orthocarbonate must polymerize after the resin system has gelled.

In an alternative mechanism postulated by Shimbo et al. (6), the orthocarbonate's expansion properties have little to do with its effectiveness. Shimbo determined that most stress is formed during the cooling stage from the glass transition temperature (T_g) to the ambient temperature (T) and that the amount of residual stress in the cured resin is directly proportional to this difference ($ds_r/d[T_g - T]$ is a constant; s_r is resin residual stress). Consequently, a reduced difference between T_g and T should result in a lower-stressed resin. Subsequently they demonstrated that this material (the orthocarbonate) consistently reduces both cured resin T_g and the residual internal stress accordingly.

In an attempt to expose the true mechanism of this phenomenon, we used Fourier transform infrared (FTIR) spectroscopy coupled with dynamic mechanical testing to investigate the spiro orthocarbonate–epoxy resin system. Basic IR transmission was used to follow the orthocarbonate's ring-opening reaction with time. The resulting reaction kinetics were compared with the rheological data obtained via dynamic mechanical spectrometry (DMS). The coordinated FTIR–DMS approach is useful to elucidate the true polymerization behavior of this system and to demonstrate the complimentary nature of the two techniques.

Experimental Details

Epoxy resin (diglycidyl ether of bisphenol A (DGEBA), Shell Epon 828) was heated to 100 °C; 10 phr (parts per hundred parts of resin) based on the weight of epoxy resin was used. Finely ground crystalline dinorbornene spiro orthocarbonate was added with vigorous magnetic stirring. The system was stirred under these conditions for 1–2 h. The resulting dispersion was then allowed to cool to room temperature. After reaching ambient temperature, 10 phr of $BF_3 \cdot$ MEA catalyst was mixed in (MEA, monoethylamine, Anchor 1115, is complexed with the highly reactive BF_3 to impart stability). The system was degassed under reduced pressure before any curing studies were initiated. Infrared samples were prepared by placing the resin described in the IR transmission cell.

At least 12 spectra were obtained at discrete intervals during three isothermal cures at temperatures of 80, 90, and 100 °C. All spectra are presented in absorbance mode and were taken with 200 scans. Sample and reference spectra were obtained at a resolution of 4 cm^{-1} on a double-beam spectrophotometer (Digilab FTS-20) equipped with a mercury–cadmium telluride (MCT) detector cooled with liquid nitrogen.

Resin prepared in an identical manner was subjected to rheological testing in a dynamic mechanical spectrometer (Rheometrics RMS-800). The parallel-plate configuration with a plate diameter of 25 cm and a gap of 2 mm was used in the dynamic cure mode to provide resin viscosity as a function of time during a 100 °C cure. Other relevant settings include a strain of 0.5%, strain rate of 10 per minute, and a sampling rate of one measurement per minute. Because of memory limitations, three sequential curves (obtained sequentially on the same sample) were pieced together to cover the entire cure range.

Results and Discussion

The IR cell, which accommodates low-viscosity liquids, was used to maintain a constant thickness of easily flowing material in the sampling beam. The resulting chronological spectra were analyzed, and a band growth at 1750 cm^{-1} was assigned to the C=O stretching mode of the carbonate formed during the double-ring-opening polymerization reaction (Figure 1).

An absorption at 1891 cm^{-1}, also seen in Figure 1, was assigned as a summation band because of the aromatic nature of the bisphenol A component of the epoxy resin and was used as an internal reference to correct for variations in sample thickness. Spectral manipulation software was employed to evaluate peak areas, and the ratio A_{1750}/A_{1891} was taken, where A is the absorbance at the wavelength indicated by the subscript. The plateau value toward the end of the reaction was assumed to be 100% conversion of the double-ring spiro orthocarbonate, and a chemical conversion profile was calculated and is shown in Figure 2. The conversion-vs.-time data were tested for first-order kinetics and found to fit the model well through 95% of the reaction data (Figure 3).

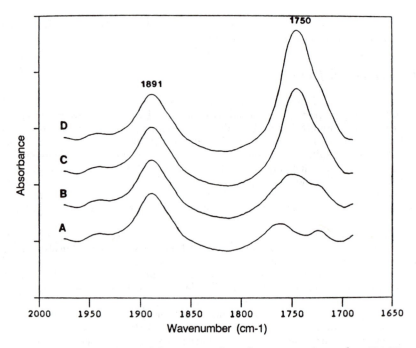

Figure 1. FTIR spectra of the spiro orthocarbonate–epoxy cured at 100 °C. Curve A, uncured: B, 30 min; C, 120 min; and D, 390 min.

Reaction rate constants (k) were calculated as the slope from the first-order plots and were used to construct an Arrhenius plot. The reaction rate constants are 1.11×10^{-2} min^{-1} at 100 °C ($R = 0.99$), 5.42×10^{-3} min^{-1} at 90 °C ($R = 0.98$), and 2.82×10^{-3} min^{-1} at 80 °C ($R = 0.99$), where R represents the correlation coefficient. From this result, an activation energy was calculated for the spiro orthocarbonate's ring-opening polymerization reaction. The resulting value of 18.1 kcal/mol compares quite favorably with those values generally accepted for most epoxy polymerizations (10–25 kcal/mol) (7) and indicates a cure compatibility of the two compounds (dinorbornene spiro orthocarbonate and epoxy resin).

A comparison with the epoxide cure data obtained at 100 °C shows that spiro orthocarbonate and epoxy have much the same reaction kinetics (Figure 4). The orthocarbonate's activation energy is in the appropriate range, a result indicating that the monomer polymerizes and expands as the epoxy resin polymerizes and shrinks. Postcuring the slurry at 100 °C for 1200 min indicated that the expansion is approximately 90–92% completed before the plateau region is reached at 360 min. It seems reasonable to ask how much of this (expansion) takes place after the epoxy had reached its gel point and could effectively combat residual stress formation.

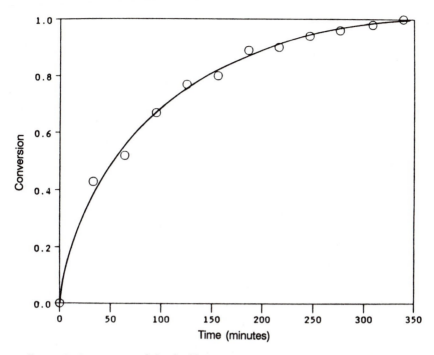

Figure 2. Conversion of the double-ring structure of spiro orthocarbonate in epoxy at 100 °C.

The rheological data presented in Figure 5 were obtained at 100 °C by dynamic mechanical spectroscopy. As plotted, the data indicate that resin viscosity becomes significant after 250 min of heat treatment and increases in a regular manner up to 700 min. During this time, the resin is a highly viscous material and is impervious to residual stress formation because it can flow in response to imposed forces.

From 700 min, the viscosity increased drastically. This behavior indicates a transition from a viscous to a more gelled state. The viscosity asymptotically approaches an infinite value that has been extrapolated to occur at 850–900 min. Although gelation is generally associated with infinite viscosity, the resin can no longer be considered viscous after 750 min. IR cure data indicate that 90–92% of the orthocarbonate's double-ring-opening reaction is complete by 360 min, leaving only 8–10% for cure beyond this time. Most likely, a major portion of the 8–10% is expended in the 360–750-min range, leaving little or none for expansion when the resin is well gelled. A predominant amount of the spiro orthocarbonate compound should react after 360 min and continue to open (ring-opening expansion) well past 750 min if it is to reduce shrinkage and to provide residual stress reduction via the expansion mechanism. Most of the orthocarbonate has been expended while

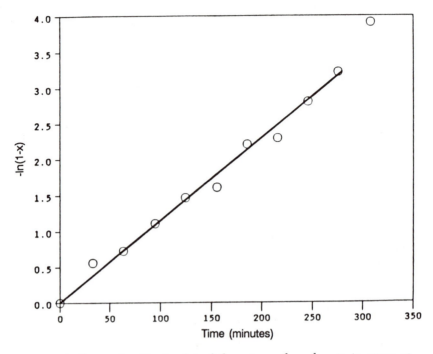

Figure 3. First-order kinetic plot of the spiro orthocarbonate in epoxy at 100 °C.

the system is viscous and is thus ineffectual in combating residual stress formation. The 8–10% that reacts after 360 min probably cannot contribute considerably to the reduction of this stress.

Conclusions

The data suggest that the orthocarbonate's expansion abilities do not act to reduce epoxy cure-induced stress formation as proposed by Bailey et al. (*1, 2, 4, 5*). Instead the experimental evidence reported here supports the theory postulated by Shimbo (*6*), in which this additivity effectively alters resin thermal properties such that less cure shrinkage is converted into residual stress. In this sense, then, the spiro orthocarbonate may not be unique; any substance that can reduce epoxy resin's glass transition temperature without altering cured resin integrity should provide for similar resin performance.

Along with helping to clarify the spiro orthocarbonate–epoxy resin interaction, this study presented a combined study of FTIR and DMS. It is hoped that the complementary nature of the rheological and IR properties and the utility of such an approach will prove useful in the future study of other systems.

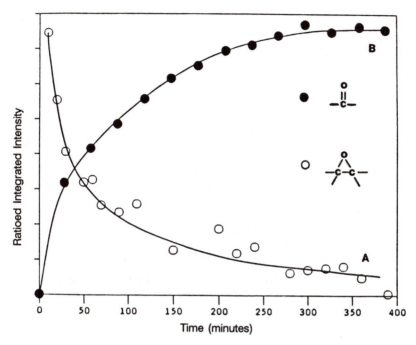

Figure 4. Normalized concentration of the epoxide ring and the carbonate during curing at 100 °C.

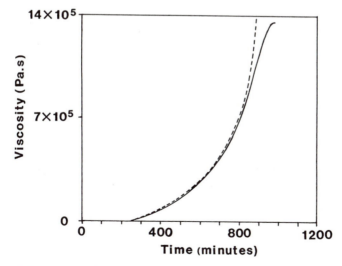

Figure 5. Viscosity of the spiro orthocarbonate–epoxy system at 100 °C.

Acknowledgments

The authors gratefully acknowledge the financial support of the Center for Adhesives, Sealants, and Coatings (CASC), Case Western Reserve University, Cleveland, Ohio.

References

1. Bailey, W. J.; Iwama, H.; Tsushima, R. *J. Polym. Sci.* **1977**, *56*, 117.
2. Bailey, W. J.; Siago, K. *Polym. Prepr.* **1980**, *21*, 4.
3. Piggott, M.; Lam, P. W.; Lim, J. T.; Woo, M. S. *Comp. Sci. Technol.* **1985**, *23*, 247.
4. Bailey, W. J.; Sun, R. L.; Katsuki, H.; Endo, T.; Iwama, H.; Tsushima, R.; Saigo, K.; Bittrito, M. M. In *Ring-Opening Polymerization;* ACS Symposium Series No. 59; American Chemical Society: Washington, DC, 1977; p 38.
5. Bailey, W. J.; Endo, T. *J. Polym. Sci.* **1978**, *60*, 17.
6. Shimbo, M.; Ochi, M.; Inamura, T.; Inoue, M. *J. Mater. Sci.* **1985**, *20*, 2965–2972.
7. Antoon, M., Ph.D. Thesis, Department of Macromolecular Science, Case Western Reserve University, Cleveland, Ohio; 1977.

RECEIVED for review February 14, 1989. ACCEPTED revised manuscript September 26, 1989.

16

Rheokinetic Measurements of Step- and Chain-Addition Polymerizations

D. Rosendale and J. A. Biesenberger

Department of Chemistry and Chemical Engineering, Stevens Institute of Technology, Hoboken, NJ 07030

The dependence of viscosity on the extent of reaction, or conversion, is information necessary for accurate modeling and control of reactive extrusion and reaction injection molding. This chapter presents rheo-kinetic data and describes a new instrument, called the rheocalorim-eter, which simultaneously measures the heat released and viscosity of a polymerizing solution and makes it possible to obtain viscosity-vs.-conversion data directly. Experimental results and a simple model show that this viscosity growth behavior is different for different polymerization mechanisms. Methyl methacrylate polymerization, styrene–acrylonitrile copolymerization, and polyurethane polymer-ization data are presented and compared to the predictions of the model.

THE EFFECT OF CONCENTRATION AND MOLECULAR WEIGHT on the viscosity of linear (non-network) polymers during their formation has long been ignored, and only recently have crude attempts been made to produce realistic models of this dependence. The effect is enormous, spanning seven or more decades of viscosity, diminishing by comparison the effects of shear rate ("shear-thinning") and even reaction temperature, which are not inconsiderable.

Most polymerizations can be viewed as falling into one of three broad categories based upon fundamentally distinguishable chain-growth mechanisms (1):

$$\text{random: } m_x + m_y \rightarrow m_{x+y} \tag{1}$$

0065–2393/90/0227–0267$06.00/0

$$\text{step addition: } m_x + m \rightarrow m_{x+1} \tag{2}$$

$$\text{chain addition: } m_x{}^* + m \rightarrow m_{x+1}{}^* \tag{3}$$

where the subscript denotes the number of monomer units in a particular molecule, and the asterisk refers to a highly reactive free radical site.

Random polymerization is characterized by chain growth occurring throughout the reaction randomly among molecules of all sizes. Examples are polycondensations and catalyzed linear urethane polymerizations. At the onset of a random reaction, monomer molecules are most likely to encounter another monomer molecule and form a dimer. The dimer may react with either a monomer or another dimer, depending on which is present in higher concentration, to form a longer molecule. Thus, in the beginning, the polymer molecules form very slowly, but as the average length of molecules in the reacting mixture increases, the chains begin growing very rapidly through each individual reaction step as long chains link with other long chains.

The random polymerization mechanism is in contrast to the addition mechanisms whereby chains grow by addition of one monomer molecule at a time. Chain addition differs from step addition because termination reactions occur only in chain addition:

$$m_x{}^* \rightarrow m_x \tag{4}$$

that make it impossible for the same molecules to continue growing throughout the reaction. Each chain lives only a very short time compared to the overall time of polymerization, and termination of fully grown chains is followed by initiation of succeeding generations of new ones:

$$m \rightarrow m_1{}^* \tag{5}$$

Free-radical polymerizations are obvious examples of the scheme illustrated by equations 3–5. By contrast, in both step-addition and random polymerization, the same molecules grow (in principle) throughout the entire course of the reaction.

These distinct features are reflected in the molecular-weight growth of the resulting polymers, giving rise to three very different degrees of polymerization (DP) versus conversion curves (1). They are illustrated in Figure 1 for number-average DP. Thus, the steep rise in DP at the very end of the random reaction stands in sharp contrast to the very large initial DP of chain-addition polymers. In fact, DP in chain-addition polymers actually declines when the kinetic parameter $\alpha_k < 1$ (1). This result can be explained only by the existence of chain termination, for if the same chains grew throughout the entire reaction it would be impossible for the DP to decline. The pa-

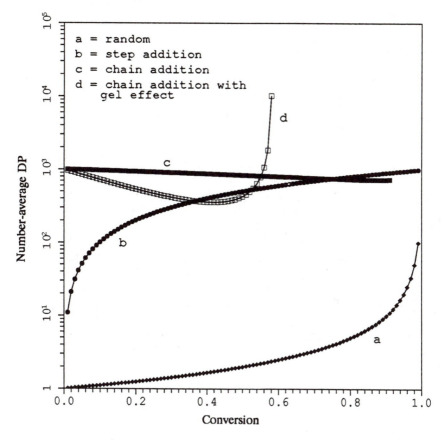

Figure 1. Number-average degree of polymerization ($\bar{x}_N$) vs. conversion for various polymerization mechanisms.

rameter α_k is defined as the ratio of the time constant for monomer conversion to that for initiator decomposition, and for a conventional chain-addition polymerization without branching or chain transfer:

$$\alpha_k = \frac{(k_t k_d)^{1/2}}{k_p (2C_0)^{1/2}} \tag{6}$$

The k_t, k_d, and k_p are the rate constants for the termination, dissociation of initiator, and propagation steps, respectively, and C_0 is the initial initiator concentration. When $\alpha_k > 1$, initiator is depleted before monomer and the reaction "dead ends" (DE).

In an attempt to predict the effect of polymer molecular weight and concentration on reaction viscosity, Malkin (2) evidently classified polymerizations in precisely the same way. Thus, our term "random" corresponds to his "condensation", and our "step-addition" to his "ionic" polymerization. It can be argued that random is a more general label, because not all reactions of the form of equation 1 are condensations, for example, linear polyurethane formation from diisocyanates and diols. Similarly, step addition is a more general term because it includes at once such extremes as very rapid ionic polymerizations and relatively slow, uncatalyzed ring-opening polymerizations. The absence of termination is the criterion for step addition, not the speed of the propagation step.

Theory

Following the approach of Malkin with some modification, we shall attempt to predict the general behavior of viscosity growth with polymer formation for the aforementioned reaction types. Anticipated viscosity growth versus time curves based upon Figure 1 have been sketched elsewhere (1a). They show chain-addition viscosity rising most rapidly, random most slowly, and step addition between them. The random curve is concave upward, whereas the addition curves exhibit downward concavity. The treatment reported here, which is more quantitative, will examine this expected behavior more closely.

We postulate that the dependence of viscosity (η) upon shear rate ($\dot{\gamma}$), temperature (T), weight fraction of polymer in solution (w_p), and weight-average DP ($\bar{x}_w$):

$$\eta = \eta(\dot{\gamma}, T, w_p, \bar{x}_w) \tag{7}$$

can be separated into the product (1b):

$$\eta = f(\dot{\gamma})\eta_0(T, w_p, \bar{x}_w) \tag{8}$$

and that the Newtonian or "zero shear" viscosity η_0 can be separated further as follows (1c):

$$\eta_0 = K(T)w_p{}^\alpha \bar{x}_w{}^\beta \tag{9}$$

where exponents α and β depend upon the "entanglement" DP in solution $(\bar{x}_w)_{es}$, and $K(T)$ is a temperature-dependent empirical constant. Thus,

$$\alpha = 1, \ \beta = 1 \quad \text{when} \quad \bar{x}_w < (\bar{x}_w)_{es} \tag{10}$$

$$\alpha = 4.7, \ \beta = 3.4 \quad \text{when} \quad \bar{x}_w > (\bar{x}_w)_{es} \tag{11}$$

The locus of values $(\bar{x}_w)_{es}$ corresponding to concentrations w_p is computed from the expression:

$$w_p(\bar{x}_w)_{es}{}^\gamma = (\bar{x}_w)_{ep}{}^\gamma \tag{12}$$

where $(\bar{x}_w)_{ep}$ is the entanglement DP for polymer melt ($w_p = 1$). Whereas values of 0.63 and 0.68 have been suggested for the empirical constant γ (3), we used 1.0 (1d).

The following expressions, when substituted into equation 9, subject to equations 10–12, predict growth of η_0 with conversion Φ:

$$w_p = 1 \tag{13}$$

$$\bar{x}_N = \frac{1}{1 - \Phi} \tag{14}$$

$$\bar{x}_w = \frac{1 + \Phi}{1 - \Phi} \tag{15}$$

where $\bar{x}_n$ is the number-average DP. These equations follow directly from those tabulated elsewhere (1e, 1f). Equation 13 reflects the fact that monomer is also counted as polymer with length of one unit in random-type polymerizations.

For step addition,

$$w_p = \Phi \tag{16}$$

$$\bar{x}_N = 1 + x_0\Phi \tag{17}$$

$$\bar{x}_w = \bar{x}_N + \frac{\bar{x}_N - 1}{\bar{x}_N} \tag{18}$$

where x_0 is the ratio of monomer to initiator in the feed. These equations also follow directly from those listed elsewhere (1e, 1g). Equation 16 reflects the fact that molar conversion of monomer is proportional to weight of polymer formed in addition polymerizations. The dispersion index $\bar{x}_w/\bar{x}_N$ approaches 2.0 at high conversions.

For chain addition with the gel effect,

$$w_p = \Phi \tag{19}$$

$$\bar{x}_N = \frac{2\alpha_k x_0\Phi}{1 - [1 + \frac{1}{2}\alpha_k G(\ln(1 - \Phi))]^2} \tag{20}$$

$$\bar{x}_w = 2\bar{x}_N \tag{21}$$

where parameter α_k indicates whether the reaction will dead end ($\alpha_k > 1$) or not ($\alpha_k < 1$), as well as whether the DP will drift upward or downward, respectively. One may incorporate the gel effect into the model by substituting for G an appropriate gel-effect model for comparison with data, for example, $G = \exp[-(B/2.0)\Phi]$, where B is an empirical constant. If no gel effect is present, then G is set equal to 1.0. Other gel-effect models may be found in ref. 1h. Equation 20 applies to all values of monomer conversion below the conversion where dead-ending occurs (Φ_{DE}), given elsewhere (1j) as:

$$\Phi_{DE} = 1 - \exp \frac{-2}{\alpha_k} \tag{22}$$

The dispersion index in equation 21 has been arbitrarily taken to be 2.

This analysis differs in several ways from that of Malkin (2). He used number-average DP ($\bar{x}_N$) instead of weight-average ($\bar{x}_w$) in equation 9, and he did not incorporate an entanglement DP, but instead used exponents in excess of 1 (5 and 3.4) throughout the polymerization. Malkin also assumed that DP remains constant throughout chain-addition polymerization, whereas we have allowed it to drift with reaction, and even to dead end.

The curves in Figure 1 are graphs of equations 14, 17, and 20. Corresponding graphs of η_0 versus conversion have been plotted in Figure 2. An

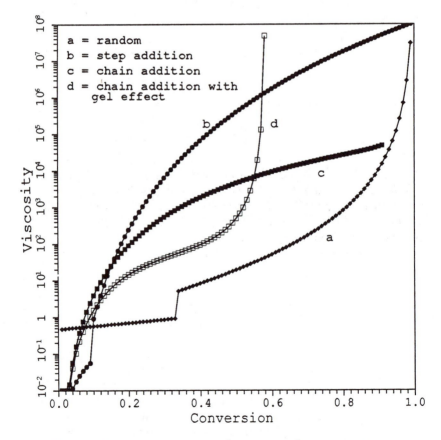

Figure 2. Viscosity vs. conversion for various polymerization types.

average value of $K(T)$ from equation 9 was computed from experimental η-versus-Φ data to produce these curves. The discontinuities in curves a and b reflect the fact that the exponents α and β in equations 9–11 change when the $\bar{x}_w$ exceeds the entanglement value. In chain-addition polymerizations, the DP begins at values higher than the entanglement x_w, and thus α and β do not change.

For all possible combinations of polymer conversion (concentration) and molecular weight for each monomer–polymer type at constant shear rate and temperature, equation 9 can be represented by a three-dimensional surface, and each curve in Figure 2 would thus correspond to a space curve on its respective surface, whose trajectory is determined by the specific reaction path followed in each case. Such surfaces and their corresponding trajectories are shown in Figures 3–5 for our three model reactions.

A typical chain-addition reaction is exhibited in Figure 3, where the DP

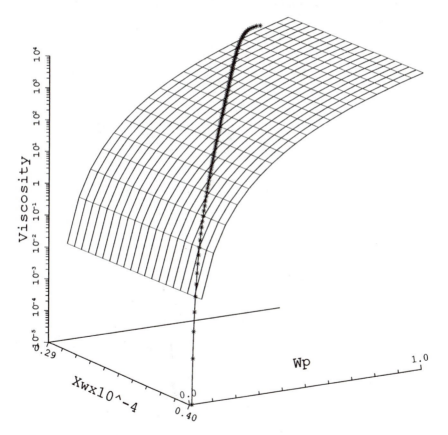

Figure 3. Viscosity–concentration (w_p)*–molecular weight* $(\bar{x}_w)$ *surface for a chain-addition polymerization* $(x_0 = 1000, \alpha_k = 0.8, B = 0)$.

initially is large and drifts downward as the monomer pool is depleted. It may also begin to drift upward at high conversions, depending upon the magnitude of α_k. Figure 4 shows that, for random polymerizations, $w_p = 1$ at all times, and the viscosity rises as the weight-average DP increases. The interval between asterisks in Figure 4 represents an equal amount of conversion and, as expected, the viscosity is increasing most rapidly at high levels of conversion.

Figure 5 shows the characteristic behavior of a step-addition polymerization with both $\bar{x}_w$ and w_p increasing as the reaction proceeds. Conventional methods for estimating reaction viscosities for polymerization and polymer-modification reactions are actually attempts to characterize the surface $\eta_0(w_p, \bar{x}_w)$ at each T, rather than the trajectory, via numerous viscosity measurements on solutions of the polymer in monomer at various concentrations and molecular weights in the absence of reaction. Concentration (conversion) and molecular weight would be subsequently related by the specific kinetic path taken during reaction, and thereby yield viscosity as a

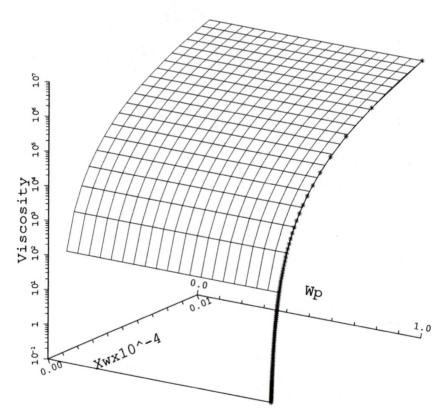

Figure 4. Viscosity–concentration–molecular weight surface for a random polymerization.

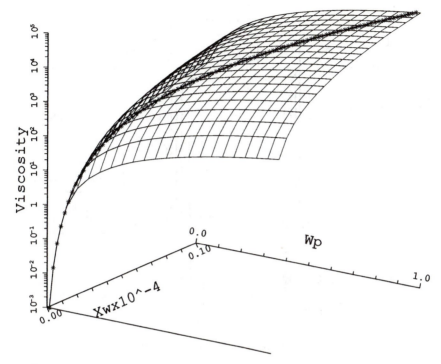

Figure 5. Viscosity–concentration–molecular weight surface for a step-addition polymerization.

function of conversion. Such measurements are often not feasible, however, because many such mixtures are not inert to reaction.

The experimental goal in this study was to simultaneously measure viscosity and conversion during reaction. Thus, in geometrical terms, we are determining viscosity directly along a specific trajectory on a system surface.

Experimental Details

A specially designed rheocalorimeter was constructed (4, 5) for the purpose of measuring reaction viscosity versus conversion, and has been used to study representative examples of some of the aforementioned reaction types in search of anticipated differences in their viscosity growth behaviors. The instrument uses Couette geometry with a very small gap to facilitate temperature control of the polymerizing solution (Figure 6). Coolant is circulated around both the inner and outer walls of the annulus at approximately 20 °C below the reaction temperature. To maintain reaction temperature, heaters, which are wrapped inside the inner wall and outside the outer wall, must operate continuously. As the reaction produces heat, the heaters must decrease their power output by the same amount to maintain isothermal conditions. The change in power output then provides a rate of heat evolution curve

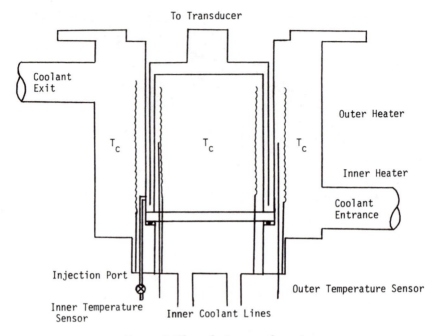

Figure 6. Rheocalorimeter schematic.

that is proportional to the rate of reaction. Concurrent to this measurement, the viscosity of the solution is monitored at relatively low shear rate (5–10 s⁻¹) in the dynamic mode.

A limitation of the rheocalorimeter is the minimum rate at which measurements can be made because, as the rate is decreased, the length of time required for each viscosity measurement increases and too much information is lost. Therefore, although zero shear viscosity is not actually being measured, but the deviation from zero shear is less than 10% of the viscosity effect seen during the reaction, and in fact will only deviate at higher viscosities when the reaction mixture becomes non-Newtonian. Steady shear viscosity can also be obtained for lower viscosity materials, but transducer overload becomes a problem as the solution becomes more viscous.

The reaction chamber itself is constructed of aluminum with an aluminum oxide hard coat for chemical inertness and electrical insulation. A stainless steel insert is fitted tightly in the annulus, which may be easily ejected when the polymer has gelled and placed in a furnace for pyrolysis of the polymer to simplify the cleaning process. Operating limits for the instrument are as follows:

> Geometry: 0.5 mm gap, 8.0 mL volume
> Temperature: 25–200 °C
> Pressure: 0–5 atm
> Steady rotational rate: 0.5–300 s⁻¹
> Dynamic shear rate: 0.1–50 rad/s
> Dynamic amplitude of rotation: 1–1000% of gap
> Transducer range: 0.2–100 g/cm
> Viscosity range: 0.001–10,000 P
> Maximum heat-evolution rate: 1.0 cal/mL

The following systems have been investigated to date: methyl methacrylate (MMA) polymerization in toluene and in bulk, styrene–acrylonitrile (SAN) copolymerization, and linear polyurethane (PU) polymerization. The reaction compositions and temperatures are given in Table I.

The MMA and SAN reactions are chain-addition polymerizations, and the PU reaction is an example of a random polymerization. MMA and SAN monomer solutions were prepared in advance and injected at room temperature into the reaction chamber that had been allowed to come to equilibrium at the reaction temperature for approximately 1 h. Upon injection, a large amount of heat is required to bring the reactants quickly to reaction temperature. This heating is typically accomplished in 20 s. As the reaction begins to release heat, the heaters decrease their power output by exactly this amount to maintain isothermal conditions.

Immediately following injection, the injection port is closed and the chamber pressurized to approximately 45 psig with nitrogen. Simultaneous to the measurement of the heat evolution, the viscosity of the reacting mixture is measured as a function of time. This measurement, done primarily in the dynamic mode, often spans 8 orders of magnitude in viscosity. The amplitude of rotation in the dynamic mode or the rate of rotation in steady shear must be decreased as the viscosity of the reaction mixture increases to prevent transducer overload. Without this adjustment the measurement of such a wide range of viscosities would be impossible.

Examples of the data obtained are given in Figures 7 and 8. Figure 7 shows the heat generation-vs.-time data, and Figure 8 exhibits the raw viscosity data for bulk methyl methacrylate at 90 °C. A huge gel effect is exhibited at approximately 50% conversion in the heat-generation curve, and a simultaneous exponential rise occurs in the viscosity curve.

The PU reactions were performed in a similar manner. The polyol, catalyst, and isocyanate were mixed manually with a stirring rod for 30 s and immediately injected into the chamber and pressurized. The viscosity of this reacting mixture typically spanned 4 orders of magnitude. Examples of the data obtained are given in Figures 9 and 10. Conversion was obtained by integrating the heat of reaction-vs.-time curve and relating this conversion to the viscosity by eliminating the time coordinate. Viscosity-versus-conversion curves for the different reactions are shown in Figures 11–13.

Table I. Reaction Compositions and Temperatures

Run Number	Reactant 1 (mol %)	Reactant 2 (mol %)	Temperature (°C)
1	60% styrene	40% AN	90.0
2	70% styrene	30% AN	100.0
3	80% styrene	20% AN	100.0
4	100% MMA	none	90.0
5	80% MMA	20% toluene	90.0
6	50% polyol	50% MDI	60.0
7	50% polyol	50% MDI	60.0
8	55% polyol	45% MDI	60.0
9	45% polyol	55% MDI	60.0

NOTE: The catalysts were 0.1 M AIBN for runs 1–5 and 0.1 wt% T-12 for runs 6–9. The inhibitor was 0.01 M BQ for runs 1–5; no inhibitor was used in runs 6–9.

ABBREVIATIONS: Polyol is Niax 16–56 (Union Carbide), MDI is diphenylmethane diisocyanate (Dow Isonate 143L), AIBN is 2,2'-azobisisobutyronitrile (Kodak), T-12 is dibutyltin dilaurate (M&T), and BQ is *p*-benzoquinone (Fisher).

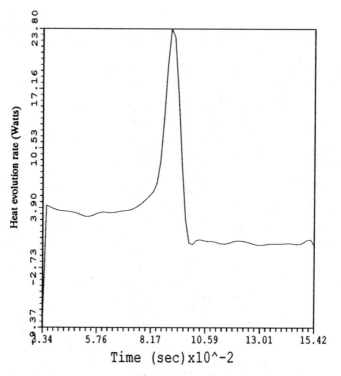

Figure 7. Heat-evolution curve for bulk MMA.

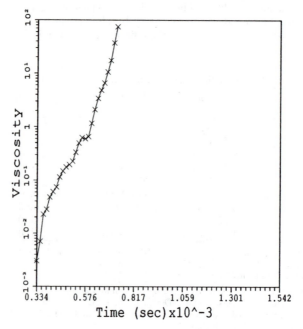

Figure 8. Viscosity-rise data for bulk MMA.

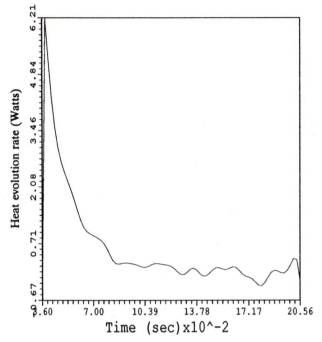

Figure 9. Heat-evolution curve for 1:1 PU (run 7).

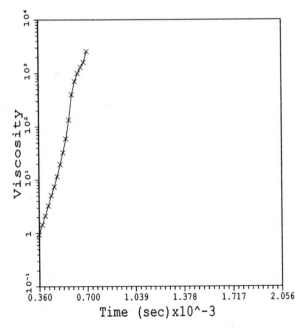

Figure 10. Viscosity-rise data for 1:1 PU (run 7).

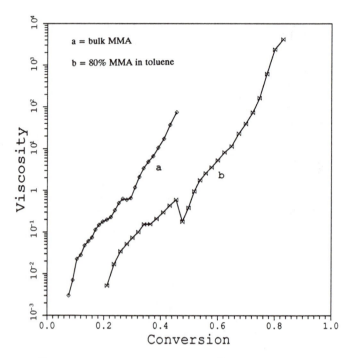

Figure 11. Viscosity vs. conversion for MMA reactions.

Results and Conclusions

Figure 11 reveals the behavior of MMA reactions. The mixture containing toluene as solvent has lower viscosities throughout, as expected. Both reactions show a gel effect where the viscosity increases exponentially. This rapid increase coincides with a large peak in the heat-generation curve. Discontinuities in the viscosity data are caused by changes in the amplitude of rotation of the viscometer. As previously mentioned, as the viscosity increases, the amplitude must be decreased to avoid transducer overload.

Figure 12 gives the curves for the SAN reactions. Curve a is highest because of the lower reaction temperature. Curves b and c represent different compositions reacted at 100 °C. Curve b, the 80:20 copolymer, is expected to have higher viscosities than curve c because of the higher concentration of styrene monomer (0.72 cP at 25 °C) than acrylonitrile (0.34 cP at 25 °C). The slightly different slopes of these curves may be due to the drift in monomer pool composition that results in polymer of different composition being formed at the end of the reaction rather than at the beginning.

Some problems were encountered in analyzing data from the PU reactions (Figure 13). A small heat of reaction causes difficulties in integrating the heat-generation data to obtain conversion. In some cases (e.g., run 9)

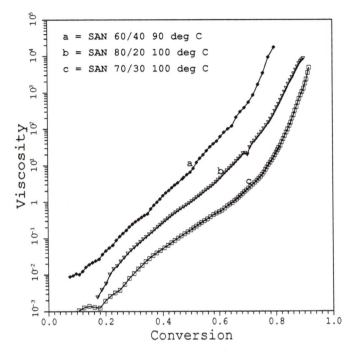

Figure 12. Viscosity vs. conversion for SAN reactions.

this difficulty produces questionable results at high levels of conversion. This problem is inherent in random reactions because of the mechanism of long chains linking at high conversions and producing huge viscosity effects but negligible thermal effects. Runs 6 and 7 (curves b and c) appear in between the other curves but do not coincide, possibly because of inaccuracies in the integration process. Run 8 (curve a), containing an excess of high-viscosity polyol, shows higher viscosities throughout, and run 9 (curve d) exhibits lower viscosities because of the excess of low-viscosity isocyanate.

Comparison of the experimental data with our simple model based on kinetic mechanisms (including the accelerating effect due to gelation) yielded good agreement (Figures 14 and 15). In these cases, experimental data were used to obtain values for $K(T)$, α, and β for use in equation 9. The value of $B/2$ in the gel-effect model used to produce Figure 14 was 3.0.

In summary, the rheocalorimeter has shed some light on the behavior of polymer viscosity growth during reaction. It has qualitatively verified the semitheoretical treatment of the behavior described herein. It is hoped that, through the use of this instrument, quantitative relationships such as equations 9–12 will become more precise and well understood. In addition, the instrument has utility in applications involving reactive extrusion, in particular for identifying the optimum point of juncture between prereactor and extruder–reactor.

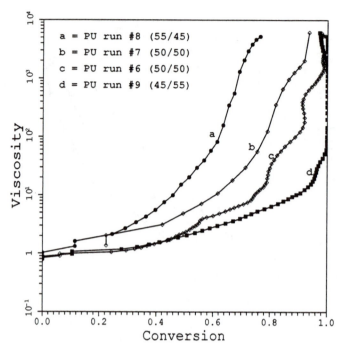

Figure 13. Viscosity vs. conversion for PU reactions.

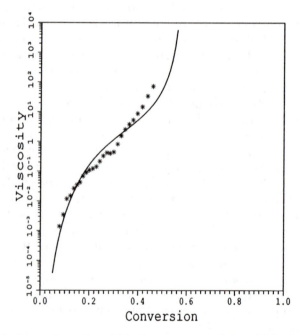

Figure 14. Comparison of model (—) to experimental data (*) for bulk MMA.

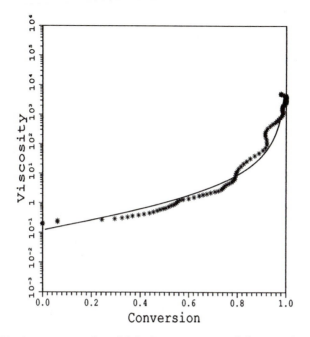

Figure 15. Comparison of model (—) to experimental data (*) for 1:1 PU reaction.

Acknowledgments

The authors thank Rheometrics, Inc., for constructing and assisting in the design of this instrument.

References

1. Biesenberger, J. A.; Sebastian, D. H. *Principles of Polymerization Engineering;* Wiley: New York, 1983; a, Figure 5.2–7; b, equation 5.2–1; c, equation 5.2–8; d, p 518; e, Table 1.5–4; f, Table 2.2–1; g, equations 1.7–9, 2.3–7; h, Section 2.9–1; i, equation 2.2–7; and j, equation 2.7–8.
2. Malkin, A. Y. *Polym. Eng. Sci.* **1980**, *20*, 1035.
3. Middleman, S. *The Flow of High Polymers;* Wiley: New York, 1968.
4. Rosendale, D., Masters Thesis, Stevens Institute of Technology, 1988.
5. Biesenberger, J. A.; Rosendale, D. "Rheokinetic Measurements of Acrylic Polymerizations;" *Tech. Pap. Soc. Plast. Eng.*, 1988.

RECEIVED for review February 14, 1989. ACCEPTED revised manuscript October 30, 1989.

SPECTROSCOPY

Probing Polymer Structures by Photoacoustic Fourier Transform Infrared Spectroscopy

Marek W. Urban*, Scott R. Gaboury, William F. McDonald, and Ann M. Tiefenthaler

North Dakota State University, Department of Polymers and Coatings, Fargo, ND 58105

This chapter presents recent developments as well as the theory of photoacoustic Fourier transform infrared (PA FTIR) spectroscopy. New techniques such as temperature photoacoustic and rheophotoacoustic (RPA) FTIR measurements and their applications to the surface analysis of fibers and cross-linking reactions of amorphous networks are discussed. In situ photoacoustic FTIR detection of cross-linking reactions permits monitoring such transitions as gelation and vitrification of the network as a function of temperature, and rheophotoacoustic FTIR spectroscopy allows one to relate the molecular deformations with external forces applied to a polymer.

T HE TERM "PHOTOACOUSTIC" (PA) refers to the generation of acoustic waves by modulated optical radiation. This effect was discovered in 1880 by Alexander Graham Bell, who observed that audible sound is produced when sunlight modulated by a chopper is incident on optically absorbing materials (*1, 2*). For almost 100 years, this 19th century concept has been overwhelmed by other spectroscopic techniques. It was rediscovered in the early 1970s with the advent of new sources of radiation. More sensitive detectors were followed by the theory of photoacoustic effect.

*Corresponding author

0065-2393/90/0227-0287$07.75/0

Photoacoustic spectroscopy (PAS) detects the acoustic signal emitted from a sample that has absorbed a modulated electromagnetic radiation. A sample is placed in a small chamber to which a sensitive microphone is attached. Upon absorption of modulated light, the sample generates heat. Its release leads to temperature fluctuations at the surface. The frequency of the temperature fluctuations is in phase with the modulation frequency. The temperature fluctuations of the sample surface cause pressure changes in a surrounding gas, which, in turn, generate acoustic waves in the sample chamber. These pressure changes of the gas are detected by a sensitive microphone.

Several processes may occur after light absorption. Depending upon the nature of detection, there are essentially four classes of PA signal: (1) PA spectroscopy that measures the amplitude of PA signal for a range of optical excitation wavelength; (2) PA spectroscopy that monitors deexcitation processes after optical excitation; various decays are possible including luminescence, photochemistry, photoelectricity, and heat that may be generated directly or through energy-transfer processes; (3) PA probing of thermoelastic or other physical properties of materials such as sound velocity, elasticity, flow viscosity, specific heat, substrate defects; and (4) PA generation of mechanical motion. Although these methods have generated various applications and have been the subject of numerous studies, the most common photothermal effect is caused by the heating of a sample after the absorption of optical energy. Other deexcitation processes besides heating may also occur. Figure 1 illustrates various paths producing the photoacoustic signal.

In its broader sense, photoacoustics is the generation of acoustic waves or other thermoelastic effects by any type of energetic radiation, including electromagnetic radiation from radio frequency to X-rays, electrons, protons,

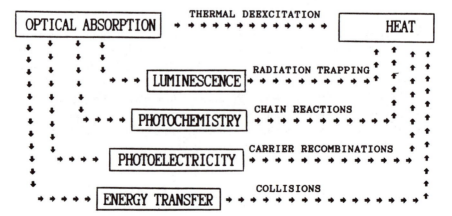

Figure 1. Various processes leading to the production of heat and generation of acoustic waves.

ions, and other particles. As a consequence, quite a substantial amount of experimental and theoretical work has been presented in the literature on applications not only in spectroscopy, but also in many other disciplines such as physics, chemistry, biology, and medicine.

Depending upon the method of PA signal generation, several criteria of classification have been established. Here, we will focus on the generation of an indirect PA signal. In direct PA generation, the acoustic waves are created within the sample where the excitation energy is absorbed, but in indirect generation, the acoustic waves are generated in a coupling medium adjacent to the sample, usually because of heat produced at the sample surface and subsequent emission of acoustic waves in the coupling medium. This form of detection is essential in photoacoustic infrared spectroscopy. The infrared (IR) region of radiation offers considerable advantages in polymer science because the energy of vibrating atoms forming chemical bonds falls in this range. As a result, an IR spectrum can be obtained. This process is schematically depicted in Figure 2. Because of the energy-conversion processes (absorption of light–emission of acoustic waves), such detection

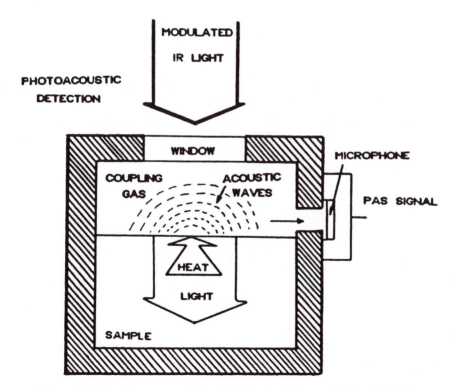

Figure 2. Schematic representation of indirect photoacoustic signal detection for condensed samples.

can be a valuable tool when the optical absorption is so strong that it prevents light passage through the sample.

Theory of Photoacoustic Effect

Figure 3 schematically depicts the generation of photoacoustic signals. The modulated IR radiation with intensity I_0 enters the sample with refractive index n and absorption coefficient β. The intensity of the IR radiation diminishes exponentially as it penetrates the sample, giving rise to the intensity at depth x:

$$I(x) = I_0(1 - n) \exp(-\beta x) \tag{1}$$

The amount of light absorbed within the thickness x is equal to:

$$E(x) = \beta I(x) = \beta I_0(1 - n) \exp(-\beta x) \tag{2}$$

The depth of optical penetration is defined as optical absorption length, L_β, and is inversely proportional to β:

$$L_\beta = \frac{1}{\beta} \tag{3}$$

In other words, L_β is the distance from the surface at which the initial IR intensity, I_0, attenuates to $(1/e)I_0$. The absorbed energy is released in a form of heat that is transferred to the sample surface. The efficiency of the heat transfer is determined by the thermal diffusion coefficient of the sample, a_s, and the modulation frequency of the incident radiation, ω:

$$a_s = \left[\frac{\omega}{2\alpha}\right]^{1/2} \tag{4}$$

where α is the thermal diffusivity [$\alpha = k/\rho C$, that is, thermal conductivity divided by (density $\times$ specific heat)]. The thermal diffusion length, μ_{th}, is related to the thermal diffusion coefficient, a_s, as

$$\mu_{th} = \frac{1}{a_s} = \left[\frac{2\alpha}{\omega}\right]^{1/2} \tag{5}$$

The amount of heat periodically transferred to the surface through the sample is then equal to

$$H(x) = E(x) \exp(-a_s x) = \beta I_0 \frac{1 - n}{\exp[(\beta + a_s)x]} \tag{6}$$

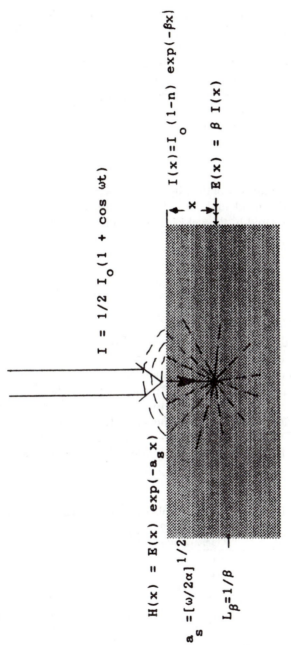

Figure 3. Generation of photoacoustic signal at a sample surface.

Applications of PA FTIR Spectroscopy

Although the theory of indirect PA signal generation has been described previously (3–5), we will briefly focus on its applications to the analysis of polymers. As suggested by Gersho and Rosencwaig (6–8), materials are classified according to their thermal and optical properties. Table I summerizes these results and demonstrates the relationship between photoacoustic intensity and modulation frequency. For example, most polymeric materials are optically thin and thermally thick. For such materials, the Rosencwaig–Gersho (RG) theory predicts that

$$PA \propto \omega^{-3/2} \tag{7}$$

where PA is the intensity of the photoacoustic signal and ω is the modulation frequency of the incident light. The modulation frequency of the FTIR instrument is related to the velocity of the moving mirror of the interferometer:

$$\text{for Michelson interferometer: } \omega = 2V\nu \tag{8}$$

$$\text{for Ganzel interferometer: } \omega = 4V\nu \tag{9}$$

where V is the mirror velocity (cm/s), and ν is the vibrational frequency. The thermal diffusion length, that is, a distance below the surface from which the generated heat can communicate with the surface, is related to the modulation frequency by equation 5. Thus, by changing the mirror velocity of the interferometer, it is possible to vary the thermal diffusion length which, in turn, is the effective penetration depth. However, the thermal diffusion length is also a function of IR wavenumber. For a typical polymer, $k = 0.0003$ cal/(m s °C); $\rho = 1.2$ g/cm^3; $C = 0.35$ cal/g °C; and the thermal diffusion length is of the order of 6.5, 6.9, 8.4, and 11.4 μm at

Table I. Dependence of Modulated Frequency on Magnitude of Photoacoustic Signal

Optical Property	Thermally Thin	Thermally Thick
Optically transparent	1. $\mu_{th} > \beta$; $\mu_{th} < L_\beta$ PA $\propto \omega^{-3/2}$	1. C $\mu_{th} \gg \beta$; $\mu_{th} > L_\beta$ PA $\propto \omega^{-1}$
	2. $\mu_{th} < \beta$; $\mu_{th} \ll L_\beta$ PA $\propto \omega^{-3/2}$	2. $\mu_{th} > \beta$; $\mu_{th} < L_\beta$ PA $\propto \omega^{-1}$
Optically opaque	1. $\mu_{th} < \beta$; $\mu_{th} > L_\beta$ PA $\propto \omega^{-3/2}$	1. $\mu_{th} \gg \beta$; $\mu_{th} \gg L_\beta$ PA $\propto \omega^{-1}$
	2. $\mu_{th} \ll \beta$; $\mu_{th} < L_\beta$ PA $\propto \omega^{-3/2}$	

NOTE: ω is modulation frequency (in hertz); β is the optical absorption coefficient of the sample (in reciprocal centimeters); $L_\beta = 1/\beta$ – optical absorption length of the sample; and μ_{th} is the thermal diffusion length.

1700, 1500, 1000, and 500 cm^{-1}, respectively, when the mirror speed of the Michelson interferometer is 0.16 cm/s. The absorption coefficient in the IR region for a typical polymeric material is of the order of 10^{-3} cm^{-1}, so the optical path length will be of the order of about 10 μm. Thus, most polymer samples of 10–50-μm thickness are optically transparent and may range from thermally thick to thermally thin. However, the spatial resolution of PA FTIR spectroscopy, that is, the smallest thickness that can be detected, is fairly high and is in the range of two Langmuir–Blodgett monolayers with a length of approximately 20 C–C bonds each (9).

Surface Depth Profiling

To demonstrate the capability of the depth profiling, Urban and Koenig (10) performed studies on a double layer of poly(vinylidene fluoride) on poly(ethylene terephthalate) (PVF$_2$ on PET). Figure 4 shows photoacoustic FTIR spectra of a 6-μm overlayer of PVF$_2$ on PET obtained with various modulation frequencies (mirror speeds). When the modulation frequency is fast, the spectrum contains primarily information from the surface (top PVF$_2$ layer). A decrease of modulation frequency causes an increase of the thermal diffusion length, and the PA spectrum consists of information from the bulk.

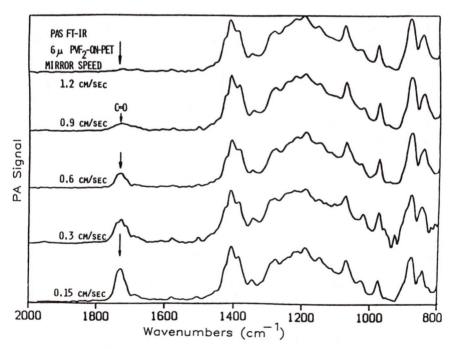

Figure 4. PA FTIR spectra of 6-μm PVF$_2$ on PET recorded with various modulation frequencies. (Reproduced with permission from reference 10. Copyright 1986.)

Figure 5 illustrates similar measurements with the 9-µm PVF$_2$ on PET and indicates the same features; the intensity of the carbonyl band at 1738 cm^{-1} decreases as the mirror speed increases. A log–log plot of the integrated intensity of the carbonyl band as a function of mirror velocity is shown in Figure 6. In both the 6- and 9-µm-thick PVF$_2$ films, a straight line is obtained. The value of the slope represents the power of the modulation frequency in the photoacoustic equation and corresponds to $-\frac{3}{2}$ power in the photoacoustic equation for thermally thick and optically thin samples (eq 7).

In addition to the capability of surface depth profiling, PAS allows the examination of almost any shape of material without special sample preparation. Oddly shaped polymeric samples are difficult to examine by other IR spectroscopic techniques because of the thickness or roughness of the sample. If the sample is cast by dissolving in a suitable solvent or melt-pressed into a film, the conformation of the polymer chains may be altered. Photoacoustic FTIR spectroscopy requires minimal or no sample preparation, and therefore it is a well-suited method for nondestructive analysis of fibers. One of the useful applications of PAS was demonstrated by Yang et al. (11), who examined cotton yarns sized with urethane coatings. Using the depth profiling capability of PA FTIR spectroscopy, the authors concluded

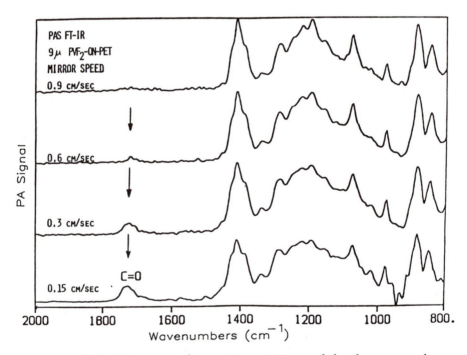

Figure 5. PA FTIR spectra of 9-µm PVF$_2$ on PET recorded with various modulation frequencies. (Reproduced with permission from reference 10. Copyright 1986.)

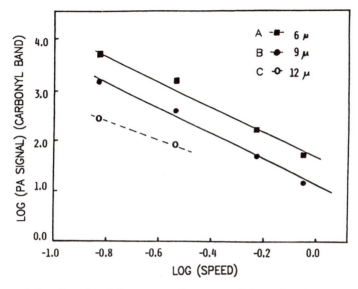

Figure 6. Log–log plot of the integrated intensity of the carbonyl band plotted as a function of the mirror velocity. (Reproduced with permission from reference 10. Copyright 1986.)

that the concentrations of polyurethane sizing agents are higher at the yarn surface than in its interior.

Surface Analysis of Fibers

Nextal ceramic fibers (3M Company) are composed of alumina (62%), boron oxide (14%), and silica (24%). Such a combination of thermally stable oxides provides a fiber with a high thermal stability. Long-term exposure to harsh environments may cause changes in the mechanical and thermal integrity of the fibers, but understanding their behavior at the early stages can help to prevent further degradation and determine their chemical stability. Because of the polycrystalline character of the fibers and their chemical composition, each phase has a different environmental stability.

PA FTIR spectroscopy was used to monitor the extent of degradation of the Nextal fiber surface upon exposure to acidic, neutral, and basic environments (12). The PA FTIR spectrum of a plain, heat-cleaned Nextal fiber is shown in Figure 7. The spectral region between 4000 and 1600 cm^{-1} does not show significant spectral features, but the region below 1500 cm^{-1} suffers from a strong overlap of the bands due to SiO_2, B_2O_3, and Al_2O_3. Although the fibers were heat-treated to eliminate the traces of hydrocarbons present due to processing, weak C–H stretching bands in the 3000-cm^{-1} region are still present.

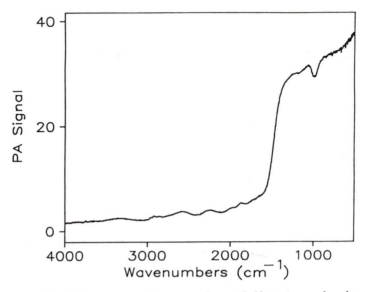

Figure 7. PA FTIR spectrum of untreated Nextal fiber. (Reproduced with permission from reference 12. Copyright 1989.)

To reveal the spectral features of fibers exposed to acidic, neutral, and basic environments, the spectrum of untreated fibers was subtracted from that corresponding to each treatment. Figure 8A illustrates a difference spectrum of untreated Nextal fiber subtracted from a fiber sample treated in water at room temperature. The negative bands at 3700 and 1640 cm^{-1} indicate a loss of water from the treated fiber surface. These bands are due to the OH stretching and bending vibrations of free hydroxyl groups and water adsorbed on the fiber. On the other hand, much more pronounced changes are observed in the low-wave number range, namely, the appearance of two strong positive bands at 1300 and 850 cm^{-1}. Similar changes are observed for acidic and basic treatments.

To gain further insight into the surface changes after the treatment, the origin of these bands must be understood. Al_2O_3 and SiO_2 have strong IR bands due to metal–oxygen vibrations at 650 and 1100 cm^{-1}, respectively, but these bands are not present in the subtraction spectrum. Borates, on the other hand, absorb in the 1300- and 850-cm^{-1} region (*13, 14*). Such observations were made earlier in studies on E-glass fibers, but the band at 1300 cm^{-1} remained unassigned (*15*). The authors were hesitant to assign the broad band to Si–O modes but failed to consider the significant amount of B_2O_3 present in the glass fibers. Because both of these bands are positive on subtraction, an increased amount of borates is observed upon treatment. Although the exact structures of newly formed species on the surface are difficult to determine, most likely a fraction of the polymeric form of B_2O_3, upon reaction with water, preferentially forms the $B(OH)_3$ phase. The band

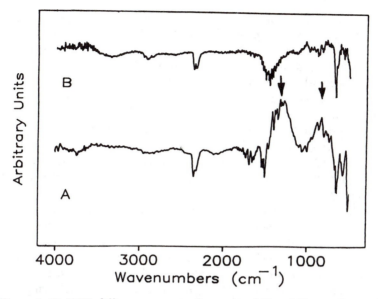

Figure 8. PA FTIR difference spectra of as-received Nextal fibers subtracted from fibers treated in H₂O at room temperature (A), and the same treatment followed by heating at 500 °C for 1 h (B). (Reproduced with permission from reference 20. Copyright 1989.)

at 1300 cm^{-1} is due to the B–O stretching vibrations, whereas the band at 850 cm^{-1} is attributed to the B–OH deformation mode (*13*).

The next question is whether the presence of aqueous media reduces the thermal performance of the fibers because of the formation of a $B(OH)_3$ phase. This aspect is important because the stability of $B(OH)_3$ will also affect the fibers' performance. However, exposure to elevated temperatures will result in conversion of boric acid to boron oxide. This process is governed by the following reaction (*16*):

$$B(OH)_3 \xrightarrow[\text{H}_2\text{O}]{\text{heat}} HBO_2 \xrightarrow[\text{H}_2]{\text{heat}} B_2O_3$$

This reaction occurs at temperatures much below the actual performance conditions of the fibers (1200 °C), namely, at about 250 °C. To confirm this hypothesis, the previously treated Nextal fibers, with the $B(OH)_3$ phase, were heated at 500 °C for 1 h. The resulting difference spectrum (Figure 8B) does not show the presence of the $B(OH)_3$ phase. This result is demonstrated by the absence of the bands at 1300 and 850 cm^{-1} observed in the previous spectra. Also, the negative peak in the C–H stretching region indicates that the entire process irreversibly removed the small hydrocarbon phase from the fibers.

A common problem encountered when incorporating ceramic or glass fibers into composites is poor adhesion of the fibers to the polymer matrix, causing low dry flexural and tensile strength, and even lower wet flexural and tensile strength (17). This problem is usually overcome by treating the fiber surface with silane coupling agents that, through their ability to bond with both organic and inorganic surfaces, give the composites increased strength.

The initial studies (18, 19) on coupling agents used PA FTIR spectroscopy to analyze orientation as a function of relative concentrations of γ-MPS, γ-GPS, and γ-APS coupling agents on the silica surface. (γ-MPS is γ-methacryloxypropyltrimethoxysilane, γ-GPS is γ-glycidoxypropyltrimethoxysilane, and γ-APS is γ-aminopropyltriethoxysilane.) These studies showed that, with appropriate modifications, PA FTIR spectroscopy provides suitable sensitivity to follow changes in coupling agents deposited on the fiber surface. In more recent studies (20), PA FTIR spectroscopy was used to analyze the thermal stability of a series of coupling agents with various functionalities. The coupling agents were deposited on Nextal ceramic fibers. Because of a high thermal stability, Nextal fibers are primarily used in high-temperature composites. However, to incorporate Nextal into high-temperature composites, the coupling agent must remain stable at elevated temperatures and function as a thermally stable interface. Because the thermal stability of potential coupling agents as a neat liquid is only a prerequisite, the analysis of the thermal stability was performed with PA FTIR spectroscopy after the coupling agent was applied to the fibers.

To spectroscopically analyze the various coupling agents deposited on the Nextal surface, a spectral region must be chosen that would be common for the studied samples and, at the same time, would represent thermal stability of the coupling agent on the Nextal surface. As illustrated in Figure 7, the PA FTIR spectrum of untreated Nextal fiber below 1600 cm^{-1} indicates the presence of heavily overlapping bands due to metal oxides forming the fiber. Therefore, the analysis is difficult. However, all the analyzed silanes contain organic groups that absorb in the 3000-cm^{-1} region of the IR spectrum because of absorption of the C–H stretching modes.

Figure 9 shows a series of PA FTIR spectra after the sample was treated at various temperatures. As the temperature of the treatment increases, the C–H stretching band decreases. Thus, this region was used for the analysis of thermal stability by plotting the percent intensity loss as a function of the treatment temperature of the sample. Figure 10 shows the results of the analysis for the coupling agents with various functionalities. The highest thermal stability is achieved for the coupling agent designated as Prosil 2107 containing a combination of epoxy and aromatic functionalities. This plot illustrates that Prosil 2107 remains on the fiber surface even after heat treatment at 550 °C. Table II summarizes the experimental procedure and the results of analysis of the coupling agents with various functionalities.

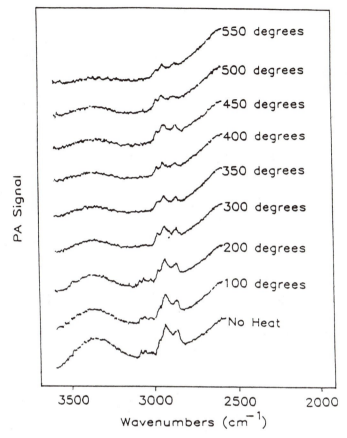

Figure 9. PA FTIR spectra of Nextal fibers treated with 1% w/w Prosil 2107 in the C–H stretching region at various treatment temperatures. (Reproduced with permission from reference 12. Copyright 1989.)

Rheophotoacoustic FTIR Measurements

As was indicated earlier, PA IR detection involves two stages: first, IR light is absorbed by the medium and, second, because of reabsorption processes, heat is produced, which, in turn, generates acoustic waves in the surrounding gas. However, the second stage can also be induced by external forces leading to deformations and conformational changes within the polymer. These movements cause an energy release that also generates acoustic waves.

A simple example of this phenomenon is the cracking of ice on a pond, which produces sounds audible to the human ear. If stress is induced in a polymer and PA measurements are performed, in addition to a "normal" PA IR spectrum obtained as a result of the reabsorption process, an acoustic signal due to deformations within the polymer will occur.

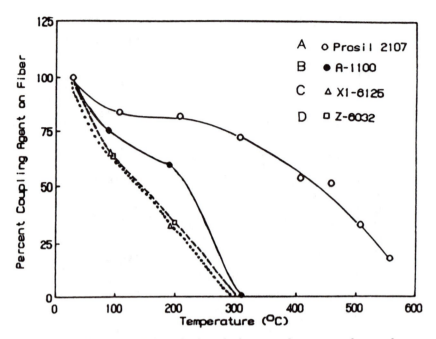

Figure 10. Thermal stability of selected silane coupling agents deposited on
the surface of Nextal fibers. The data are presented as a function of the original
absorbance using 3000-cm⁻¹ region bands.

This example can help explain the deformations of poly(p-phenylene
terephthalamide) (PPTA) fibers and analyze the spectral changes occurring
when external forces are applied. For that purpose, we designed (21) a
rheophotoacoustic (RPA) cell. It is a simple stretching device incorporated
inside the sample compartment of the photoacoustic cell. Figure 11 illustrates the cell design and each element of the cell. The cell design is universal
and permits the measurements of various samples such as fibers, composites,
and films.

Figure 12 illustrates a series of PA FTIR spectra of PPTA fibers in the
4000–2500-cm⁻¹ region recorded as a function of elongation. With increasing
elongation, the intensity of the N–H stretching vibrational mode increases.
Because of extensive N–H• • •O = C– associations, a similar behavior is observed for the C=O band at 1656 cm⁻¹. To further relate the molecular
deformations in PPTA fibers to their elongation, the integrated intensities
of both bands were plotted as a function of percent elongation. As seen in
Figure 13, the intensities of both bands increase as the fiber is stretched.
The intensity decreases, however, when the sample breaks, a result indicating that both bands are sensitive to the shear forces involved when the
fiber is elongated. These plots are similar to the load–elongation curve
(Figure 13C), a result indicating that the changes of thermal properties with
the fiber elongation do not necessarily contribute to the intensity changes.

Table II. Silane Coupling Agents and Their Preparation Parameters

Commercial Name	Structure	Manufacturer	Amount of Coupling Agent (g)	Amount of Water (g)	Acid Type and Amount (g)	Alcohol Type and Amount (g)	Stability Range (°C)
A-1100	γ-aminopropyltriethoxysilane	Union Carbide	0.25	49.75	—[a]	—	200–300
Y-9576	phenylaminoalkyltrialkoxysilane	Union Carbide	0.25	48.175	acetic, 1.325	—	200–300
X1-6125	50% N,β-aminoethyl-γ-aminopropyltrimethoxysilane and 50% phenyltrimethoxysilane	Dow Corning	0.25	49.0	formic, 0.09	methanol, 0.75	200–300
X1-6100	alkoxysilanes	Dow Corning	0.25	49.5	formic, 0.09	methanol, 0.25	200–300
X1-6106	organotrimethoxysilanes	Dow Corning	0.25	49.75	acetic, pH 3.8	—	200–300
Z-6032[a]	N,2-(vinylbenzylamino)ethyl-3-aminopropyltrimethoxysilane monohydrogen chloride	Dow Corning	0.59	49.38	acetic, 0.03	—	200–300
P-0320	phenyltriethoxysilane	Petrarch	0.25	49.5	formic, 0.05	methanol, 0.25	300–350
Prosil 2107	epoxy and methoxy functional silanes	PCR	0.25	49.5	acetic, pH 3.2	isopropyl, 0.25	>550
Prosil 2212[b]	polyimide	PCR	50	—	—	—	200–300
Prosil 9214[c]	trialkoxysilyl acid silane	PCR	0.29	49.5	acetic, pH 3.2	isoropyl, 0.46	100–200
Prosil 9102	10–20% p-tolyltrimethoxysilane and 70–80% p-(chloromethyl)-phenyltrimethoxysilane	PCR	0.25	49.5	acetic, pH 3.2	isopropyl, 0.50	200–300

NOTE: The surfactant was octoxynol (Triton X-100, Rohm and Haas), 0.09 g with X1-6125, 0.085 g with X1-6100, and 0.05 g with P-0320. No surfactant was used with the other coupling agents.

[a] 40% coupling agent in methanol, prepared as a prehydrolyzate solution as per company literature.
[b] Supplied prehydrolyzed at 0.43% by weight.
[c] Supplied 85% w/w in methanol.

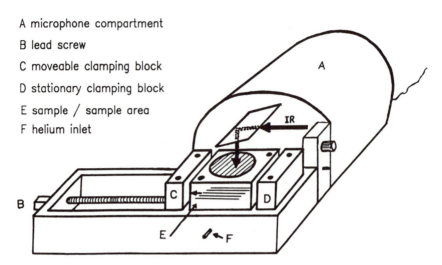

A microphone compartment

B lead screw

C moveable clamping block

D stationary clamping block

E sample / sample area

F helium inlet

Figure 11. Rheophotoacoustic (RPA) FTIR cell. (Reproduced with permission from reference 21. Copyright 1989.)

The polymer backbone is affected by the external shear forces, as can be seen by examining the band due to the C–C stretching at 1408 cm^{-1} and the C–N in-plane bending at the 1261-cm^{-1} vibrational modes. Figure 14 plots the integrated intensities of both bands as a function of elongation and indicates that the intensity changes are observed only when elongation exceeds 0.85%. In contrast, the N–H and C=O groups are sensitive virtually throughout the entire elongation process.

A comparison of the N–H and C=O bands with the aromatic C–C and the C–N bands is important because of the different contributions of each group to the chemical structure of the PPTA fiber. The structure of the PPTA fiber shows that the N–H and C=O species are the side groups and therefore are capable of taking part in hydrogen bonding with the amide carbonyl and amide N–H groups of neighboring chains. On the other hand, the C–C aromatic stretching band arises from the aromatic ring of the polyimide

PPTA fiber

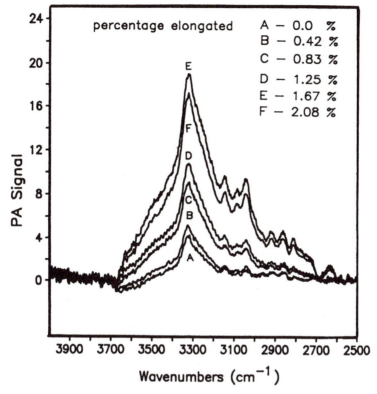

Figure 12. RPA FTIR spectra in the N–H stretching region recorded as a function of fiber elongation. (Reproduced with permission from reference 21. Copyright 1989.)

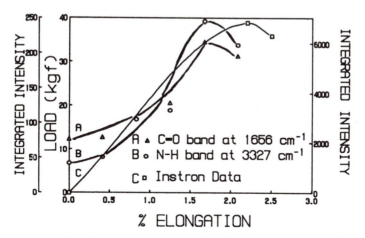

Figure 13. Integrated intensities of the C=O (curve A) and N–H (curve B) stretching bands plotted as a function of elongation; Curve C, rheometric data. (Reproduced with permission from reference 21. Copyright 1989.)

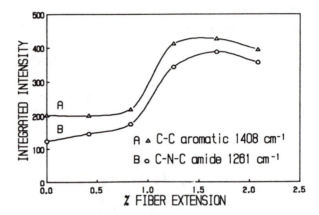

Figure 14. Integrated intensities of the C–C (A) and C–N–C (B) vibrational modes plotted as a function of elongation. (Reproduced with permission from reference 21. Copyright 1989.)

backbone. The aromatic ring is not capable of participating in strong inter-molecular interactions such as hydrogen bonding, although it does interact with the neighboring chains through Π–Π interactions. Likewise the aromatic C–C groups and, even more so, the C–N part of the backbone will not participate in intermolecular interactions.

In summary, rheophotoacoustic FTIR spectroscopy is a novel technique that, to our best knowledge, has not been reported in the literature. On the other hand, rheo-optical FTIR spectroscopy has been around for over a decade (22, 23). A distinct difference and advantage of rheophotoacoustic FTIR spectroscopy comes from the fact that it can be used for the deformation studies of all materials, regardless of the shape or optical properties. Moreover, our initial studies also indicate that this approach can be used to monitor interfacial failure between two phases (24). Rheo-optical FTIR spectroscopy is limited only to optically transparent polymers.

Temperature PA FTIR Studies

According to the Rosencwaig–Gersho (RG) theory (6), elevated temperatures ought to diminish the intensity of the PA spectrum. The relationship between temperature and PA intensity is

$$PA \propto \frac{(\kappa' p)^{1/2}}{T} \tag{10}$$

where κ' is the gas thermal conductivity, p is the pressure in the PA cell, and T is the temperature of the gas.

To examine the temperature effect on PA spectra, we designed and built

(25) a PA cell. Figure 15 shows a schematic of a custom-built temperature sample compartment. Photoacoustic FTIR spectra of silica were obtained at temperatures ranging from 5 to 103 °C. Figure 16, trace A, shows the spectrum recorded at 5 °C. Two intense bands at 1100 and 3500 cm^{-1} are observed. Because the amount of the surface-adsorbed species is temperature dependent, and consequently will affect the band intensities in the IR spectrum, a distinction must be made between the modes associated with molecules adsorbed on the surface and those due to the bulk. The band at 1100 cm^{-1} is assigned to the Si–O–Si lattice mode (bulk), whereas the 3500-cm^{-1} band is due to water molecules adsorbed on the silica surface.

Because of the high thermal stability of silica between 5 and 103 °C, the Si–O–Si lattice mode can be used as a probe to examine the effect of temperature on the intensity of PA signal generated from the bulk. Figure 16, traces B–G, illustrates PA FTIR spectra recorded as a function of temperature, and Figure 17 depicts integrated intensities of the Si–O–Si and O–H stretching vibrations. Apparently, the intensity of the Si–O–Si band remains constant in the examined temperature range. This result is somewhat surprising, considering that the RG theory predicts that the temperature increase should lead to a decrease of the PA signal (eq 10). However, at elevated temperatures, a pressure builds up in the cell. Thus, according to equation 10, both temperature and pressure effects cancel each other out,

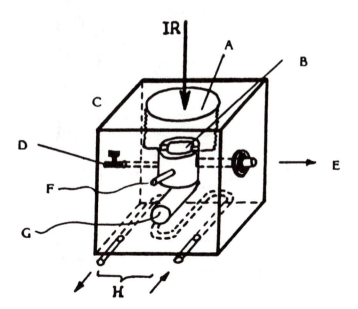

Figure 15. High-temperature PA FTIR cell: A, KBr window; B, sample compartment; C, brass cube; D, gas inlet; E, microphone coupling; F, thermocouple; G, heater; and H, cooling system. (Reproduced from reference 25. Copyright 1988 American Chemical Society.)

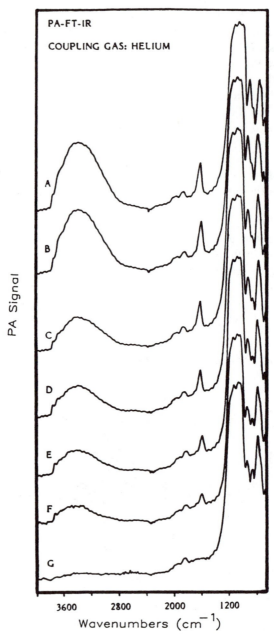

Figure 16. PA FTIR spectra of silica recorded as a function of temperature: A, 5; B, 21; C, 33; D, 44; E, 55; F, 88; and G, 103 °C. (Reproduced from reference 25. Copyright 1988 American Chemical Society.)

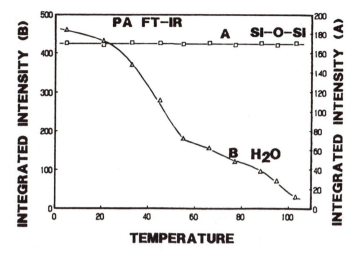

Figure 17. Integrated intensities of the Si–O–Si and O–H stretching vibrational modes in silica plotted as a function of temperature. (Reproduced from reference 25. Copyright 1988 American Chemical Society.)

and therefore a net PA signal of the Si–O–Si lattice vibrations does not change. The situation will be different, however, if the solid or liquid sample has phase transitions in the studied temperature range.

The cross-linking process of amorphous polymers is considered to be the formation of a polymer network as it undergoes transition from liquid to solid. During molecular weight build-up, thermal properties also change. Because photoacoustic detection is a two-stage process, and the second stage involves heat propagation to the surface, the changes of thermal properties during the cross-link formation will significantly influence photoacoustic intensity.

To determine how the second stage of the detection process can affect the PA signal, we will compare the cross-linking process of hydroxyl-terminated poly(dimethylsiloxane) (PDMS) detected by transmission and PA FTIR spectroscopy (26). Figure 18 shows a series of transmission FTIR spectra recorded as a function of curing time. In spite of the cross-link formation, no changes are detected in the spectra.

One of the essential problems in monitoring cross-linking reactions by transmission FTIR spectroscopy is the small number of cross-links compared to the number of other bonds in the system. In addition, during the cross-linking process of this particular system (cross-linker: tetraethoxysilane or TES), the Si–OH bonds of PDMS and H_5C_2–O–Si of TES break to form the Si–O–Si network and ethanol. Thus, the simultaneous cleavage and formation of energetically similar bonds result in a heavy spectral overlap of strongly absorbing bands. If the same process is monitored photoacoustically, the spectral changes appear to be quite significant. This result is

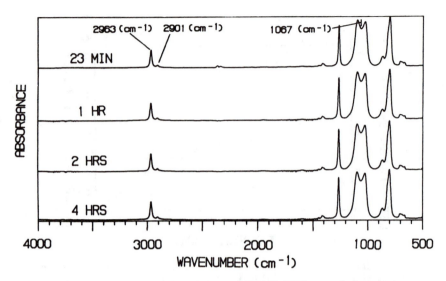

Figure 18. Transmission FTIR spectra of PDMS–TES recorded at various stages of the cross-linking process. (Reproduced from reference 26. Copyright 1989 American Chemical Society.)

demonstrated in Figure 19, which illustrates that the 1100-cm^{-1} region is highly sensitive to the cross-linking process.

The next question is why photoacoustic detection provides a sensitive tool in detecting the cross-linking reactions whereas transmission measurements show no sensitivity. Two approaches can be taken to answer this question. First, because ethanol is produced during the reaction, the decreasing intensities may be related to the escape of ethanol from the system. Second, as the cross-linked network is being formed, the thermal properties of the system (heat capacity and thermal conductivity) may also change. Because only selected IR bands are affected by the cross-linking reactions, the property changes play some role in this system, but they are not fully responsible for the observed changes. As a matter of fact, examination of the intensity changes depicted in Figure 19 indicates that the bands that decrease during the reaction correspond to ethanol. Thus, the observed changes can probably be attributed to the removal of ethanol from the reaction mixture to the surrounding gas phase. Although we hoped that the changes of thermal conductivity or heat capacity during cross-linking could have been detected, the evaporation of ethanol from the mixture submerged the changes in thermal properties. The usefulness of these data was demonstrated by correlating the intensity changes with the viscosity measurements (26).

In the cross-linking process of polyester–styrene, no byproducts are produced. The polyester–styrene system cross-links by a free-radical mechanism. Figure 20 illustrates a series of PA FTIR spectra recorded as a function

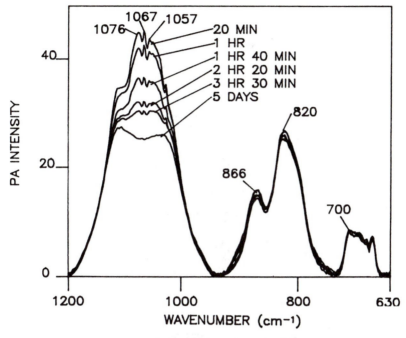

Figure 19. PDMS–TES cross-linking process monitored by PA FTIR. (Reproduced from reference 26. Copyright 1989 American Chemical Society.)

of time. Several IR bands, namely at 991, 910, 775, and 694 cm^{-1}, diminish by about 90% of their original intensity. Similar changes are observed in the C–H stretching region. To reveal the nature of the detected changes, the origin of these bands must be understood. Apparently, all diminishing bands are due to normal vibrational modes of styrene. Because styrene is a highly volatile monomer, the intensity changes were expected to result from the loss of monomer. However, parallel to the PA FTIR spectroscopy, weight-loss experiments indicated that only 4% of the styrene monomer evaporates.

The photoacoustic responses of gas-phase and fluid samples are different. The photoacoustic intensities of the gas-phase samples are usually 100 times greater than those of the fluid samples. Thus, these changes are possibly attributed to the styrene vapor above the polyester–styrene sample. However, the amount of styrene vapor in equilibrium with polyester–styrene is thermodynamically related to the extent of cross-linking. Hence, the intensities of styrene monomer bands can be semiquantitatively related to the cross-linking process.

The intensity changes of the C–H bending mode of styrene as a function of time were examined. Figure 21 illustrates two cross-linking reactions conducted at 15 and 30 °C. The cross-linking process at 15 °C shows steady and linear decrease of intensity, and after 6 h, the slope of the line changes.

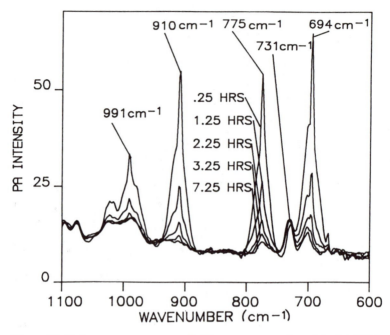

Figure 20. Polyester–styrene cross-linking process monitored by PA FTIR.

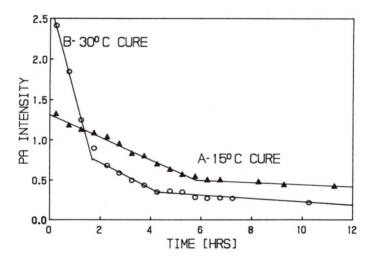

Figure 21. Integrated intensity of the C–R bending mode at 775 cm⁻¹ plotted as a function of time.

On the other hand, the same process conducted at 30 °C indicates two break points. To provide further understanding of the cross-linking process, we conducted isothermal experiments covering the temperature range from 15 to 40 °C and plotted temperatures of the break points as a function of time. This plot is illustrated in Figure 22. The two curves observed are attributed to gelation and vitrification processes.

At the initial stages of cross-linking, the process is kinetically controlled until molecular weight builds up and leads to gelation. Then the process become diffusion controlled. Depending on the cure temperature and glass transition temperature of the system, one or two transitions are observed. The second transition occurs at higher temperatures and is attributed to the vitrification process, upon which a sol–gel glassy state is produced. If the process is conducted at lower temperatures, the liquid undergoes transitions directly to the glassy phase without gelation. This spectroscopic evidence is the first to show a time–temperature–transformation diagram. These diagrams are known for epoxy systems and were obtained by using torsional braid analysis (27).

Conclusions

The use of PA FTIR spectroscopy for in situ studies of cross-linking can provide further insight into the chemical and physical processes involved in

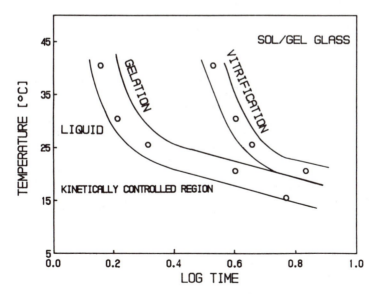

Figure 22. Time–temperature–transition diagram for the polyester–styrene system.

these complex reactions. Apparently, PA FTIR has sufficient sensitivity to monitor phase transitions occurring during the cross-linking processes. Novel RPA FTIR measurements demonstrate that it is possible to monitor molecular deformations in PPTA fibers imposed by external forces. The inter- and intramolecular interactions detected photoacoustically can be correlated to the load–fiber elongation curves. At the initial stages of application of external stress on the fiber, hydrogen bonding plays a key role, whereas later on, the polymer backbone is affected.

Acknowledgments

The authors are thankful to 3M Company and the National Science Foundation (EPSCoR Program) for partial support of this work. Alan P. Bentz (R & D Center, U.S. Coast Guard, Groton, CT) is acknowledged for providing an FTS-10M spectrometer.

References

1. Bell, A. G. *Am. J. Sci.* **1880**, *20*, 305.
2. Bell, A. G. *Philos. Mag.* **1881**, *5*, 11, 510.
3. Urban, M. W. *Prog. Org. Coat.* **1989**, *16*, 32l–353.
4. Urban, M. W. *J. Coat. Technol.* **1987**, *59(745)*, 29.
5. Graham, J. A.; Grim, M. W. III; Fateley, W. G. In *Fourier Transform Infrared Spectroscopy: Applications to Chemical Systems;* Vol. 4, Ferraro, J. R.; Basile, L. J., Eds.; Academic: New York, 1986.
6. Rosencwaig, A. *Photoacoustics and Photoacoustic Spectroscopy;* Wiley: New York, 1980.
7. Gersho, A.; Rosencwaig, A. A. *Appl. Phys.* **1976**, *47*, 64.
8. Gersho, A.; Rosencwaig, A. *Science* **1975**, *190*, 556.
9. Chatzi, E. G.; Urban, M. W.; Ishida, H.; Koenig, J. L.; Laschewski, A.; Ringsdorf, H. *Langmuir,* **1988**, *4*, 846.
10. Urban, M. W.; Koenig, J. L. *Appl. Spectrosc.* **1986**, *40(7)*, 994.
11. Yang, C. Q.; Bresee, R. R.; Fateley, W. G. *Appl. Spectrosc.* **1987**, *41(5)*, 889.
12. Tiefenthaler, A. M.; Urban, M. W. *Composites,* **1989**, *20(2)*, 145.
13. Colthup, N. B.; Daly, L. H,; Wiberly, S. E. *Introduction to Infrared and Raman Spectroscopy;* Academic: New York, 1975, pp 335–338; 431–433.
14. Gerrard, W. *The Organic Chemistry of Boron;* Academic: London, 1961.
15. Ikuta, N.; Sakamoto, T.; Kouyama, T.; Abe, I.; Hirashima, T. *Sen-I Gakkaishi* **1987**, *43*, 313.
16. Cotton, F. A,; Wilkinson, G. *Advanced Inorganic Chemistry;* Wiley: New York, 1972, 232–233.
17. Arkles, B.; Peterson, W. *Proc. Annu. Conf. Reinf. Plast. Compos. Inst. 35th;* Society for Plastics Industry, **1980**, 20-A, 1-20-A, 2.
18. Urban, M. W.; Koenig, J. L. *Appl. Spectrosc.* **1986**, *40*, 513.
19. Urban, M. W.; Koenig, J. L. *Appl. Spectrosc.* **1985**, *39*, 1051.
20. Tiefenthaler, A. M.; Urban, M. W. *Composites,* **1989**, *20(6)*, 585.
21. McDonald, W. F.; Goettler, H.; Urban, M. W. *Appl. Spectrosc.* **1989**, *43(8)*, 1387.
22. Siesler, H. W. *Adv. Polym. Sci.* **1984**, *65*, 1.

23. Bretzlaff, R. S.; Wool, R. P. *Macromolecules,* **1983,** *16,* 1907.
24. McDonald, W. F.; Urban, M. W., *J. Adhes. Sci. Technol.,* in press.
25. Urban, M. W.; Koenig, J. L. *Anal. Chem.* **1988,** *60(21),* 2408.
26. Urban, M. W.; Gaboury, S. R. *Macromolecules* **1989,** *22,* 1486.
27. Enns, J. B.; Gillham, J. K. In *Polymer Characterization: Spectroscopic, Chromatographic, and Physical Instrumental Methods;* Advances in Chemistry Series 203, Craver, C. D., Ed.; American Chemical Society: Washington, DC, 1983, p 27.

RECEIVED for review February 14, 1989. ACCEPTED revised manuscript November 28, 1989.

18

Characterization of Oriented Surfaces

Polarized Refractometry and Polarized Attenuated Total Reflection Techniques

Randy E. Pepper[1] and Robert J. Samuels*

School of Chemical Engineering, Georgia Institute of Technology, Atlanta, GA 30332–0100

This chapter examines and compares two nondestructive techniques, polarized refractometry and polarized attenuated total reflection (ATR) spectroscopy, for the three-dimensional characterization of polymer surfaces. Both techniques depend on measurement at the critical angle for three-dimensional analysis, and they share a similar molecular model for orientation description. Surface orientation data from both techniques is presented for both thin films and thick sheets of isotactic polypropylene.

T HE FABRICATION PROCESSES used for producing polymer films, sheets, and moldings frequently lead to products with surface structures that differ from the bulk structures. Furthermore, the top surface may differ from the bottom surface, and either or both surfaces may be oriented. Such differences can strongly affect the properties of the fabricated product; therefore, these surfaces must be characterized. This chapter compares two surface orientation characterization techniques, polarized refractometry and polarized attenuated total reflection (ATR) spectroscopy. Series of oriented isotactic polypropylene films and sheets were used as test samples. This study is part of an ongoing effort to develop rapid nondestructive techniques for the

---

*Corresponding author

[1]Current address: Phillips Petroleum Company, Phillips Research Center, Bartlesville, OK 74004

0065–2393/90/0227–0315$06.00/0
© 1990 American Chemical Society

characterization of molecular orientation in fabricated polymer systems (1, 2).

The driving force for this study was the discovery of an ATR surface orientation technique that shared two important similarities with polarized refractometry. First, accurate data from both techniques are available only when the measurements are done at or near the critical angle. Second, both techniques can describe their orientation results by using a fractional orientation parameter derived from very similar orientation models.

Theory

Polarized Refractometry.   A brief review of the polarized refractive index technique is given here. More detailed theoretical descriptions can be found elsewhere (3–6).

Anisotropic samples, such as oriented polymer films, split incoming unpolarized light rays into two polarized rays, the O and E rays (Figure 1). The electric vectors of these two rays, which vibrate perpendicular to the rays' propagation direction, are oriented perpendicular to each other. At the critical angle in an Abbe refractometer, the O ray's electric vector vibrates through, or normal, to the sample plane, while the E ray's electric vector vibrates parallel to the sample plane. If a piece of polarizing film is placed in the eyepiece of the Abbe refractometer, the O and E rays are separated

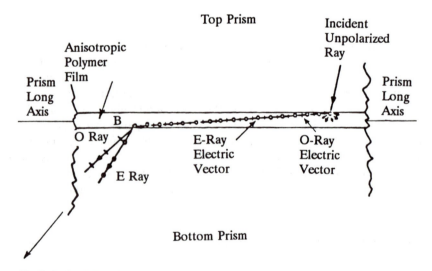

Figure 1. Diagram showing the film position between the prisms of the Abbe refractometer. The O and E rays are refracted separately at B.

at their individual critical angles. This separation allows the refractive indices parallel and normal to the sample plane to be measured.

Three principal refractive indices are required for surface characterization: N_Z, the principal refractive index parallel to the optical symmetry axis (or primary orientation direction in the sample plane); N_Y, the principal refractive index perpendicular to the optical symmetry axis in the sample plane; and N_X, the principal refractive index perpendicular to the optical symmetry axis and normal to (or through) the sample plane. For unidirectional stretch, the sample's optical symmetry axis (or the Z axis) will lie along the stretch direction. For biaxially oriented samples, the exact direction of the optical symmetry axis cannot be determined without optical measurement (3).

N_Z and N_Y are measured by using the E-ray electric vector vibrating in the sample plane. N_X is measured by using the through-the-plane O-ray electric vector. These three measured principal refractive indices yield the following information about a sample's surface: its crystallinity, axiality, three-dimensional birefringences, and, most importantly for this study, its fractional orientation.

The model for the fractional orientation, $f(N_i)$ (where i = X, Y, or Z), is given in Figure 2. In this figure, a sample's molecular chain orientation

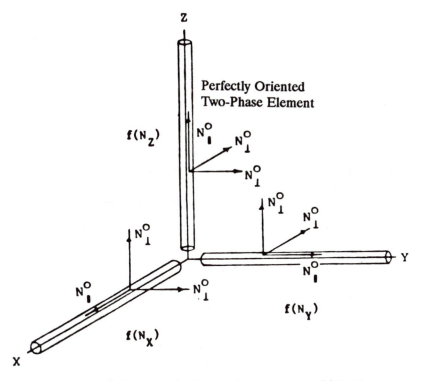

Figure 2. Refractive index fractional orientation model (3, 4).

is represented by a distribution of perfectly oriented elements along each principal axis. The number of perfectly oriented polymer elements lying along any given principal axis is proportional to the amount of polymer chain orientation in that direction.

For example, if the sample is perfectly oriented along the Z axis (optical symmetry axis), then all of the perfectly oriented elements lie along the Z axis. The "fraction", $f(N_Z)$ (*see* Figure 2), of the total number of chains lying along the Z axis is then 1.0 (i.e., $f(N_Z) = 1.0$, and $f(N_X) = f(N_Y) = 0$). Similarly, if all of the polymer chains are randomly distributed perpendicular to the Z axis, then $f(N_Z) = 0$, and $f(N_X) = f(N_Y) = 0.5$.

The equations for the fractional orientation are

$$f(N_X) = \frac{N_X - N_\perp^0}{N_\parallel^0 - N_\perp^0} \tag{1a}$$

$$f(N_Y) = \frac{N_Y - N_\perp^0}{N_\parallel^0 - N_\perp^0} \tag{1b}$$

$$f(N_Z) = \frac{N_Z - N_\perp^0}{N_\parallel^0 - N_\perp^0} \tag{1c}$$

where the $N_i (i = X, Y, \text{or } Z)$ are the three principal refractive indices from the polarized refractive index measurement, and $N_\parallel^0$ and $N_\perp^0$ are the refractive indices parallel and perpendicular, respectively, to the perfectly oriented polymer chains. $N_\parallel^0$ and $N_\perp^0$ are intrinsic material constants, and reference 4 gives a procedure for determining them.

If a polymer is semicrystalline, that is, composed of both crystalline and noncrystalline phases, then $N_\parallel^0$ and $N_\perp^0$ become functions of the intrinsic refractive indices for both the crystalline and noncrystalline phases as follows:

$$N_\parallel^0 = V_c N_{\parallel c}^0 + (1 - V_c) N_{\perp nc}^0 \tag{2a}$$

$$N_\perp^0 = V_c N_{\perp c}^0 + (1 - V_c) N_{\perp nc}^0 \tag{2b}$$

where V_c is the sample's volume fraction of crystallinity, and the subscripts c and nc refer to the crystalline and noncrystalline phases, respectively. Thus, for semicrystalline polymers, $N_\parallel^0$ and $N_\perp^0$ represent average orientation descriptors over both the crystalline and noncrystalline phase orientations.

$N_{\parallel,p}^0$ and $N_{\perp,p}^0$ (where $p = c$ or nc) are the intrinsic refractive indices either parallel or perpendicular, respectively, to a perfectly oriented crystalline or noncrystalline polymer chain, and can also be determined by the procedures given in reference 4. The intrinsic refractive indices for semicrystalline isotactic polypropylene were reported previously (3, 4).

From equations 1a–1c and 2a–2b, then, if all the orientation is directed along the Z axis as mentioned previously, $N_Z = N_{\parallel}^0$, and $f(N_Z) = 1$. Similarly, if all the orientation is perpendicular to Z, then $N_Z = N_{\perp}^0$, and $f(N_Z) = 0$. In the general case, the measured refractive indices lie between $N_{\parallel}^0$ and $N_{\perp}^0$, and $f(N_i)$ lies between 0 and 1.

As these examples have indicated, the three fractional orientations have a sum of 1, such that:

$$f(N_X) + f(N_Y) + f(N_Z) = 1 \qquad (3)$$

This orthogonal property of the fractional orientation will be exploited later as a triangular diagram, which gives very fast and easy visual information about the degree and type of the sample's (or sets of samples') orientation (3–7).

For semicrystalline polymers, one useful result of the definition of $N_{\parallel}^0$ and $N_{\perp}^0$ through V_c is that the effects of crystallinity are normalized out. Samples of different crystallinities can be compared strictly on the basis of their anisotropy, independent of any differences in crystallinity. Thus, although the sheet series and the two film series measured in this study are all of different crystallinities, their fractional orientations are comparable.

Polarized ATR Spectroscopy.   This technique was developed by Mirabella (8–10) for surface analysis as a modification of Kissin et al.'s (11) transmission infrared (IR) orientation technique. Mirabella developed this technique in response to two deficiencies he perceived in current ATR surface orientation techniques: (1) the lack of an available direct three-dimensional orientation measurement, and (2) the surface contact problems inherent in the mounting–demounting of samples for dichroic ratio determination. Because his technique also describes the orientation data in terms of a fractional orientation similar to that obtained from refractive indices, we sensed a possible correspondence of Mirabella's IR orientation data with fractional orientation data measured in the visible spectrum with polarized refractometry. The following is a brief description of how the fractional orientation method addressed Mirabella's two concerns about conventional ATR surface orientation analysis.

Figure 3 shows the experimental setup. An ATR crystal is pressed firmly against the sample, and an IR beam is introduced along the x_1 direction of the crystal at an angle, Θ, which is just greater than the critical angle. At this angle the beam totally reflects off the crystal–sample interface, but a weak evanescent wave, with electric vectors vibrating in all three orthogonal directions ($\mathbf{E}_{x_1}$, $\mathbf{E}_{x_2}$, and $\mathbf{E}_{x_3}$), is propagated inside the sample. The absorption

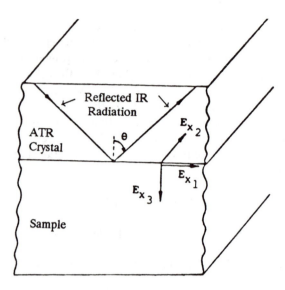

Figure 3. Experimental ATR spectroscopy setup. E_{x_1}, E_{x_2}, and E_{x_3} are the three evanescent wave electric vectors present in the sample for unpolarized radiation.

due to this evanescent wave is detected by the IR detector after the beam exits from the ATR crystal.

In conventional polarized ATR spectroscopy (10), the incoming IR beam is polarized parallel to the sample plane, and the resultant evanescent wave electric vector, E_{x_2}, is used to measure the two absorptions parallel (A_Z) and perpendicular (A_Y) to the sample optical symmetry axis. For A_Z, the sample optical symmetry axis is oriented along the x_2 axis of the prism (*see* Figure 3). A_Y is measured by demounting the sample and orienting its optical symmetry axis along the x_1 axis of the prism. This mounting and demounting raises serious questions about sample–prism contact and whether the sample is seeing the same amount of energy for each measurement (9).

Also, in conventional polarized ATR spectroscopy, the IR beam is not usually polarized normal to the sample surface because the resultant polarized wave is not purely one electric vector (E_{x_2}) as in horizontal polarization, but is a mixture of E_{x_3} and E_{x_1}. Thus, the resulting absorptions are mixed, and the normal absorption, A_X, cannot be purely extracted.

Mirabella, however, using Harrick's equations (10), found that E_{x_1} decreased dramatically with respect to E_{x_3} as Θ approaches the critical angle, Θ_c. Experimentally, Mirabella found that no information was available just at the critical angle, but good information was available at an angle just 0.4° above Θ_c (8). Furthermore, Harrick's equations showed that, at 0.4° above Θ_c, the ratio of the electric vector amplitudes, $E_{x_3}:E_{x_1}$, was 90:10, and A_X was very nearly a pure normal absorption. Thus, all three principal absorptions, A_X, A_Y, and A_Z, were available experimentally for a three-dimensional orientation description of the sample, provided measurements were taken near the critical angle of the sample.

The second problem with conventional surface ATR orientation techniques, which use dichroic ratios from a single band, is that the sample must be demounted after the first measurement has been made parallel to the sample's optical symmetry axis. After a 90° rotation, the sample is remounted for the perpendicular measurement. Serious questions arise as to the similarity of crystal–sample contact for both measurements and its effect on the dichroic ratio data. To avoid this problem, Sung (*12*) developed a rotating ATR crystal mount; however, that solution was found unsatisfactory (*13*).

Mirabella's fractional orientation technique avoids all contact problems by taking the necessary absorption information from a single spectrum. To do so, he uses the ratio of two different absorption bands in the same spectrum (a single mounting), rather than taking the ratio of the data for one absorption band from two different spectra (which requires demounting and then remounting the sample).

Because only one spectrum is used, two different absorption bands are required for Mirabella's technique. One band must have a parallel transition moment vector, $\mathbf{M}$, that is parallel to the molecular chain axis. The other band must have a perpendicular transition moment vector, $\mathbf{M}_\perp$, that is perpendicular to the molecular chain axis (*see* Figure 4). Following Kissin et al., Mirabella used the 841-cm^{-1} parallel band and the 809-cm^{-1} perpendicular band for isotactic polypropylene surface characterization (*9, 11*).

The ratio of the parallel band absorption to the perpendicular band absorption results in a fractional orientation descriptor, P_i. The physical model for P_i is very similar to the refractive index fractional orientation model, except that transition moment vectors, $\mathbf{M}$ and $\mathbf{M}_\perp$, which are oriented parallel and perpendicular to the molecular chain axis, are used in place of the parallel and perpendicular intrinsic refractive indices, $N_\parallel^0$ and $N_\perp^0$ (cf. Figures 4 and 2). The three equations for the ATR fractional orientation along all three axes are

$$P_X = \frac{R_X}{R_X + N} \tag{4a}$$

$$P_Y = \frac{R_Y}{R_Y + N} \tag{4b}$$

$$P_Z = \frac{R_Z}{R_Z + N} \tag{4c}$$

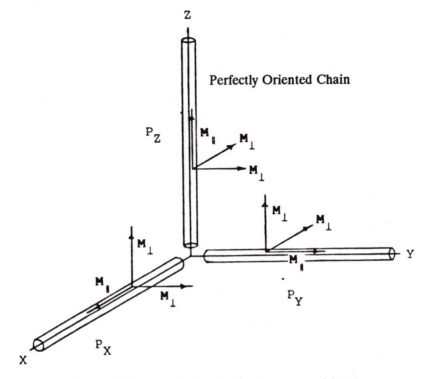

Figure 4. Kissin et al.'s fractional orientation model (11).

where

$$R_i = \frac{A_{i,841}}{A_{i,809}} \qquad i = X, Y, \text{ or } Z \tag{5}$$

The A_i are absorptions at the wavenumbers indicated, and the i subscripts in equation 5 denote polarization of the IR electric vector parallel to axis i. N in equations 4a–4c is a constant that includes the absorption coefficients and transition moment vector amplitudes. N is derived from measurements run on isotropic samples (9).

The fractional orientation from polarized IR (either in transmission or ATR) also is orthogonal, so that (9, 11)

$$P_X + P_Y + P_Z = 1 \tag{6}$$

The strong similarity of polarized ATR spectroscopy and polarized refractometry, through both their common dependencies on the critical angle and their common orientation description as a fractional orientation, provided the impetus for the comparison study described here.

Experimental Details

Samples. Three sets of oriented polypropylene samples were measured in this study. Two thin film series, homogeneous through their cross sections, were measured as a control set. One series, labeled as the A series, is a set of individual cast polypropylene films that were unidirectionally drawn at 110 °C to different extensions. Original cast film thicknesses were chosen to achieve the same final thickness after drawing. Morphological characteristics for these films are given in reference 14. Of the seven members in this series, A-2, A-5, and A-6 were measured in this study.

The second thin film series, labeled the F series, was processed similarly to the A series films, but at a temperature of 127 °C. Morphological data for these films can be found in reference 15, where they are referred to only by number. Of the six members in this series, F-2, F-3, F-4, and F-5 were used in this study.

The oriented thick sheet samples were supplied by F. M. Mirabella. These samples, labeled here as the N series, consist of an original cast sheet that was then unidirectionally drawn on a tenter frame at 149 °C. Thicknesses of these samples varies from 63 mils at 0% extension to 13 mils at 500% extension.

Samples from each film and sheet were measured with polarized refractometry, polarized ATR spectroscopy, and, where possible, birefringence. The birefringence measurement was run both as a check on the polarized refractometry results and as a comparison of the refractive index surface orientation to the microscope bulk sample orientation.

Polarized Refractometry. A Carl Zeiss Company model A Abbe refractometer with a rotatable polarizing film in the eyepiece was used for all refractive index measurements. A constant prism temperature of 23 ± 0.1 °C was maintained by circulating temperature-controlled water through the prisms' water jackets.

Polarized ATR Spectroscopy. IR measurements were made on a Nicolet 60SX Fourier transform spectrometer equipped with an ATR attachment manufactured by Harrick Scientific. The parallelogram ATR crystal was made of KRS5 with entrance and exit faces specially cut at 40° to allow the IR beam to totally reflect in the crystal at angles close to the 40° critical angle (based on $n_{KRS5} = 2.37$ and $n_{sample} = 1.50$; n is the refractive index). All ATR measurements were done at an IR beam incident angle ca. 1° above the critical angle.

Birefringence. Birefringence measurements were made on each film and sheet with a Carl Zeiss Universal R Pol microscope coupled with a Carl Zeiss calcite Ehringhaus compensator.

Results and Discussion

In a previous paper (5) on the N sheet series, the individual surface orientation data from polarized refractometry (data from our laboratory) and polarized ATR (outside laboratory data) were related by using an INNER and OUTER label. The INNER surface and OUTER surface referred to the inner and outer sides of the curvature present in the samples due to the roll on which they were collected. Subsequent refractive index and ATR analyses run only in our laboratory revealed some inconsistencies in this labeling

method, and some confusion arose as to which was the "inner" and which
was the "outer" surface. Another method, which is more internally consistent, is used here. When the sheets were measured for their refractive indices
in the Abbe refractometer, one surface of the sheets gave a very clear, sharp
interface, and the interface of the opposite surface was slightly hazier. As a
result, the surface data from both polarized refractometry and polarized ATR
spectroscopy for the N series thick sheets is indicated in this chapter using
the terms "clear" and "hazy".

For the A and F thin film series, a notch in one corner of the sample
is used to distinguish which surface of the sample is being measured. Therefore, the terms NR (notch located to the right of the sample's centerline)
and NL (notch to the left of the centerline) are used to distinguish results
from the two different surfaces. However, unlike the N sheet series, where
all samples were produced from the same cast sheet, the A and F series
films were drawn from different cast films. Therefore, there is no relationship
between the NR and NL sides of the samples in these film series.

Figure 5 presents the fractional orientation data from both polarized
refractive index and polarized ATR spectroscopy plotted versus percent extension for the N series sheets. Figure 5 presents the data for the clear and
hazy surfaces, respectively. Again, the Z direction is chosen to lie along the
optical symmetry axis in the sample plane, the Y axis is perpendicular to
the optical symmetry axis in the sample plane, and the X axis is perpendicular
to the optical symmetry axis and normal to the sample plane. The sheets
were unidirectionally stretched, and Figure 5 shows that the orientation
increased in the Z direction with increased extension.

Three curves were drawn in both figures at the average of all refractive
index and ATR data. As the fractional orientation curves in the Y and X
directions show, the two low orientation members in the series, N-000 and
N-100, show definite biaxiality; that is, the three fractional orientations
(either $f(N_i)$ or P_i) are different. However, the four more highly drawn sheets
show uniaxial orientation, that is, $f(N_X) = f(N_Y) f(N_Z)$.

The fractional orientation as measured by both techniques agrees very
well on both surfaces. The correspondence is somewhat better on the hazy
surface than the clear surface. However, if the Z direction fractional orientation data from refractive index and ATR spectroscopy ($f(N_Z)$ and P_Z) is
compared, the average orientation difference between the two methods is
only 0.02 for all points in Figure 5 (less than 3% average deviation).

Figure 5 illustrates the desirability of taking data for both surfaces in
order to understand the effect of processing on the polymer surface orientation. These figures show that the clear surface oriented much faster than
the hazy surface. We know few details about how these sheets were processed, but we can speculate that the clear surface was the surface that saw
a chill roll downstream of the drawing oven (tenter frame). The orientation
at this surface was frozen in quickly. During the time it took for the surface

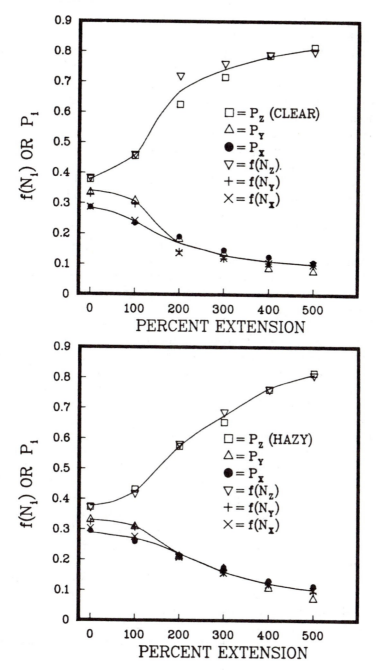

Figure 5. Fractional orientation data at each extension for the N series sheets: top, clear surface; and bottom, hazy surface.

on the far side of the sheets to cool, the hazy surface orientation relaxed. The thicker the sample, the longer the cooling time, and the greater the orientation differences between both sides as shown by the Z direction orientation of the N-200, N-300, and N-400 sheets. [Both surfaces were identical for the N-500 (500% extended) sheet.]

The N series sheets are thick, and, as shown by Figure 5, very inhomogeneous between their surfaces. Therefore, a control set of thin, homogeneous films was chosen to more finely investigate the degree of correspondence between polarized refractometry and polarized ATR spectroscopy. Figures 6 and 7 present the results of this study for the A and F series, respectively. The format of these figures is identical to Figure 5. NR and NL refer, as mentioned before, to the position of an identifying notch placed on the samples in order to uniquely identify each surface. Again, however, unlike the N series, the films series samples were drawn from different cast films, and we can give no physical correspondence to either surface among different samples.

As expected, the refractive index measurements showed no differences between the two film surfaces. Therefore, the $f(N_i)$ data are the same in Figures 6 and 7 for the NR and NL surfaces. The P_i data do suggest some small surface differences, but the average deviation between the $f(N_z)$ and P_Z data is only 0.03 (6% average deviation). This deviation is somewhat higher than the average deviation for the more nonhomogeneous N series sheets, but the data from both techniques still correspond very well. Again, as with the N sheet series, the greatest deviations tend to occur for those samples with low to moderate orientation levels.

As already mentioned, the refractive index measurements for all of the thin films showed no orientation differences between the surfaces, but small orientation differences were indicated by the ATR data for several films. To resolve the issue of orientation homogeneity in the thin films, microscope birefringences were run on both film series to compare their bulk orientation with the surface orientation data from the refractometer. Figure 8 gives the results. Except for two samples, the correspondence between the microscope bulk birefringences, Δ_T, and the refractometer surface birefringences, ΔZ_{ZY}, is excellent. These data show that the orientation is homogeneous through the films' cross sections.

Close examination of Figures 5–7 reveals that the ATR data appear to scatter more than the refractometer data. To test this observation, the thick sheet and thin film data for each method were combined on separate triangular diagrams as shown in Figures 9 and 10. These diagrams exploit the orthogonal property of the fractional orientation as given in equations 3 and 6.

In these diagrams, the dashed lines represent uniaxial orientation with respect to a given axis. Using the Z axis as an example, the point at the apex, $Z_\parallel$, represents a sample that is perfectly oriented in the Z direction

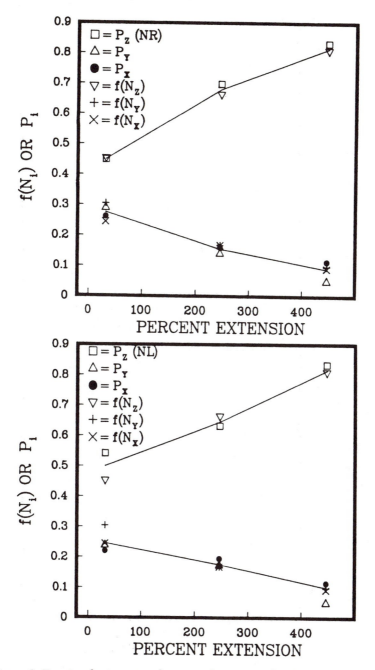

Figure 6. Fractional orientation data at each extension for the A series films: top, notch right (NR) surface; and bottom, notch left (NL) surface.

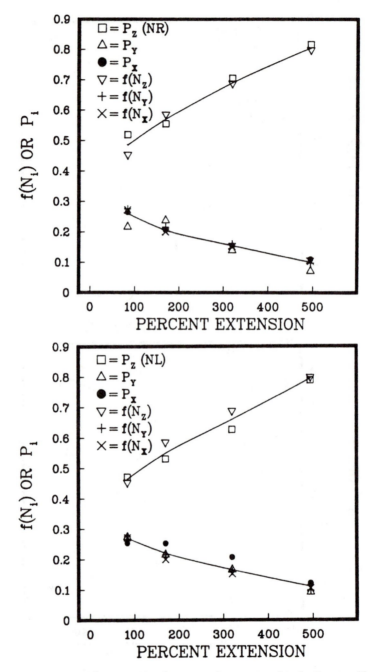

Figure 7. Fractional orientation data at each extension for the F series films: top, notch right (NR) surface; and bottom, notch left (NL) surface.

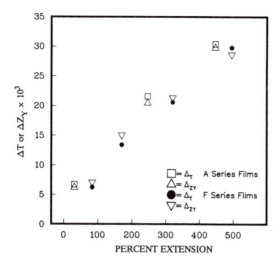

Figure 8. Comparison of the microscope (Δ_T) *and Abbe refractometer* (ΔZ_{ZY}) *birefringences for the A and F series films.*

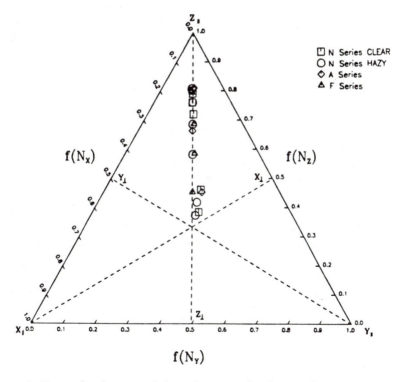

Figure 9. Triangular diagram of the refractive index fractional orientation data of all samples studied.

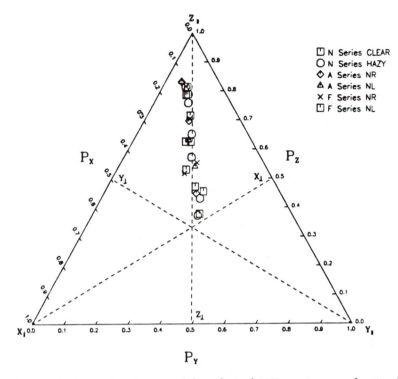

Figure 10. Triangular diagram of the polarized ATR spectroscopy fractional orientation data of all samples studied.

$(f(N_Z) = 1; f(N_X) = f(N_Y) = 0)$. At the other extreme, the point $Z_\perp$ represents a sample that is oriented perfectly perpendicular to the Z axis $(f(N_Z) = 0;$ $f(N_X) = f(N_Y) = 0.5)$. Data that plot at the intersection of all three dashed lines are isotropic $(f(N_X) = f(N_Y) = f(N_Z) = 0.333)$.

Figures 9 and 10 show general uniaxial orientation for all samples, but the refractive index data in Figure 9 clearly show much less scatter.

Finally, the polarized ATR technique used in this study yields only three-dimensional fractional orientation information about the sample surface being measured, but the polarized refractometry technique yields much more information. This information includes the three principle refractive indices, the average refractive index (and hence the density), the crystal volume fraction, the three birefringences, the symmetry axis direction, and the axiality (isotropic, uniaxial, biaxial) of each surface (3, 4).

Conclusions

In conclusion then, the data from this study on both thick sheets and thin films show excellent agreement between the surface fractional orientation results from both polarized refractometry and polarized ATR spectroscopy. This finding confirms our earlier suggestion that, because both methods require that measurements be made close to the critical angle, and both methods share similar physical models, their fractional orientation results should correspond.

Furthermore, the orientation data from polarized ATR spectroscopy are subject to somewhat greater scatter than similar refractive index data.

Finally, the additional surface structure information gained from the polarized refractometry technique strongly recommends its use for such studies.

References

1. Samuels, R. J. *Polym. Eng. Sci.* **1984**, *23*, 1293.
2. Samuels, R. J. *Polym. Eng. Sci.* **1988**, *28*, 852.
3. Pepper, R. E.; Samuels, R. J. "Refractometry", in the *Encyclopedia of Polymer Science and Engineering;* Vol. 14, 2nd ed., Kroschwitz, J. I., Ed.; Wiley: New York, 1988, pp 261–298.
4. Samuels, R. J. *J. Appl. Polym. Sci.* **1981**, *26*, 1383.
5. Pepper, R. E.; Samuels, R. J. *Annu. Tech. Conf. Soc. Plast. Eng.* **1987**, *33*, 556.
6. Pepper, R. E.; Samuels, R. J. *Annu. Tech. Conf. Soc. Plast. Eng.* **1988**, *34*, 884.
7. Collier, L. W., IV; Samuels, R. J. *Annu. Tech. Conf. Soc. Plast. Eng.* **1983**, *29*, 390.
8. Mirabella, F. M., Jr. *J. Polym. Sci. Part A-2* **1984**, *22*, 1283.
9. Mirabella, F. M., Jr. *J. Polym. Sci. Part A-2*, **1984**, *22*, 1293.
10. Mirabella, F. M., Jr. *Appl. Spectrosc. Rev.* **1985**, *21*, 45.
11. Kissin, Y. V.; Lekaye, I. A.; Chernova, Y. A.; Davydova, N. A.; Cherkov, N. M. *Vysokomol. Soyed* **1974**, *A16*, 677.
12. Sung, C. S. P. *Macromolecules* **1981**, *14*, 591.
13. Mirabella, F. M., Jr. *Appl. Spectrosc.* **1988**, *42*, 1258.
14. Samuels, R. J. *Structured Polymer Properties;* Wiley: New York, 1974.
15. Samuels, R. J. *J. Macromol. Sci. Phys.* **1973**, *B8*, 41.

RECEIVED for review February 14, 1989. ACCEPTED revised manuscript August 8, 1989.

Infrared Studies on the Grafting Reactions of Poly(vinyl alcohol)

Neal J. Earhart[1], Victoria L. Dimonie, Mohamed S. El-Aasser, and John W. Vanderhoff

Emulsion Polymers Institute, Departments of Chemistry and Chemical Engineering, Lehigh University, Bethlehem, PA 18015

The emulsion copolymerization of vinyl acetate and butyl acrylate with poly(vinyl alcohol) of different degrees of hydrolysis as the sole emulsifier was studied. The grafting reactions of the poly(vinyl alcohol) and the vinyl acetate in the aqueous phase affect the rates of individual monomer consumption and the overall polymerization kinetics. The copolymer products of this grafting reaction were analyzed by Fourier transform infrared (FTIR) spectroscopy. A limiting value for the amount of vinyl acetate grafting to the poly(vinyl alcohol) to maintain water solubility was determined.

POLY(VINYL ALCOHOL) WITH DIFFERING DEGREES of hydrolysis is often used as the sole emulsifier and stabilizer in the emulsion polymerization of vinyl acetate (*1*). The kinetic characteristics of the polymerization process as well as the surface, colloidal, and bulk properties of the resulting latex seem to be determined by the grafting reaction that involves the poly(vinyl alcohol) (PVA) and the vinyl acetate (VAc) in the aqueous phase (*2*).

The degree of hydrolysis of the poly(vinyl acetate) (PVAc) greatly affects the polymerization kinetics. Okamura and Yamashita (*3*) found a greater degree of grafting on fully hydrolyzed PVA than on the partially hydrolyzed PVA. During the emulsion polymerization process, the water-soluble potassium persulfate $(K_2S_2O_8)$ initiator radicals are generated in the aqueous

[1]Current address: Ciba-Geigy Corporation, Additives Division, Ardsley, NY 10502–2699

phase. The radicals react with the aqueous-phase vinyl acetate monomer and the PVA to create vinyl acetate radicals and macro PVA and PVAc radicals, which can then undergo grafting reactions. The PVA–PVAc copolymer products of this grafting reaction are thought to be the loci of particle nucleation (4).

Vinyl acetate and butyl acrylate emulsion copolymerizations using anionic surfactants have been extensively studied in terms of their copolymerization kinetics, the mode of monomer addition (as reflected by the copolymer composition and particle morphology), and the particle nucleation mechanism. The goal of the research work reported here was to determine the effect of poly(vinyl alcohol) on the mechanism and kinetics of the copolymerization of the monomers, with emphasis placed on examination of the grafting reactions with the poly(vinyl alcohol) (5–8).

Experimental Details

Vinyl acetate and butyl acrylate, commercial monomers, were purified by distillation. The poly(vinyl alcohol)s, Vinol 205 and Vinol 107, were provided by Air Products and Chemicals, Inc. Sodium bicarbonate ($NaHCO_3$), potassium persulfate ($K_2S_2O_8$) (Fisher Scientific), and tert-dodecyl mercaptan (Pennwalt) were reagent grade. Distilled–deionized water was used throughout the experiments.

The poly(vinyl alcohol) solutions were prepared by dispersing the PVA into room temperature distilled–deionized water. The PVA dispersion was then heated to 85 °C for the partially hydrolyzed Vinol 205 and to 95 °C for the fully hydrolyzed Vinol 107 for 30 min to allow the PVA to completely dissolve.

A standard latex recipe was used throughout the experiments:

Component	Weight (g)
Vinyl acetate (VAc)	50.00
Butyl acrylate	50.00
Distilled–deionized water	300.00
Poly(vinyl alcohol) (PVA)	10.00
Sodium bicarbonate ($NaHCO_3$)	0.10
t-Dodecyl mercaptan	0.10
Potassium persulfate ($K_2S_2O_8$)	0.10

Polymerizations were carried out with constant stirring in a four-neck round-bottom flask equipped with a sampling port, reflux condenser, and thermometer. The polymerization temperature was held constant at 60 ± 0.5 °C by a water bath. Samples were withdrawn periodically to study the kinetics of the reaction. The fractional monomer conversions were determined by gas chromatography with a Hewlett-Packard model 5890A gas chromatograph. The overall conversion was determined by gravimetric analysis.

The samples withdrawn for the kinetic study were centrifuged at 18,000 rpm at 4 °C for 2 h to separate the aqueous-phase serum from the polymer and monomer phases. The solids content of the aqueous phase was determined by gravimetric analysis.

Fourier transform infrared (FTIR) spectroscopy was used to determine the increase in the acetate character of the polymer chains in the aqueous-phase serum resulting from the grafting between the poly(vinyl alcohol) and the vinyl acetate. Pure PVA solutions before polymerization were compared to the latex serum. Potassium thiocyanate (KSCN) was used as the reference standard for the FTIR spectra and was added quantitatively to both the pure PVA solution and the latex serum. KSCN was selected as the reference standard because it is water-soluble and the peak wavelength of the nitrile stretch (2052 cm^{-1}) is isolated from the other functional peak wavelengths (Figure 1).

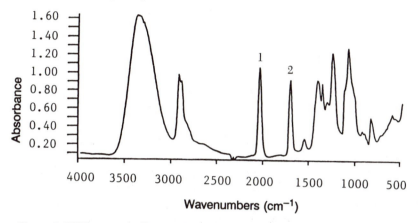

Figure 1. FTIR spectra of latex 205–10 serum with KSCN added as the spectral standard. Peak 1: 2052 cm^{-1}, –C≡N; Peak 2, 1740 cm^{-1}, >C = O.

Films of the samples were cast onto zinc selenide (ZnSe) IR cells. FTIR spectra were obtained for each sample with a Mattson Sirius 100 FTIR spectrometer. From the IR spectra, specific peak absorbances were measured, and the increase in the absorbance due to grafted vinyl acetate was recorded.

To determine the poly(vinyl acetate-*co*-butyl acrylate) copolymer composition by FTIR, the copolymer was dissolved in dimethyl sulfoxide (DMSO), and a film was cast onto the ZnSe IR cells. The peak absorbances were recorded for the –C–O– stretch of VAc at 1248 cm^{-1} and for butyl acrylate (BA) at 1167 cm^{-1}. A calibration curve plotting the absorbance ratio of PVAc–PBA versus the weight ratio of PVAc–PBA was generated. The copolymer composition of the sample was then calculated by using the calibration curve.

Results

The research plan was to examine the polymerization kinetics of identically prepared latex systems by using poly(vinyl alcohol)s that have a similar molecular weight (M_n and M_w), but a different degree of hydrolysis (% OH) as given in Table I.

Table I. Characteristics of Poly(vinyl alcohols)

Sample	OH Content (%)	M_n	M_w
Vinol 205	87.0–89.0	18,470	34,940
Vinol 107	98.0–98.8	15,570	26,250

NOTE: M_n and M_w were determined by aqueous-phase gel permeation chromatography.

When polymerizations were carried out with the partially hydrolyzed Vinol 205 PVA, the initial rates of polymerization (R_p) as well as the overall conversion are faster than the polymerizations using the fully hydrolyzed Vinol 107 (*see also* Figure 2):

PVA Sample	R_p, initial (g/L-min)
Vinol 205 (partially hydrolyzed)	5.76×10^{-3}
Vinol 107 (fully hydrolyzed)	2.30×10^{-3}

This difference in the kinetics results from the higher degree of hydrolysis of the Vinol 107, which decreases its effectiveness as an emulsifier. Grafting has to occur to a greater extent on the fully hydrolyzed Vinol 107 chains in order for the graft PVA-*co*-PVAc to function as a locus for particle nucleation.

The grafting reaction can be further quantified by examining the percent solids and the composition of the aqueous-phase serum during the polymerization (Table II).

In the latex prepared with the partially hydrolyzed PVA (Vinol 205) the percent solids of the aqueous-phase serum decreases from the initial amount. However, in the latex prepared with the fully hydrolyzed PVA (Vinol 107) the solids content of the aqueous-phase serum increases from the initial content. This result suggests that, while grafting is taking place in the aqueous phase, the fully hydrolyzed PVA chains are capable of a similar, if not greater, degree of grafting with the vinyl acetate and can still maintain their water solubility. The partially hydrolyzed PVA chains, in contrast, attain

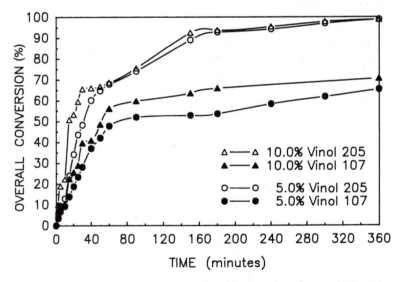

Figure 2. Overall conversion versus time data for 50:50 (weight ratio) VAc–BA copolymer latices prepared with 10.0% and 5.0% PVA concentrations at a 0.10% $K_2S_2O_8$ concentration.

Table II. Solids Content and Composition of the Latex Serum

Reaction Time (min)	Solids Content (%)	Overall Degree of Hydrolysis[b]	PVAc Grafted[a] (%)	PVA Original[a] (%)
		Fully hydrolyzed Vinol 107		
0.0	3.23	98.0	0.00	100.00
5.0	3.25	92.2	5.85	94.15
15.0	3.36	89.1	9.88	90.12
120.0	3.38	82.0	16.00	84.00
360.0	3.40	81.9	16.05	83.95
		Partially hydrolyzed Vinol 205		
0.0	3.42	88.0	0.00	100.00
5.0	2.98	83.0	5.02	94.98
15.0	2.80	81.9	6.11	93.89
40.0	2.72	81.8	6.18	93.82
360.0	2.71	81.8	6.20	93.80

[a] Determined by FTIR spectroscopy.
[b] PVOH.

a certain degree of grafting with the vinyl acetate and become water insoluble, and thus, they are not present in the aqueous phase. FTIR analysis of the serum for composition showed that for the fully hydrolyzed Vinol 107 samples, the aqueous-phase serum does contain more grafted vinyl acetate than the samples prepared with the partially hydrolyzed Vinol 205 samples.

In both cases, the amount of vinyl acetate grafted to the poly(vinyl alcohol) becomes constant during the polymerization. This constant value is a limiting value for the graft PVA-co-PVAc polymer to remain water soluble. From the initial number of acetate groups present on the PVA chains and from the data regarding serum composition obtained by FTIR, the limiting value for acetate content on the PVA chains to keep its water solubility can be calculated to be approximately 18.0%.

This finding suggests that both types of poly(vinyl alcohol) can be grafted with vinyl acetate until 18.0% of the polymer chains are acetate groups or an overall 82.0% degree of hydrolysis. The partially hydrolyzed PVA (Vinol 205) with an initial 88.0% degree of hydrolysis can graft with the VAc until 82.0% overall degree of hydrolysis is reached, or what represents an increase of 6.0% acetate character on the copolymer chain. The fully hydrolyzed PVA (Vinol 107) has an initial 98.0% degree of hydrolysis and can also graft with the vinyl acetate until an 82.0% overall degree of hydrolysis is reached, or what represents an increase of 16.0% acetate character.

The grafting process of the vinyl acetate with the poly(vinyl alcohol) chains can be further demonstrated by examining the fractional conversion for the vinyl acetate and butyl acrylate monomers during the course of the polymerization. Figures 3A–3C show the fractional conversion of each monomer at the same initiator concentration. In both cases where the PVA is present as the emulsifier (Figures 3B and 3C), the initial consumption of the vinyl acetate is greater than that of the butyl acrylate monomer. This trend is not predicted from the copolymerization reactivity ratios of the two monomers ($r_{1 VAc}$ = 0.04, $r_{2 BA}$ = 5.5) (9, 10).

The data in Figure 3A show that in the latex prepared solely with sodium lauryl sulfate (SLS) as the emulsifier (no PVA present), the fractional conversion of each monomer reflects the difference in the reactivity ratios, as expected. Butyl acrylate was consumed at a much faster rate than the vinyl acetate until it was exhausted; after that point the vinyl acetate was consumed

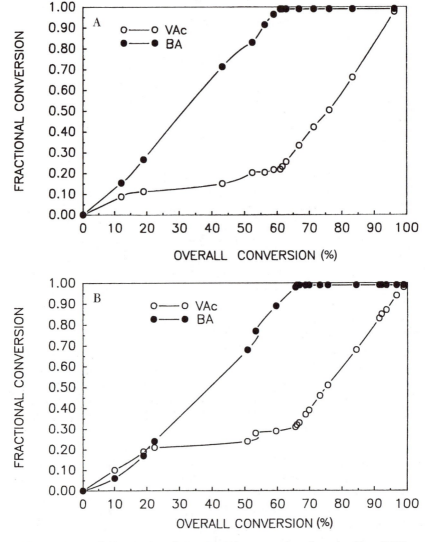

Figure 3. Copolymerization of VAc–BA (A) using sodium lauryl sulfate (SLS), and (B) using 10% based on monomer of Vinol 205 (88% degree of hydrolysis).
Continued on next page.

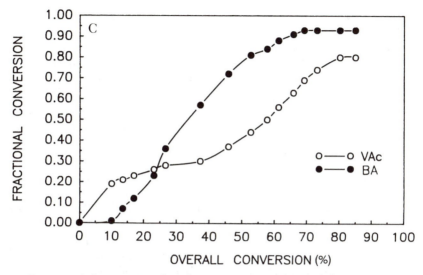

Figure 3. (C) Using 10% based on monomer of Vinol 107 (98% degree of hydrolysis).

at an increased rate. The higher rate of vinyl acetate consumption in the initial stages of the polymerization (less than 35% overall conversion) in the presence of PVA suggests that the vinyl acetate is involved in another reaction besides the copolymerization with the butyl acrylate monomer.

The polymer fraction remaining from the centrifugation of the latex was analyzed for composition by FTIR to determine the poly(vinyl acetate)–poly(butyl acrylate) ratio in the copolymer. The data obtained from FTIR for the copolymer composition (Table III) show less poly(vinyl acetate) present in the copolymer than the data obtained by gas chromatography. Because gas chromatography is measuring the total amount of monomer consumed and FTIR spectroscopy is determining the composition of the copolymer formed, it can be assumed that this difference also shows that

Table III. Copolymer Composition Determined by Gas Chromatography and Fourier Transform Infrared Spectroscopy

Latex Sample	Reaction Time (min)	Overall Conversion	Ratio of PVAc to PBA	
			GC	FTIR
Vinol 107	5.0	9.44	5.33	4.26
	15.0	22.48	0.58	0.47
	120.0	60.36	0.28	0.23
	360.0	71.03	0.42	0.34
Vinol 205	5.0	18.89	1.25	1.11
	15.0	50.73	0.31	0.27
	40.0	65.98	0.33	0.30
	360.0	98.96	1.00	0.90

the vinyl acetate is involved with another reaction in addition to that with the butyl acrylate during the initial stages of polymerization.

In the latex prepared with the partially hydrolyzed PVA (Vinol 205) the difference between the PVAc–PBA ratio obtained by gas chromatography and by FTIR spectroscopy is less than the difference between the same PVAc–PBA ratios calculated for the latex prepared with the fully hydrolyzed PVA (Vinol 107). These data also show that the fully hydrolyzed PVA can graft to a greater extent than the partially hydrolyzed PVA and consume more of the vinyl acetate in that process.

Conclusions

The experimental copolymerization kinetic data showed that the vinyl acetate monomer was involved in another reaction besides the emulsion copolymerization with butyl acrylate. The vinyl acetate was shown to react in the aqueous phase with the poly(vinyl alcohol) emulsifier. This aqueous-phase grafting reaction between the poly(vinyl alcohol) and the vinyl acetate causes a faster initial consumption of the vinyl acetate, which was not predicted by the copolymerization reactivity ratios. According to the prediction of the reactivity ratios, the butyl acrylate, being the more reactive monomer, should have reacted faster then the vinyl acetate. This faster reaction of vinyl acetate was illustrated when sodium lauryl sulfate was used as an emulsifier in the copolymerization of VAc and BA. This grafting process serves as one mechanism for particle nucleation in this emulsion polymerization system.

The particle nucleation mechanism can be illustrated as follows: The poly(vinyl alcohol) chains are acting as the nucleation site. The potassium persulfate ($K_2S_2O_8$) is initiating the vinyl acetate monomer that is present in the aqueous phase (vinyl acetate solubility in water is 2.50 g/100 mL of H_2O). The vinyl acetate radicals then can graft with the water-soluble PVA chains. The grafting process continues until the water-soluble PVA chains begin to exhibit a hydrophilic–hydrophobic balance due to the grafted water-insoluble vinyl acetate.

At this point the butyl acrylate, present as monomer droplets dispersed into the aqueous phase, begins to diffuse from the droplets into the hydrophobic regions of the highly grafted PVA–PVAc copolymer, and particles begin to form. The grafting reaction has to occur first, which causes the faster initial consumption of the vinyl acetate monomer. As the particles then begin to form, the more reactive butyl acrylate polymerizes faster than the vinyl acetate until it is completely consumed.

References

1. Finch, C. A. *Polyvinyl Alcohol: Properties and Uses;* Wiley: New York, 1973.
2. Gavat, I.; Dimonie, V.; Donescu, D. *J. Polym. Sci. Polym. Symp.* **1978,** *64,* 125.

3. Okamura, S.; Yamashita T. *Kobunshi Kagaku* **1958**, *15*, 165.
4. Heublein, G.; Meissner, H. *Acta Polym.* **1984**, *35*, 744.
5. Chugo, K.; et al. *J. Polym. Sci. Part C* **1969**, *27*, 321.
6. Makgawinata, T., Ph.D. Dissertation, Lehigh University, Bethlehem, PA, 1981.
7. Misra, S. C.; Pichot, C.; El-Aasser, M. S.; Vanderhoff, J. W. *J. Polym. Sci. Polym. Lett. Ed.* **1979**, *17*, 567.
8. Pichot, C.; Makgawinata, T.; El-Aasser, M. S.; Vanderhoff, J. W. *Polym. Prepr.* **1979**, *19*, 870.
9. Delgado, J., Ph.D. Dissertation, Lehigh University, Bethlehem, PA, 1986.
10. Pichot, C.; Llauro, M. F.; Pham, Q. T. *J. Polym. Sci. Chem. Ed.* **1981**, *19*, 2619.

RECEIVED for review February 14, 1989. ACCEPTED revised manuscript September 6, 1989.

Fourier Transform Infrared Spectroscopic Method for Evolved-Gas Analysis

Cheng-Yih Kuo and Theodore Provder

The Glidden Company, 16651 Sprague Road, Strongsville, OH 44136

Qualitative and quantitative Fourier transform infrared (FTIR) spectroscopic methodologies were developed for monitoring the gases evolved during chemical cure. The evolved-gas profile can be obtained as a function of temperature (nonisothermal dynamic scan) or time (isothermal). The major gas evolved can be identified readily by comparison of the experimental spectra with known spectra in a vapor-phase library. The gases coevolved also can be identified after spectral subtraction. Computer programs written for thermal analysis were adapted to obtain nth-order reaction kinetics parameters from the evolved-gas profile. Application of evolved-gas analysis (EGA)–FTIR to the thermal deblocking and chemical cure of model coatings systems is shown. Results are compared with those obtained from other techniques, such as thermal gravimetric analysis and thin-film FTIR studies.

CURE REACTIONS have been studied by various techniques, for example, thermal gravimetric analysis (TGA), differential scanning calorimetry (DSC), and dynamic mechanical analysis (DMA). By combining information generated from DSC and DMA methods with the information generated from thin-film Fourier transform infrared (FTIR) spectroscopic studies, progress has been made in the basic understanding of cure reactions in terms of the buildup of physical properties and the changes of chemical functionalities. This approach was successfully applied to the study of the deblocking and subsequent cure of blocked isocyanate-containing coatings (1–3) by using

0065–2393/90/0227–0343$06.00/0

FTIR spectroscopy to follow the changes in the reactive functionalities in a thin film during cure and correlating this information with that generated from thermal analysis (DSC or DMA).

However, important facts yet to be determined are the types and quantities of volatile gases given off during the cure reaction. TGA has been used to answer the question of quantity by following the weight changes as a function of temperature or time, but identification of the evolved gases is not clear, especially when two or more volatile gases evolve simultaneously. Coupling FTIR spectroscopy to TGA will enable the in-line identification of the volatile gases evolved from the TGA and provide information on which and what quantity of the volatile gases are given off during cure. This type of information will be extremely useful for the further elucidation of cure mechanisms.

TGA–FTIR spectroscopy also can be used to provide valuable information on the identities of thermal degradation byproducts. Although EGA–FTIR spectroscopic results obtained from experimental coupling of TGA and FTIR spectroscopy were reported in the literature (4–12) as early as 1980, dedicated instruments were not commercially available until 1988 (13, 14). This chapter describes straightforward instrumentation and methodology for FTIR spectroscopic monitoring of the gases evolved during chemical cure and illustrates the methodology with some cure reaction kinetics applications.

Experimental Details

Instrumentation. Figure 1 shows a schematic diagram of the experimental instrumentation. The sample, in the form of either a viscous liquid or a coated aluminum rectangular strip, is placed in $\frac{1}{4}$- $\times$ 2-in. stainless steel tubing that is then connected to an empty column in a Varian 3700 gas chromatography (GC) oven with temperature programming. For a dynamic temperature scan, the heating rate was 10 °C/min. The gases evolved from the sample passed through a gold-coated light pipe and were monitored with a mercury–cadmium telluride (MCT) detector cooled with liquid nitrogen. A Digilab FTS-15E FTIR spectrometer was used. The infrared data were collected with standard GC–FTIR software. The total amount of gas evolved is displayed as a Gram–Schmidt response in real time as a function of time or temperature. The instantaneous absorbance spectrum as a function of time or temperature also is obtained.

Materials. Blocked –NCO cross-linkers were prepared from trimerized isophorone diisocyanate (IPDI) (Veba-Chemie T-1890), which was used as received to synthesize blocked derivatives. In a typical blocking reaction, 100 g of T-1890 (0.15 mol) was dissolved in 100 g of dry ethyl acetate, and 0.1 g of dibutyltin dilaurate (DBTDL) was added. Methyl ethyl ketoxime (MEKO), 48.9 g (0.187 mol), was added over 2 h, during which time the temperature rose to 40 °C. Aliquots were removed, and the amount of unreacted –NCO functionality was determined by infrared absorbance. When the reaction of the –NCO was complete, the mixture was cooled and precipitated into diethyl ether. The product was isolated, dissolved in toluene, and reprecipitated three times. The final product was isolated and dried.

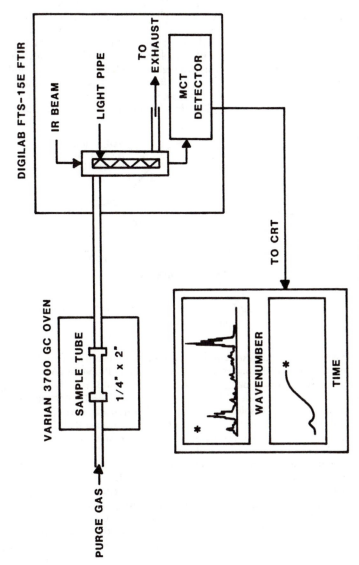

Figure 1. Instrumentation for EGA–FTIR spectroscopic analysis.

Methodology. *Qualitative: Evolved-Gas Profile.* The evolved-gas profile is represented by the Gram–Schmidt response (GSR). The GSR is reconstructed from the interferograms by the Gram–Schmidt orthogonalization method (*15*), which compares the interferograms of the background and the sample. When a sample is not present in the light pipe, the interferogram shows no difference from that of the background, and a flat base line is the result. When a sample is present in the light pipe, the orthogonalization yields a nonzero difference and deviates from the base line. The magnitude of the deviation is a function of the concentration of the sample in the light pipe. Figure 2 shows the GSR as a function of temperature of a hydroxyl (–OH) functional acrylic resin cured with a melamine cross-linker. If only one gas is evolving during cure, the GSR is the evolved-gas profile of that gas. If multiple gases are evolving simultaneously, then the GSR is the resultant profile of all gases. The evolved-gas profile of the individual gas can be reconstructed by looking for the characteristic band of interest, either directly or through spectral subtraction.

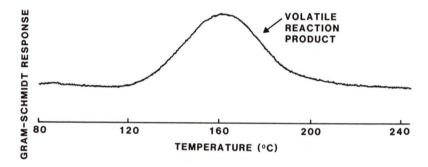

Figure 2. Evolved-gas profile for the melamine cure of a hydroxyl functional acrylic resin.

The identity of evolved gases can be determined by comparing and matching the experimental spectra with spectra in the Environmental Protection Agency vapor-phase library. In this study, about 30 spectra were generated, usually as a function of time or temperature, for each GSR. Each spectrum is a result of coadding about 20 scans together at 8-cm^{-1} resolution. The spectral search was performed with the program HITS, which compares the experimental spectra against the library spectra. The goodness of the match is represented by the HITS quality index (HQI), which is a normalized sum of squares of the difference between the absorbances of the experimental and library spectra as a function of wavelength. A lower HQI value indicates a better match. HITS allows the user to analyze the search results by displaying or plotting the spectrum of the experimental sample and the spectra of the several best-matched library spectra. Figure 3 shows a plot of the search results of the gas evolved in the reaction of the –OH functional acrylic resin with the melamine cross-linker. The low value of HQI = 0.11 and the visual comparison of that searched spectrum with the experimental spectrum clearly indicate that the evolved gas is methanol.

Quantitative: Cure Kinetics. The evolved-gas profile of Figure 2 is analogous in shape to an exothermic DSC thermogram and can be treated by the differential method of kinetics analysis. It also can be transformed into a fractional conversion curve and treated by the integral method of kinetics analysis.

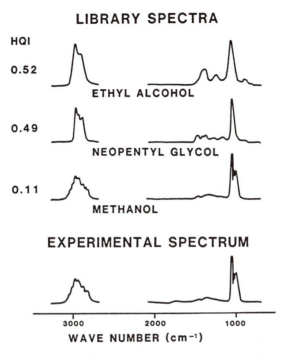

LIBRARY SPECTRA

HQI

0.52

ETHYL ALCOHOL

0.49

NEOPENTYL GLYCOL

0.11

METHANOL

EXPERIMENTAL SPECTRUM

3000   2000   1000

WAVE NUMBER (cm⁻¹)

Figure 3. Comparison of library spectra with experimental spectrum for EGA–FTIR spectroscopic volatile reaction product identification.

Differential Method.   Cure kinetics are obtained by a differential method based on the work of Borchardt and Daniels (*16*) with a single dynamic temperature scan. This method is based on the general nth-order rate equation

$$\frac{\mathrm{d}F(t,T)}{\mathrm{d}t} = k(T)[1 - F(t,T)]^n \tag{1}$$

where $F(t,T)$ is the fractional extent of conversion, n is the order of reaction, $k(T)$ is the rate constant, t is time (s), and T is the absolute temperature (K). DSC measures the heat flow in or out of the sample. A schematic GSR for a cure reaction is shown in Figure 4, where $\mathrm{d}G/\mathrm{d}t$ is the Gram–Schmidt response of the sample, G_{tot} is the total area under the GSR, and $G(t,T)$ is the partial area of the evolved GSR up to a specific time and temperature. The fractional extent of conversion is defined as

$$F(t,T) = \frac{G(t,T)}{G_{tot}} \tag{2}$$

and ranges from 0 to 1.0, as is shown in Figure 4. The Arrhenius equation gives the temperature dependence of the rate constant

$$k = A \exp\left(\frac{-E}{RT}\right) \tag{3}$$

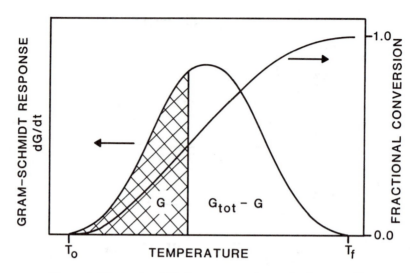

Figure 4. Illustrative GSR for cure reaction evolved-gas profile.

where A is the Arrhenius frequency factor (s^{-1}), E is the activation energy (kcal/mol), and R is the gas constant (kcal/mol K). In the differential method, the change in GSR with increasing temperature is used to obtain a working expression in terms of observable experimental parameters:

$$\ln\left[\frac{1}{G_{tot}}\frac{d\,G(t,T)}{dt}\right] = \ln A - \left[\frac{E}{RT}\right] + n\ln\left[\frac{G_{tot} - G(t,T)}{G_{tot}}\right] \tag{4}$$

Equation 4 has a form suitable for multiple regression analysis:

$$Z = a + bx + cy \tag{5}$$

where Z is equal to the left side of equation 4, $a = \ln A$, $b = -E/R$, and $c = n$. The reaction kinetics parameters n, E, and $\ln A$ are obtained simultaneously from this expression with a multiple regression analysis. The validity of the nth-order reaction kinetics model for describing the reaction mechanism that generates the evolved-gas profile is verified by comparing experimental and calculated data in an Arrhenius plot of $\ln k$ vs. $1/T$. Such a plot is shown in Figure 5 for the deblocking of T-1890–MEKO. The vertical dotted bars indicate the temperature range of the data used in the multiple regression analysis and generally corresponds to 90–95% of the area under the GSR. A detailed description of the multiple regression analysis method and its statistical validity can be found elsewhere (17–19).

Integration of equation 1 provides a generalized nonisothermal expression for the fractional conversion in terms of the kinetics parameters n, E, and A.

$$F(t,T) = 1 - \left[(n - 1)\,A\int_0^t\left[\frac{-E}{RT(t)}\right]dt + 1\right]^{1/(1-n)} \tag{6}$$

The kinetics parameters n, E, and A in conjunction with equation 6 allow the prediction of fractional conversion for a specific temperature–time profile experienced

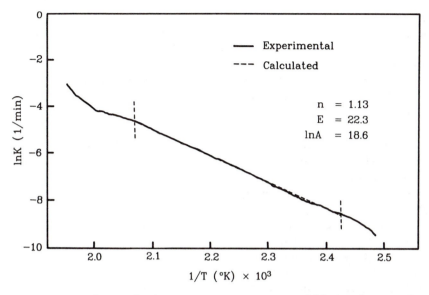

Figure 5. Arrhenius plot for EGA–FTIR spectroscopic deblocking kinetics of T-1890–MEKO.

by a coated part as the part equilibrates to the oven temperature. The most common temperature–time profile is exponential in nature and is given by the expression

$$T = T_0 + (T_f - T_0) \left[1 - \exp\left(\frac{-t}{\tau}\right) \right] \tag{7}$$

where T_0 is the initial temperature of the part, T_f is the temperature of the part equilibrated at the oven temperature, and τ (s) is the time constant governing the rate of equilibration from T_0 to T_f. A simple numerical method, the incremental isothermal-state method, for calculating the fractional conversion for any temperature–time profile was developed by Provder et al. (20).

Integral Method. To obtain the kinetics parameters, equation 6 must be solved for n, E, and A by a procedure known as the integral method, which requires the use of a Nelder–Mead simplex minimization algorithm (21–23) in conjunction with numerical procedures for Gauss–Legendre integration. A unique objective function is used in the numerical procedure to solve for n and E simultaneously. The experimental values of $F(t,T)$ are compared to the calculated values based upon initial guesses of n and E. The values of n and E are adjusted by the Nelder–Mead simplex algorithm until the objective function is minimized. The value of $\ln A$ is then determined from an algebraic expression. A detailed description of the numerical analysis procedure was provided by Koehler et al. (24).

Results and Discussion

Evolved-gas analysis will allow the identification and quantification of the amount of volatile gases evolved as a function of time or temperature. This information is quite useful for the study of polymer degradation in terms of degradation kinetics and the identification of thermal degradation byproducts. It is especially useful for the elucidation of cross-linking chemistry, kinetics, and mechanism, both qualitatively and quantitatively, as well as being complementary to physical methods for studying cure behavior. Some examples of the use of EGA–FTIR spectroscopy for coating cure characterization are as follows.

Comparative Cure of Acrylic–Melamine Coatings.   The rate of cure for –OH-containing acrylic resins with melamine cross-linkers is important information for relating chemical reactivity to film formation and ultimate film properties. In this example the rate of cure of an acrylic coating containing melamine A capped with methanol is compared to the same coating containing melamine B capped with a higher boiling alcohol. In this experiment the oven temperature was programmed from 80 to 240 °C at a rate of 10 °C/min and then held at 240 °C. The Gram–Schmidt reconstructed evolved-gas profile for the acrylic polymer cured with melamine A is shown in Figure 2. The profile shows that the reaction started at around 115 °C and peaked out at 160 °C. The EGA–FTIR spectroscopic profile for the acrylic coating containing melamine B run under the same experimental conditions shows that the reaction started at the same temperature but peaked at 10 °C higher than the acrylic coating containing melamine A.

The fractional conversion curves for the two systems are shown in Figure 6. The slower evolved-gas rate of the melamine B system did correlate with the cured film having fewer visual defects. The faster evolved-gas rate of the melamine A system caused the film viscosity to increase at too fast a rate; the result was that sufficient gas bubbles were trapped in the film to lead to unacceptable observable defects. Reaction kinetics parameters were obtained from the fractional conversion curves in Figure 6 by using the integral method and are the following:

Coating System	n	E	ln A
Melamine A	2.01	34.8	35.6
Melamine B	2.10	29.3	28.2

where E is in kilocalories per mole.

The reaction order for gas evolution for both melamine systems was second order. Berge et al. (25), Blank (26), and Bauer and Budde (27) favor an S_N1 reaction mechanism, whereas Holmberg (28), Lazzara (29), and Santer and Anderson (30) prefer the S_N2 reaction mechanism. The second-order reaction obtained from the EGA–FTIR results fits better with the S_N2 mechanism. The melamine cure is a very complex reaction, and no single mechanism could fully describe the cure kinetics. Mixed modes of reaction are known to occur.

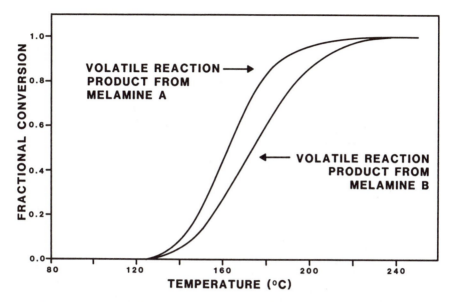

Figure 6. Fractional conversion curves for melamine cure of hydroxyl acrylic resin from EGA–FTIR spectroscopic analysis.

Cure of Reactive Diluent Coatings.   Figure 7 shows the reconstructed evolved-gas absorbance profiles of a coating containing a reactive diluent. In this case the reactive diluent and volatile reaction product of cure have noninterfering absorbance bands that allow the reconstruction of the individual evolved-gas absorbance profiles. The temperature of the oven was programmed from 60 to 200 °C and then held at 200 °C for 6 min. The evolved-gas absorbance profiles show that, at lower temperatures, the loss of the diluent predominates over the cure reaction because the cure volatile gases do not evolve until the temperature reaches 120 °C. Once the cure reaction commences, it proceeds quite rapidly. To avoid the loss of reactive diluent, a higher boiling diluent would be required or the cure reaction would have to be catalyzed to occur at lower temperatures.

Deblocking of Blocked Isocyanate-Containing Coatings.   As demonstrated previously (*1–3*), deblocking kinetics for blocked –NCO cross-linkers can be obtained from thin-film FTIR kinetics analysis methods by monitoring the appearance of free –NCO functionality with time or temperature. With the EGA–FTIR spectroscopic technique, deblocking kinetics can be obtained by following the evolved-gas profile of the volatilized blocking group. Figure 8 shows the Gram–Schmidt reconstructed evolved-gas absorbance profile of MEKO from the thermal deblocking of T-1890–MEKO. The small peak eluting before the main peak was shown to be composed of residual solvents used in sample preparation. Examination of the FTIR spec-

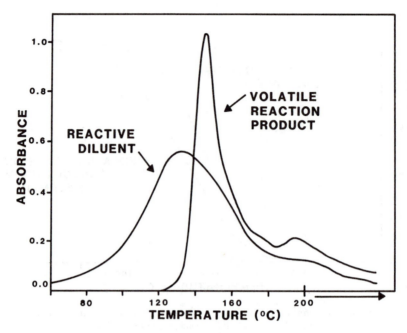

Figure 7. *Evolved-gas absorbance profile for a reactive diluent containing coating.*

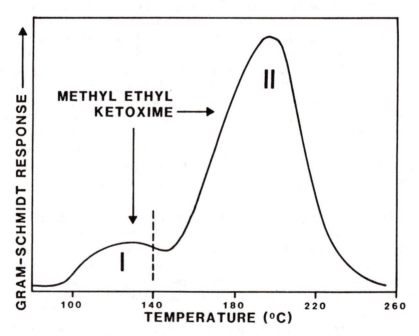

Figure 8. *Evolved-gas absorbance profile for the thermal deblocking of T-1890–MEKO.*

tra during the early stages of deblocking shows that MEKO was not present in the evolved-gas profile until the temperature exceeded 120 °C.

Evolved-gas profile data above 140 °C were used to construct a normalized differential curve and a fractional conversion curve, which were then subjected to kinetics analysis by the differential and the integral methods, respectively. The fractional conversion curves of the integral method for the experimental and fitted data are shown in Figure 9. Figure 10 illustrates the percent variation between the experimental and calculated conversion curves over the data range, which, for T-1890–MEKO deblocking kinetics, is better than ±0.5%. Table I summarizes deblocking kinetics data for T-1890–MEKO obtained by EGA–FTIR spectroscopy, thin-film FTIR spectroscopy, and isothermal TGA experiments from the literature (*31*). Given that the error in the order of reaction *n* is typically ±0.1, the agreement among the three different methods is excellent. The EGA–FTIR spectroscopic studies showed that both qualitative and quantitative kinetics information can be obtained from evolved volatile gases during the cure of various coatings systems. The EGA–FTIR spectroscopic method can corroborate chemical kinetics information obtained from thin-film FTIR spectroscopic methods. At times EGA–FTIR spectroscopy may be more suitable than thin-film FTIR spectroscopic methods for providing chemical kinetics information because of band interferences that can occur in thin-film studies.

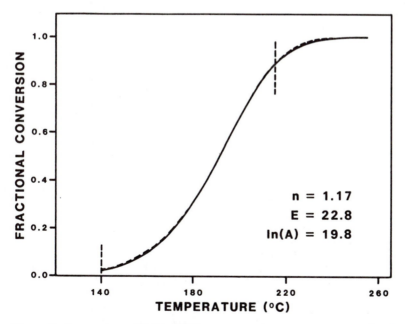

Figure 9. Comparison of EGA–FTIR spectroscopic conversion curves for T-1890–MEKO from integral method kinetics. Key: —, experimental curve, and - - -, calculated curve.

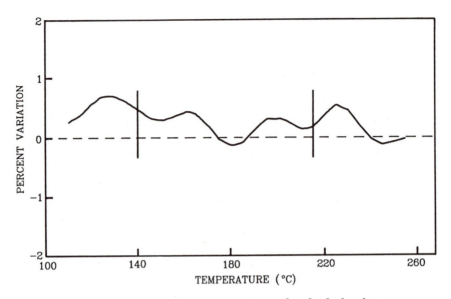

Figure 10. *Percent variation between experimental and calculated conversion curves for T-1890–MEKO deblocking kinetics.*

Table I. Deblocking Kinetics Parameters for T-1890–MEKO

Method	n	Et[a]	ln A
Thin-film FTIR spectroscopy			
2256 cm[-1] (NCO)	1.0	23.2	20.6
EGA–FTIR spectroscopy:			
Integral method	1.17	22.8	19.8
Differential method	1.13	22.3	
Kordomenos et al. (TGA)[b]	1.00	23.9	23.2

[a] In kilocalories per mole.
[b] n = 1.0 assumed (*see* reference 31).

References

1. Provder, T.; Neag, C. M.; Carlson, G. M.; Kuo, C.; Holsworth, R. M. In *Analytical Calorimetry;* Gill, P. S.; Johnson, J., Eds.; Plenum: New York, 1984; Vol. 5, p 377.
2. Carlson, G. M.; Neag, C. M.; Kuo, C.; Provder, T. In *Advances in Urethane Science and Technology;* Frisch, K. C.; Klempner, D., Eds.; Technomic Publishing: Lancaster, PA, 1984; Vol. 9, pp 47.
3. Carlson, G. M.; Neag, C. M.; Kuo, C.; Provder, T. In *Fourier Transform Infrared Characterization of Polymers;* Ishida, H., Ed.; Plenum: New York, 1987; pp 197.
4. Lephardt, J. O.; Fenner, R. A. *Appl. Spectrosc.* **1980**, *34*, 174.
5. Lephardt, J. O.; Fenner, R. A. *Appl. Spectrosc.* **1981**, *35*, 95.
6. Cody, C. A.; DiCarlo, L.; Fauleit, B. K. *Am. Lab.* **1981**, *13(1)*, 93.
7. Roush, P. B.; Luce, J. M.; Totten, G. A. *Am. Lab.* **1983**, *15(10)*, 90.

8. Mittleman, M. L.; Compton, D. A. C.; Engler, P. In *Proceedings of the 13th North American Thermal Analysis Society Conference;* Philadelphia, PA, 1984.
9. Nerheim, A. G. In *Fourier Transform Infrared Spectroscopy;* Ferraro, J. R.; Basile, L. J., Eds.; Academic: New York, 1985; Vol. 4.
10. Khorami, J.; Chauvette, G.; Lemieux, A.; Menard, H.; Jolicoeur, C. *Thermochim. Acta* **1986,** *103,* 221.
11. Carangelo, R. M.; Solomon, P. R.; Gerson, D. J. *Prepr. Am. Chem. Soc. Div. Fuel Chem.* **1986,** *31(1),* 152.
12. Sanchez, L. A. *Appl. Spectrosc.* **1988,** *41(6),* 1019.
13. Compton, D. A. C. *Internat. Labmate,* **1987,** *12(4),* 37.
14. Wieboldt, R. C.; Adams, G. E.; Lowry, S. R.; Rosenthal, R. J. *Am. Lab.* **1988,** *20(1),* 70.
15. de Haseth, J. A.; Isenhour, T. L. *Anal. Chem.* **1977,** *49,* 1977.
16. Borchardt, H. J.; Daniels, F. *J. Am. Chem. Soc.* **1957,** *41,* 79.
17. Provder, T.; Holsworth, R. M.; Grentzer, T. H.; Kline, S. A. In *Polymer Characterization: Spectroscopic, Chromatographic, and Physical Instrumental Methods;* Advances in Chemistry Series No. 203; Craver, C. D., Ed.; American Chemical Society: Washington, DC, 1988; p 233.
18. Kah, A. F.; Koehler, M. E.; Grentzer, T. H.; Niemann, T. F.; Provder, T. In *Computer Applications in Applied Polymer Science;* ACS Symposium Series No. 197; Provder, T., Ed.; American Chemical Society: Washington, DC, 1982; p 297.
19. Neag, C. M.; Provder, T; Holsworth, R. M. *J. Thermal Anal.* **1987,** *32,* 1833.
20. Provder, T.; Malihi, F. B.; Neag, C. M.; Holsworth, R. M.; Koehler, M. E. *Am. Chem. Org. Coat. Plast. Chem. Prepr.* **1982,** *46,* 493.
21. Smith, J. M. *Mathematical Modeling and Digital Simulation for Engineers and Scientists;* Wiley: New York, 1977.
22. Olsson, P. M. *J. Qual. Technol.* **1974,** *6,* 53.
23. Olsson, P. M.; Nelson, L. S. *Technometrics* **1975,** *17,* 45.
24. Koehler, M. E.; Kah, A. F.; Neag, C. M.; Niemann, T. F.; Malihi, F. B.; Provder, T. In *Analytical Calorimetry 5;* Gill, P. S.; Johnson, T. F., Eds.; Plenum: New York, 1984; p 361.
25. Berge, A.; Kvaeven, B.; Ugelstad, J. *Eur. Polym. J.* **1970,** *6,* 981.
26. Blank, W. J. *J. Coat. Technol.* **1982,** *54(687),* 26.
27. Bauer, D. R.; Budde, G. F. *J. Appl. Polym. Sci.* **1983,** *28,* 253.
28. Holmberg, K. *J. Oil Colour Chem. Assoc.* **1978,** *61,* 359.
29. Lazzara, M. G. *J. Coat. Technol.* **1984,** *56(710),* 19.
30. Santer, J. O.; Anderson, G. J. *J. Coat. Technol.* **1980,** *52(667),* 33.
31. Kordomenos, P. I.; Dervan, A. H.; Kresta, J. *J. Coat. Technol.* **1982,** *54(687),* 43.

RECEIVED for review February 14, 1989. ACCEPTED revised manuscript October 19, 1989.

Simultaneous Differential Scanning Calorimetry and Infrared Spectroscopy

Francis M. Mirabella, Jr.

Quantum Chemical Corporation, Rolling Meadows, IL 60008–4070

The simultaneous measurement of thermal response and infrared spectra on specimens of microscopic dimensions was demonstrated for polymer melting and crystallization of polymer blends. The measurement technique was achieved by combining a differential scanning calorimeter (DSC) having provision for microscopic observation of the specimen with an infrared microsampling accessory (IRMA) fitted into a Fourier transform infrared (FTIR) spectrometer. This combined technique was shown to provide important insight into structural changes associated with thermal response for a wide range of systems including polymers. In particular, the structural changes of polypropylene during melting were observed, and the separate crystallization of polypropylene and polyethylene in two-part blends was observed and correlated with the crystallization exotherms of the blends.

SIMULTANEOUS MEASUREMENTS OF THERMAL PROPERTY RESPONSE and infrared spectra on specimens of microscopic dimensions can provide important insight into structural changes associated with thermal response for a wide range of systems. The simultaneous measurement of the thermal response in a differential scanning calorimeter (DSC) and the infrared spectra in a Fourier transform infrared (FTIR) spectrometer can provide this capability. In particular, the use of a DSC cell adapted for microscopic observation during thermal treatment, combined with an infrared microsampling accessory (IRMA) in an FTIR spectrometer, permits the simultaneous collection of such data. The range of applications of such a technique is undoubtedly large and varied. The various phase transitions and thermal responses of

0065-2393/90/0227–0357$06.00/0

crystalline and amorphous polymers, polymer blends, copolymers, and liquid-crystalline polymers, as well as small molecules, could be correlated with the structural information derivable from the simultaneously collected infrared spectra. This approach would permit studies of reaction and crystallization kinetics, oxidation, efficacy of additives, etc.

Experimental Details

A Nicolet 6000 FTIR spectrometer was used. All spectra were recorded at 4-cm^{-1} resolution. Acceptable signal-to-noise ratios (S/N) were obtained by the coadding of 10 spectra at a mirror velocity of 0.586 cm/s. A Spectra-Tech IR-Plan III transmittance–reflectance microscope with a 250- × 250-μm mercury–cadmium telluride (MCT) detector was used to focus the infrared radiation through the DSC cell. The DSC cell used was a Mettler FP84 thermal analysis microscopy cell. A Mettler FP80 central processor with recorder was used to control the temperature in the FP84 DSC cell. Typical DSC thermograms and infrared spectra could be simultaneously recorded with this combined apparatus. A schematic diagram of the IRMA is shown in Figure 1.

A schematic diagram of the DSC microscopy cell is shown in Figure 2. The DSC microscopy cell was placed on the stage of the IRMA. The infrared beam was required to pass through a 2.5-mm hole in the bottom of the DSC cell, travel 3.0 mm (thereby passing through the sample and sample cup), and exit from the DSC cell through a 3.0-mm hole on the way to the MCT detector. Thus, the infrared beam must pass through a cylinder of 2.5-mm diameter and 3.0-mm length. This situation requires a careful alignment procedure prior to operation.

To permit both microscopical observation and transmission infrared spectroscopy, sample cups of potassium bromide were used to hold the sample. For simultaneous optical microscopy and DSC, sapphire sample cups were used. The thermal conductivity of KBr is 1.3 g-cal/s-cm-°C, that of sapphire is 0.6 g-cal/s-cm-°C. Thus, KBr is a reasonably good conductor of heat, but it is also transparent to visible light and infrared radiation in the wavelength range of 0.35–15 μm. It is, therefore, an excellent material for use in the present application.

The combination of temperature program rates in the DSC and spectral collection rates in the FTIR spectrometer offers a vast range in which to collect data simultaneously. Of course, the FTIR spectrometer has the capability of rapidly collecting spectra over the entire DSC thermogram. Spectra can be collected at the rate of about one per second and thereby a large number of spectra can be accumulated over a 15–20-min DSC experiment. These spectra may be important for full elucidation of the thermal response under investigation. For the purposes of this work and for ease of presentation, fewer spectra, collected over longer time, will be presented. The DSC was programmed at 10 °C/min. The FTIR spectrometer collected 10 scans in 10 s at the aforementioned conditions.

Polymer blends were prepared from Quantum USI PP 8001-LK polypropylene (melt flow rate 5 g/10 min) and NHD 6180 high-density polyethylene (melt index 1.15 g/10 min and density 0.96+ g/cm^3). Blends were prepared by repetitive melt blending in a hot press. Four blends at 20:80, 40:60, 60:40 and 80:20 wt% of polypropylene–polyethylene were prepared.

Results and Discussion

The combined DSC–microscopy–FTIR technique was applied to the study of polymer melting and crystallization. Thermal analysis of polymers is a

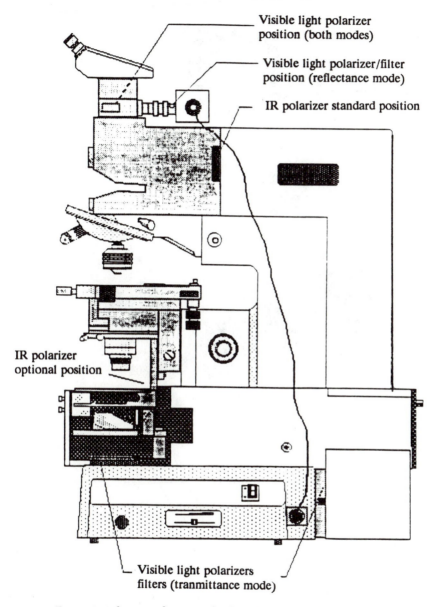

Visible light polarizer
position (both modes)

Visible light polarizer/filter
position (reflectance mode)

IR polarizer standard position

IR polarizer
optional position

Visible light polarizers
filters (tranmittance mode)

Figure 1. Schematic diagram of infrared microsampling accessory.

useful technique for obtaining qualitative and quantitative information. However, the thermal analysis technique is, by its nature, limited by the fact that polymers do not have well-defined melting points or heats of fusion. Real polymer crystals are virtually never in thermodynamic equilibrium but rather are formed on the basis of kinetic factors during crystallization. For this reason polymer crystals may exhibit a large range of melting tempera-

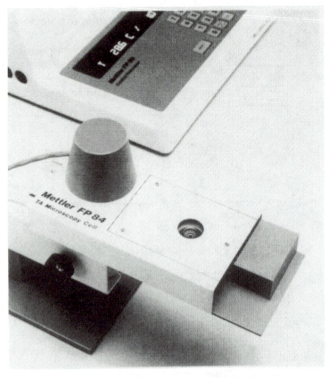

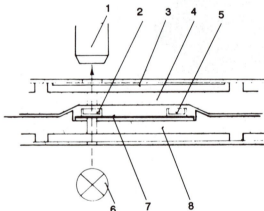

1 Microscope objective
2 Sapphire sample crucible
3 Heat protection filter
4 Metal plate with heating wires
5 Sapphire reference crucible

6 Microscope light source
7 DSC measuring sensor
8 Metal plate with heating wires
   and Pt 100 resistance sensor

Figure 2. Photograph and schematic diagram of DSC microscopy cell. (Reproduced with permission from ref. 1. Copyright 1986 Society for Applied Spectroscopy.)

tures and heats of fusion as a result of the thickness and perfection of crystal lamellae produced under the kinetic conditions of crystallization. These facts may compromise the quantitative value of thermal analysis data and, furthermore, may inhibit the use of such data for qualitative purposes because different polymers may exhibit similar melting endotherms when subjected to particular thermal histories.

Typically, these considerations are realized, and thermal analysis is quite useful in spite of these limitations. However, this situation may be significantly improved by the simultaneous recording of the DSC thermogram and the corresponding infrared spectra as a function of temperature. The melting of polypropylene is a particularly interesting case because of the drastic changes in the infrared spectrum as this transition occurs. The changes in the infrared spectrum have been explained on the basis of particular bands arising from a particular phase, crystalline or noncrystalline (2), or on the basis of particular bands arising from a critical minimum number of monomer units arranged helically and necessary for observation of the band (3, 4).

The melting of a 1-mg sample of polypropylene that had been slowly cooled from the melt over a 5.5-h period was studied by the simultaneous DSC microscopy–FTIR technique. The DSC thermogram is shown in Figure 3, and some of the infrared spectra collected over this temperature range are shown in Figure 4. Changes in the infrared spectra can be first observed in the 110 °C spectrum. This spectrum was taken over the temperature interval from 100 to 110 °C. These changes correspond to the first evidence of heat uptake observed in the DSC thermogram in Figure 3. Large changes in the infrared spectra can be observed in Figure 4 as the melting endotherm is traversed. Careful study of the thermograms and infrared spectra with the use of the simultaneous DSC microscopy–FTIR technique may aid in the

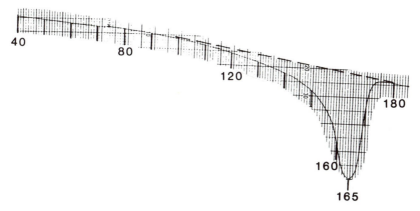

Figure 3. DSC thermogram of polypropylene cooled slowly from the melt over 5.5 h. The program rate was 10 °C/min. (Reproduced with permission from ref. 1. Copyright 1986 Society for Applied Spectroscopy.)

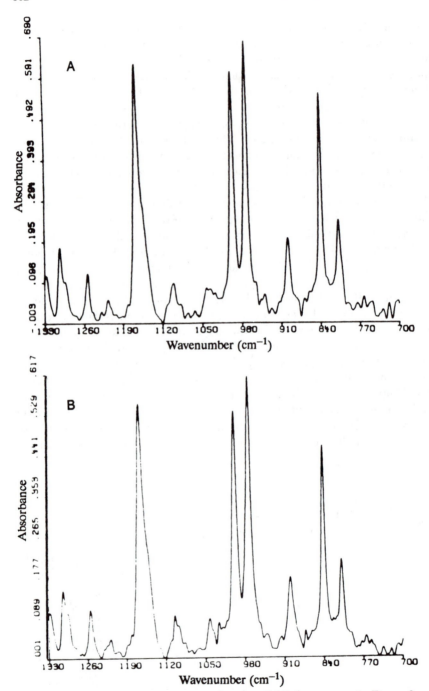

Figure 4. Selected infrared spectra taken over DSC thermogram in Figure 3 at the following temperatures (°C): A, 25; B, 82; C, 110; D, 155; E, 160; and F, 165. (Reproduced with permission from ref. 1. Copyright 1986 Society for Applied Spectroscopy.)

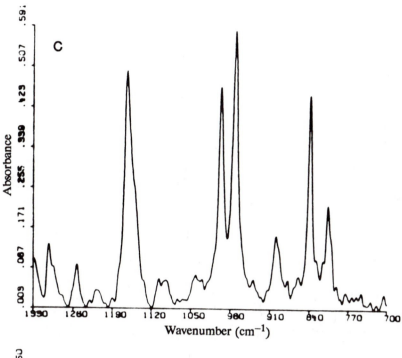

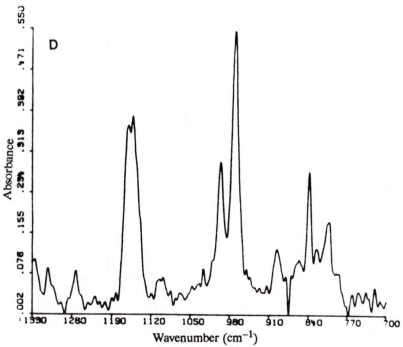

Figure 4. Continued.

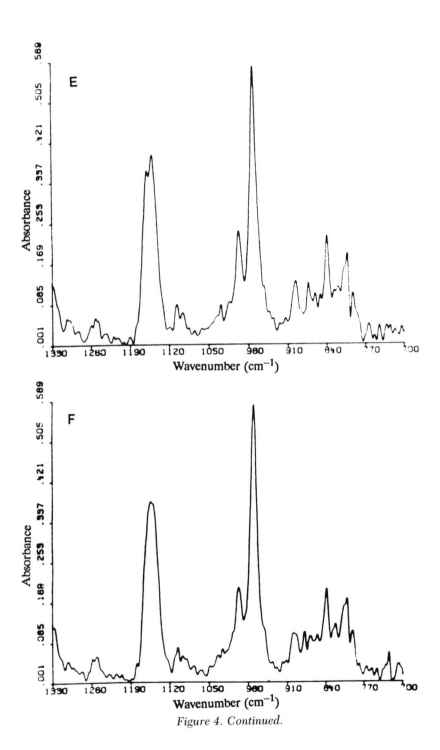

Figure 4. Continued.

explanation of the structural changes that occur in polypropylene during melting.

Polymer blend studies are particularly appropriate for the application of the simultaneous DSC microscopy–FTIR technique. The difficulty in understanding what is happening during crystallization of two blended polymers is especially evident in blends of polyethylene and polypropylene. Figures 5 and 6 show the crystallization exotherms of the polyethylene and polypropylene homopolymers, respectively. The peak crystallization temperatures are similar, and the crystallization exotherm envelopes are largely superimposed on one another.

The crystallization exotherms of the 20:80, 40:60, 60:40 and 80:20 wt% blends of polypropylene–polyethylene are shown in Figures 7–10, respectively. These figures show a single peak for each blend, a result indicating simultaneous crystallization of the two polymers in the blends. Therefore, it is not possible to understand the separate crystallization processes of each polymer in the blends by using the DSC data alone. All of the blends are initially assumed to involve no cocrystallization of polypropylene and polyethylene because these polymers are highly incompatible (5). However, the simultaneous DSC microscopy–FTIR apparatus was used to record infrared spectra over each of the crystallization exotherms in Figures 5–10. Spectral quality was excellent, as shown in Figures 11 and 12. Three spectra are

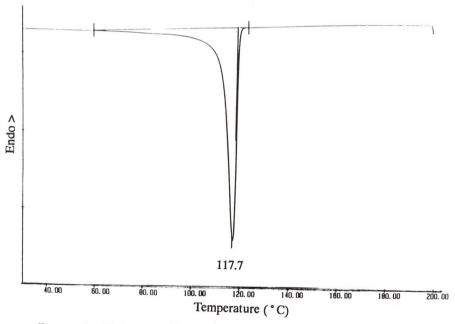

Figure 5. DSC crystallization exotherm for a 0:100 wt% polypropylene–polyethylene blend.

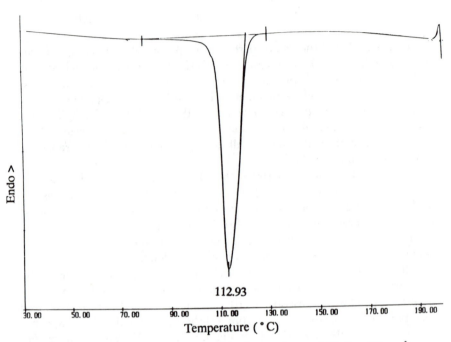

Figure 6. DSC crystallization exotherm for a 100:0 wt% polypropylene–polyethylene blend.

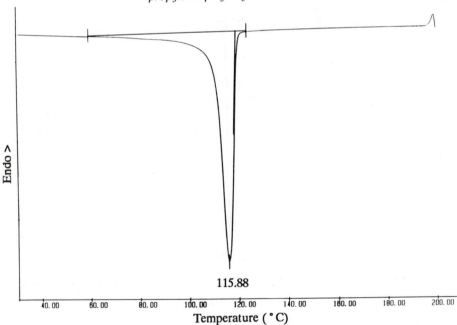

Figure 7. DSC crystallization exotherm for a 20:80 wt% polypropylene–polyethylene blend.

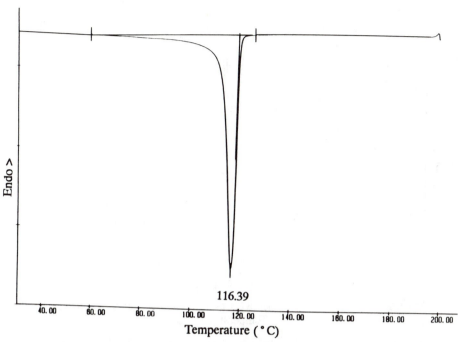

Figure 8. DSC crystallization exotherm for a 40:60 wt% poly-propylene–polyethylene blend.

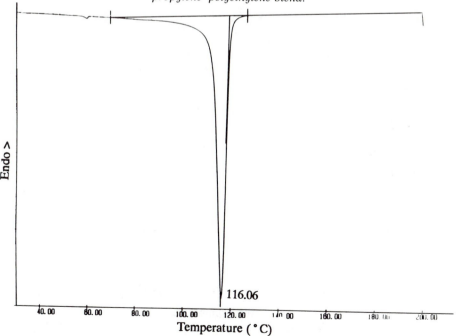

Figure 9. DSC crystallization exotherm for a 60:40 wt% poly-propylene–polyethylene blend.

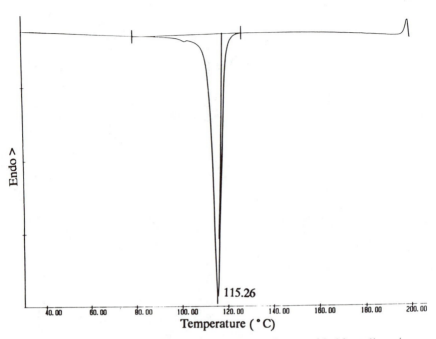

Figure 10. DSC crystallization exotherm for an 80:20 wt% poly-propylene–polyethylene blend.

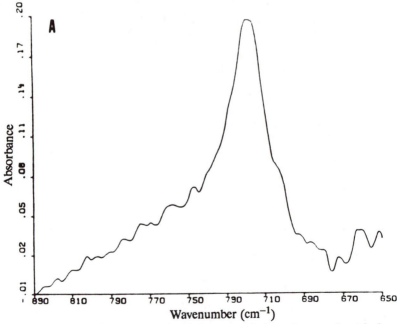

Figure 11. Infrared spectra of polyethylene recorded simultaneously with the DSC crystallization exotherm in Figure 5 at the following temperatures (°C): A, 170; B, 115; and C, 90.

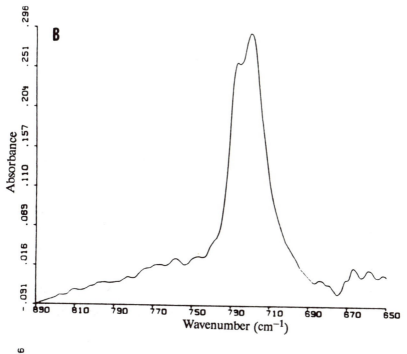

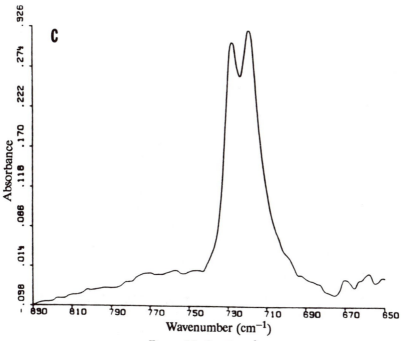

Figure 11. Continued.

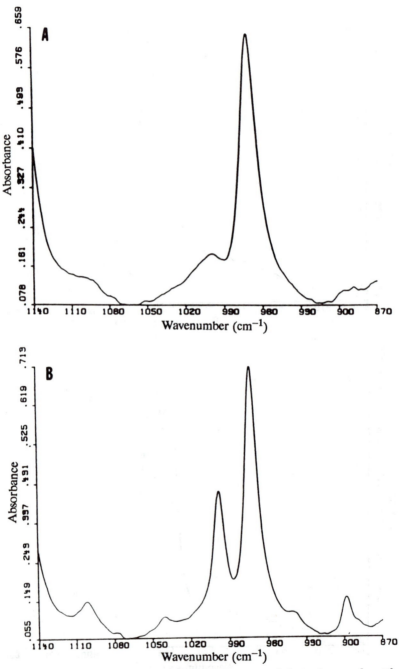

Figure 12. Infrared spectra of polypropylene recorded simultaneously with the DSC crystallization exotherm in Figure 6 at the following temperatures (°C): A, 180; B, 120; and C, 80.

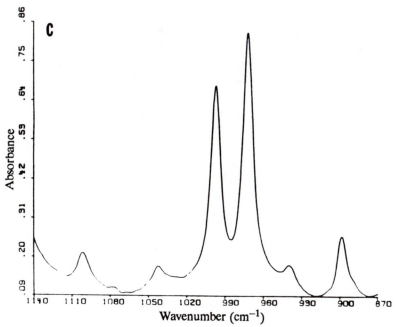

Figure 12. Continued.

shown in each figure during the crystallization of polyethylene and poly-propylene homopolymers at 170, 115, and 90 °C and 180, 120, and 80 °C, respectively.

It is not practical to show the large number of spectra recorded over each one of the crystallization exotherms in Figures 5–10. However, a wealth of information can be gleaned from these spectra, especially concerning the comparison of the crystallization of each polymer separately, as compared to crystallization of each polymer in the presence of the other in the blends. One of the methods used to analyze the data was to identify a crystalline- and noncrystalline-phase infrared band for polyethylene and polypropylene. The bands chosen are as follows:

Polymer	Infrared Band (cm^{-1})	Phase
Polyethylene	731	crystalline
	720	noncrystalline
Polypropylene	998	crystalline
	973	noncrystalline

The ratios of the 731- to 720-cm^{-1} bands in polyethylene and the 998–973-cm^{-1} bands in polypropylene were used as measures of the relative proportion of crystalline and noncrystalline material at any particular temperature. The infrared absorbance ratios A_{730}/A_{720} and A_{998}/A_{973} for polyethylene and polypropylene, respectively, are plotted in Figure 13 for the crystallization of each polymer, as shown in Figures 5 and 6, respectively. As each of these ratios increases in magnitude, the proportion of crystalline material is increasing. The ratios are seen in Figure 13 to increase rapidly

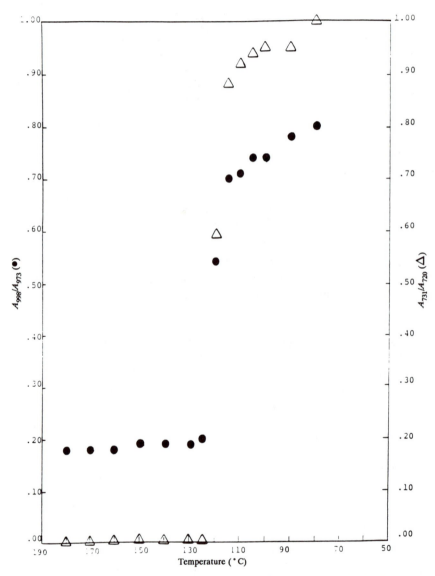

Figure 13. Absorbance ratios for polypropylene ($\bullet$, A_{998}/A_{973}) and polyethylene ($\triangle$, A_{731}/A_{720}) versus temperature during crystallization of each polymer (corresponding to Figures 6 and 5, respectively).

between 115 and 120 °C for polyethylene and polypropylene, in agreement with the peak crystallization temperatures in Figures 5 and 6, respectively.

Similar plots can be drawn for each of the blend compositions in which the absorbance ratios for each polymer can be plotted as a function of temperature. These plots are shown in Figures 14 and 15 for polypropylene and polyethylene, respectively. Figure 14 is a plot of each of the blends and 100% polypropylene showing the A_{998}/A_{973} ratio versus temperature, which

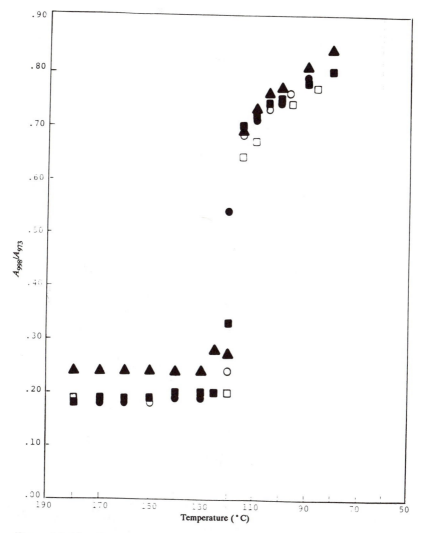

Figure 14. Absorbance ratios for polypropylene (A_{998}/A_{973}) versus temperature for the crystallization of blends of polypropylene with 0 (Δ), 20 ($\blacktriangle$), 40 ($\square$), 60 ($\blacksquare$), and 80 ($\bigcirc$) % polyethylene.

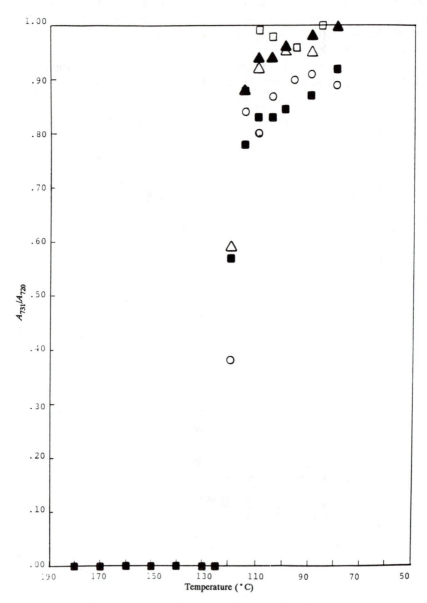

Figure 15. Absorbance ratios for polyethylene (A_{731}/A_{720}) versus temperature for the crystallization of blends of polypropylene with 0 (●), 20 (○), 40 (■), 60 (□), and 80 (▲)% polypropylene.

indicates the crystallization of polypropylene. Figure 15 is a plot of each of the blends and 100% polyethylene showing the A_{731}/A_{720} ratio versus temperature, which indicates the crystallization of polyethylene. Figures 14 and 15 indicate that one of the polymers has no effect on the crystallization of the other polymer in the blends. That is, each polymer crystallizes in the same 115–120 °C temperature range whether the polymer is in a blend or not. A slower cooling rate and more detailed analysis of the crystallization in this narrow temperature range might yield more subtle effects in the blends. Furthermore, comparisons of this type can yield separate crystallization rates and crystallization temperature profiles for each polymer in the blends.

Acknowledgments

I am grateful to Timothy Minvielle for collecting the simultaneous DSC–FTIR data and to Marlene Moellenkamp for typing the manuscript.

References

1. Mirabella, F. M. *Appl. Spectrosc.* **1986**, *40*(*3*), 417.
2. Samuels, R. J. *Makromol. Chem. Suppl.* **1981**, *4*, 241.
3. Kissin, Y. V.; Rishina, L. *Eur. Polym. J.* **1976**, *12*, 757.
4. Glotin, M.; Rahalkar, R. R.; Hendra, P. J.; Cudby, M. E. A.; Willis, H. A. *Polymer* **1981**, *22*, 731.
5. Krause, S. In *Polymer Blends;* Paul, D. R.; Newman, S., Eds.; Academic: New York, 1978; "Polymer–Polymer Compatibility," Chapter 2.

Received for review February 14, 1989. Accepted revised manuscript August 14, 1989.

22

Characterization of Crystalline Polymers by Raman Spectroscopy and Differential Scanning Calorimetry

Leo Mandelkern

Department of Chemistry and Institute of Molecular Biophysics, Florida State University, Tallahassee, FL 32306

Semicrystalline polymers formed by flexible chain molecules display complex structures and morphology. A set of independent variables governed by the molecular constitution and the crystallization conditions has been identified. These variables control microscopic as well as macroscopic properties. To understand the broad range of properties of such crystalline systems, a wide variety of experimental techniques has to be used to measure these structural variables and to comprehend the wide range in values that these systems are known to assume. Two such techniques are differential scanning calorimetry and Raman spectroscopy. In this chapter we illustrate, by a selected set of examples, how these two techniques can be used to establish melting temperatures of thermodynamic significance, to quantitatively describe the phase structure, and to elucidate the nature of the interlamellar regions of semicrystalline polymers.

POLYMERS CAN CRYSTALLIZE IN A REASONABLE TIME only at temperatures that are well below their melting points. In addition, the crystallization process is rarely, if ever, complete. Hence, a nonequilibrium state develops. Consequently, when a polymer system crystallizes from the pure melt, a morphologically complex polycrystalline state is formed. Both the macroscopic and microscopic properties of this semicrystalline state will depend on the structural and morphological features that define the state. Therefore, the key independent structural variables that describe the crystalline state

0065–2393/90/0227–0377$06.00/0

must be identified and described. A set of independent structural variables, previously identified (1), either individually or in particular combinations contribute to and control a specific property. The variables that have been identified so far in this connection are the level, or degree, of crystallinity; the structure of the residual noncrystalline region; the crystallite thickness distribution; the structure and relative amount of the interfacial region; the crystallite structure; the supermolecular structure; and the melting temperature.

Because the transformation from the liquid to the crystalline state is not complete, the system can be characterized by the degree of crystallinity or extent of transformation that exists under given conditions. Understanding the detailed nature of the phase structure that has developed thus becomes a very important matter. After the initial discovery of the lamellarlike crystallite habit, the view was expounded that the concept of the degree of crystallinity was invalid. Instead, the system was proposed (2–5) to consist of a crystalline matrix in which defects were randomly interspersed. More modern theoretical and experimental studies have banished this concept to obscurity. As will be shown, the concept of the degree of crystallinity can be placed on a rigorous and quantitative basis.

The noncrystalline portions of a system comprise two major regions. The interfacial region is associated with the basal plane of the lamellar crystallites, and the interlamellar region reflects that portion of the system where the chain units connect crystallites. It is now recognized that the interfacial region is not made up of regularly folded chains with an occasional, rare deviation from this regularity. Nor does this region result from the nucleation of regular folded chains that grow to mature crystallites of the same form (6). The problem and structural details that are primarily concerned with the interfacial structure involve the distribution of units that return to the crystallite of origin and the amount and extent of this region. Quantitatively defining this structure is very important because it defines many properties. Theoretical advances have, however, recently been made in resolving this problem (7–11). In the interlamellar region, the chain units that connect the crystallites are in random conformation so that the region is isotropic. Within the general concept of isotropy, however, a variety of detailed structures can exist. These structures need to be quantitatively specified in the future.

Although lamellar crystallites are the typical form for homopolymers over a wide range of molecular weights (12–15) as well as for random copolymers of surprisingly high co-unit content (16), other factors of crystallite structure are of importance. These include the chain tilt, the curvature of the lamellae, and their lateral extent. The supermolecular structure represents the arrangement of the lamellar crystallites into some type of higher organization. Spherulites, for example, are one type of supermolecular structure. The nature of the superstructure that forms depends on many factors, including the molecular weight, the crystallization temperature, and the chain structure (17–21). One important conclusion that has resulted from studies of the supermolecular structure under controlled conditions is that spherulites are not a universal mode of polymer crystallization. This point has been demonstrated from studies of linear polyethylene (17), branched polyethylene (18, 19), polyethylene copolymers (20), polyethylene oxide (21), and isotactic polypropylene (22, 23). The crystallite thickness distribution involves the determination of the crystallite size in the chain direction.

Because of the wide variety of structural variables that are pertinent to the problem, many different experimental techniques are needed to understand properties.

This chapter will focus attention on two such techniques, namely Raman spectroscopy and differential scanning calorimetry. The examples taken to illustrate various aspects of the problem will be from the polyethylenes. These polymers, with their wide range of molecular weights and structures, serve as very good models for crystalline systems of flexible chains.

Differential scanning calorimetry (DSC) measurements give the melting temperature, the enthalpy of fusion, which leads to the degree of crystallinity, the glass temperature, and, of course, the specific heat. The uses of Raman spectroscopy for particular applications are as follows:

- Internal modes for phase structure (24–26)
- Longitudinal acoustical mode (LAM) for crystallite thickness distribution (27–34)
- D-LAM for long-range conformation disorder (35–38)

The internal modes, which give us the phase structure, lie in the region of approximately $900\text{–}1500 \text{ cm}^{-1}$. The LAM (longitudinal acoustical mode), which gives us the crystallite thickness distribution, lies in the range of about $5\text{–}20 \text{ cm}^{-1}$. The D-LAM, which gives us a measure of the long-range conformational disorder, is in the range of 200 cm^{-1}.

We shall apply these two techniques to several selected examples. Many more examples could be discussed even when restricting the discussion to just these two methods. Raman spectroscopy and differential scanning calorimetry represent only a portion of the experimental techniques required to completely describe the properties and structure of semicrystalline polymers.

Discussion

Melting Temperature.   The first example of these techniques is the determination of the melting temperature (39), which is very often a straightforward process: determining the endothermic peak in the DSC. However, many cases present the problem of melting–recrystallization. Very often this phenomenon is not detected, so that the differential scan gives a very misleading melting temperature that results in major problems in a detailed analysis. An example of this problem is given in Figure 1, which is a thermogram for a dilute solution formed crystal of polyethylene (39). From the peak in the endotherm, the melting temperature is expected to be close to 130 °C. However, in this case, there is a strong suspicion that this value is too high because the melting endotherm is found to be independent of crystallite thickness. Hence, a melting–recrystallization process is probably occurring (39, 40).

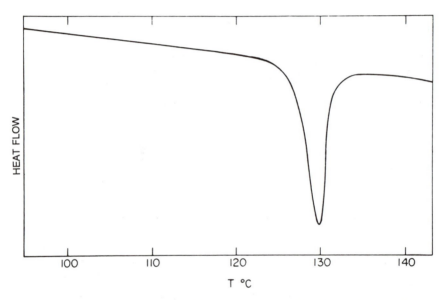

Figure 1. DSC thermogram for dilute solution crystallized polyethylene (M_w = 1.66 × 10⁵). Conditions: heating rate, 5 K/min; range, 2 mcal/s; and sample mass, 0.72 mg. (Reproduced with permission from reference 39. Copyright 1986 John Wiley & Sons.)

To ascertain if this conjecture is correct, in that we are not observing the melting point characteristic of the initial sample, we studied the distribution of crystallite thicknesses by monitoring the Raman LAM as a function of temperature. Figure 2 gives a normalized plot of the ordered sequence lengths against the length as derived from the spectra (39). For the initial sample, and for those heated to 112 °C, the size distribution stays constant, is very narrow, and is centered at about 130 Å. As the annealing temperature is raised to higher temperatures, 120.5 °C and above, major changes are observed in the LAM, as is reflected in the curves given in Figure 2. After heating at 120.5 °C, a bimodal crystallite thickness distribution develops.

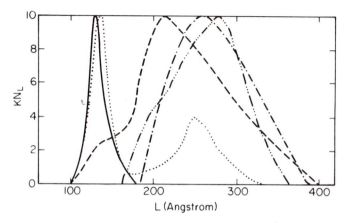

Figure 2. Normalized plot of number of ordered sequences, KN_L, of length L against L as derived from the spectra of sample illustrated in Figure 1. Key: —, original sample; • • •, T_a = 120.5; --, T_a = 123.5; –•–, T_a = 125; –• • •–, T_a = 150 °C. (T_a is the annealing temperature.) (Reproduced with permission from reference 39. Copyright 1986 John Wiley & Sons.)

One population corresponds to the initial crystallite size distribution, the other to crystallites that were formed from the pure melt. These results lend themselves to the straightforward interpretation that partial melting and recrystallization from the melt has taken place. After the sample is annealed at 123.5 °C, the resulting size distribution indicates that almost all of the sample has melted and recrystallized under this condition. Only a very small concentration of the original crystallites remains. After the sample is annealed at 125 °C, the size distribution is essentially the same as after crystallization from the pure melt, that is, from 150 °C. Therefore, the melting of the original crystallites is complete at 125 °C, and the fusion process is well under way at 120.5 °C. The melting temperature of this sample is in the restricted range of 123.5–125 °C and is not at 130 °C, as might have been interpreted from the initial thermogram. We must therefore be careful that the thermogram does indeed represent the melting of the crystallites that are initially formed if we are interested in melting temperatures of thermodynamic significance.

Phase Structure. The phase structures of semicrystalline polymers are of primary importance in understanding their properties. Three distinct regions are involved:

1. the ordered crystalline region, which, for polyethylene, represents the orthorhombic unit cell structure

2. the liquidlike, disordered region having an isotropic structure

3. the interfacial region, which involves chain units that connect these two different structural regions

Thus, a chain can traverse all three regions or be restricted to the crystalline and interfacial ones. Flory pointed out in 1949 (*41*) that a sharp demarcation between the ordered crystalline region and the random liquidlike region cannot in general take place because of the spatial requirements of chains emerging from the crystallite. This concept has been theoretically developed in more detail (*7–11, 42*).

Several different kinds of measurements can be used to describe the phase structure:

Descriptive Variable	Definition	Type of Measurement
$(1 - \lambda)_d$	degree of crystallinity	density
$(1 - \lambda)_{\Delta H}$	degree of crystallinity	enthalpy of fusion
α_c	degree of crystallinity	Raman internal modes
α_a	degree of liquidlike material	Raman internal modes
$\alpha_a + \alpha_c \neq 1$ $\alpha_b \equiv 1 - (\alpha_a + \alpha_c)$	degree of interfacial region	

Those of concern here are the density, the enthalpy of fusion, and the Raman internal modes. The analysis of the Raman internal modes was originally given by Strobl and Hagedorn (*24*). Here, α_c is the core degree of crystallinity because the only contributions to this quantity come from the structure of the orthorhombic unit cell. It is calculated from the CH_2 bending band at 1416 cm^{-1} and is the component of this vibration that is split by the crystal field. The degree of liquidlike material, α_a, is obtained from the twisting region at 1303 cm^{-1}. The total integrated intensity of the twisting region, 1295 to 1303 cm^{-1}, is independent of the phase structure (*24*). Methods of analyzing the spectra were given in detail by Strobl and Hagedorn (*24*) as well as reports from this laboratory (*25, 26*). It is found experimentally that the sum of $\alpha_a + \alpha_c$ does not equal 1. Put another way, as has been pointed out by Strobl and Hagedorn, the spectra of the completely liquidlike and the completely crystalline polymers cannot be superposed to represent any actual spectra. A partially ordered, primarily trans, anisotropic region must also be included. This contribution has been defined as the interfacial region. With this description of the phase structure, we can proceed with analyzing experimental data.

Figures 3 and 4 represent an extensive compilation of data for linear and branched polyethylene, respectively. In each case the degree of crystallinity determined from density, $(1 - \lambda)_d$, is plotted as a function of the corresponding quantity, $(1 - \lambda)_{\Delta H}$, as determined from the enthalpy of fusion. Each of these quantities was obtained in the conventional manner. Although the degree of crystallinity determined by these methods is a quantitative concept and the different methods display the same functional behavior with respect to molecular constitution and crystallization conditions, comparison of Figures 3 and 4 shows that there are significant quantitative differences between the two methods. The compilation of data demonstrates quite unequivocally that $(1 - \lambda)_d$ is always greater than $(1 - \lambda)_{\Delta H}$. In contrast, the compilation in Figure 5 shows that the core crystallinity α_c is equal to $(1 - \lambda)_{\Delta H}$. The points for the linear polymers in this plot fall on the 45° straight line over an accessible range in crystallinity level of 0.4–0.9. Thus, very good agreement is indicated between these two methods in measuring the crystallinity level.

On the other hand, for the branched polymers and copolymers, α_c is

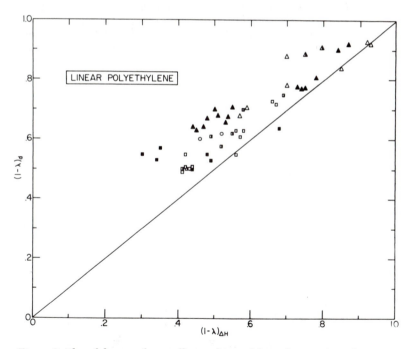

Figure 3. Plot of degree of crystallinity obtained from density, $(1 - \lambda)_d$, against value obtained from the enthalpy of fusion, $(1 - \lambda)_{\Delta H}$, for linear polyethylene. (Reproduced with permission from reference 1. Copyright 1985 Society of Polymer Science, Japan.)

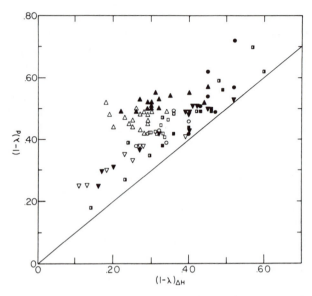

Figure 4. Plot of degree of crystallinity obtained from density, $(1 - \lambda)_d$, against value obtained from the enthalpy of fusion, $(1 - \lambda)_{\Delta H}$, for copolymers and branched polyethylene. (Reproduced with permission from reference 1. Copyright 1985 Society of Polymer Science, Japan.)

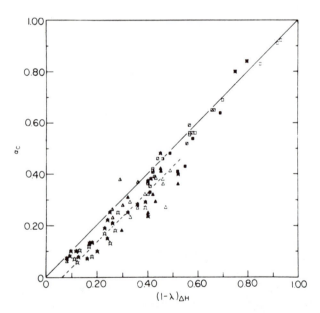

Figure 5. Plot of degree of crystallinity, α_c, as determined from the Raman internal modes, against $(1 - \lambda)_{\Delta H}$ for linear (—) and branched (--) polyethylene and ethylene copolymers. (Reproduced with permission from reference 1. Copyright 1985 Society of Polymer Science, Japan.)

about 5% less than $(1 - \lambda)_{\Delta H}$. This small disparity can be attributed to the broad melting range of the structurally irregular chains because $(1 - \lambda)_{\Delta H}$ includes the contribution of a small amount of crystallinity that has already disappeared at room temperature. Because α_c is measured at ambient temperature, this contribution is not included in the internal mode.

The basis for the discrepancy between the density and enthalpy of fusion determination of the level of crystallinity can be found in the results of Figure 6. Here $(1 - \lambda)_d$ is plotted against the sum $(\alpha_c + \alpha_b)$ and all the data fall on the 45° straight line, a result indicating a one-to-one correspondence between the two quantities. Because α_c measures the core crystallinity and α_b the interfacial content, the density measures both the core crystallinity and the partially ordered anisotropic interfacial region. On the other hand, the enthalpy of fusion measures only the core crystallinity.

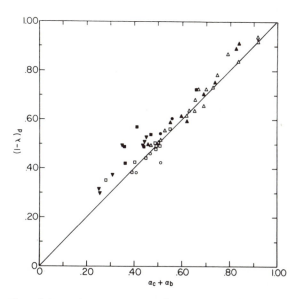

Figure 6. Plot of $(1 - \lambda)_d$ against sum of $\alpha_c + \alpha_b$ for linear and branched copolymers and ethylene copolymers. (Reproduced with permission from reference 1. Copyright 1985 Society of Polymer Science, Japan.)

The next logical question is how the phase structure is influenced by the molecular constitution. The level of crystallinity, as determined from either density or enthalpy of fusion, for linear polymers is very dependent on molecular weight (*43, 44*). For example, for linear polyethylene the level of crystallinity systematically varies from about 80% to 30% with increasing molecular weight and changes in the crystallization conditions. This rather strong influence of molecular weight on the level of crystallinity is also found in poly(ethylene oxide), poly(tetramethylphenylenesiloxane) (TMPS) and *cis-*

and *trans*-polyisoprene. It can be considered to be a general characteristic of homopolymers.

The interfacial content (α_b) of linear polyethylene is also very dependent on chain length, as is indicated in Figure 7 for rapidly crystallized molecular-weight fractions (45). At low molecular weights, the interfacial content is relatively small. However, an appreciable interfacial content, on the order of 15–20%, is observed at much higher molecular weights. This change in interfacial content with molecular weight parallels the variation in the interfacial free energy that is associated with the basal plane of the lamellae (44). This result is important in explaining certain properties (46). This result, as well as the changes in the crystallinity level with molecular weight, makes quite obvious the fact that the crystallite structure cannot involve regularly folded chains.

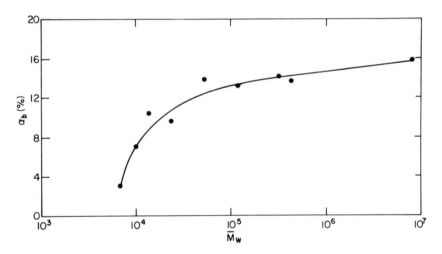

Figure 7. Plot of interfacial content, α_b, against weight-average molecular weight for rapidly crystallized fractions of linear polyethylene. (Reproduced with permission from reference 45. Copyright 1988 Elsevier Science Publishers.)

The phase structure of random copolymers depends not only on the molecular weight but also on the co-unit content and conceivably on the chemical nature of the co-unit itself. Figure 8 is a plot of the core crystallinity, as determined from the Raman internal modes, against the mole fraction of branch points for a series of random ethylene copolymers. The copolymers represented here are either compositional and molecular-weight fractions or samples of narrow composition distribution having a most probable molecular-weight distribution. The co-units include ethyl, vinyl acetate, butyl, and hexyl branches (47, 48). These types of branches do not enter the crystal lattice (47, 49). This figure clearly shows that the introduction of the non-

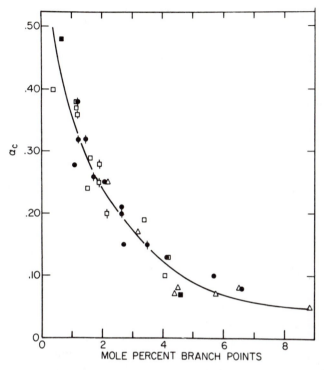

Figure 8. Plot of degree of crystallinity, α_c, calculated from Raman internal modes, against mole percent branch points. Key: $\triangle$, hydrogenated polybutadiene; •, ethylene–vinyl acetate; □, ethylene–butene; ⧫, ethylene–hexene; □, ethylene–octene. (Reproduced from reference 48. Copyright 1989 American Chemical Society.)

crystallizing co-units into the chain leads to a very rapid and continuing decrease in the crystallinity level with increasing side-group content. The level of the core crystallinity can vary from about 48% for 0.5 mol% of branches to about 7% for 6 mol% of branches. The chemical nature of these specific branch types has virtually no influence on the crystallinity value for a given co-unit content.

For those cases where the branches enter the lattice, such as CH_3 and Cl, higher levels of crystallinity will be observed. The samples studied in Figure 8 were chosen to represent essentially the same molecular-weight range. A detailed study (48) previously showed that the values $(1 - \lambda)_{\Delta H}$ and α_c are very close to each other, as would be expected from the previous discussion of linear polymers and copolymers. The changes in α_c with molecular weight follow a pattern that is similar to that of homopolymers, although the level of crystallinity is much reduced (50).

The interfacial content of random copolymers at a fixed co-unit content is essentially independent of molecular weight, as is indicated in Figure 9

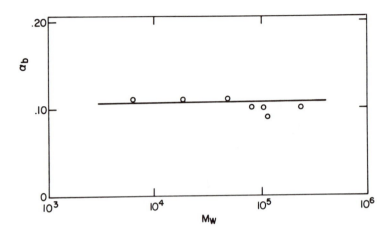

Figure 9. Plot of the fraction of the interfacial region, α_b, against the weight-average molecular weight for ethylene–hexene copolymers. (Reproduced from reference 48. Copyright 1989 American Chemical Society.)

for a series of ethylene–hexene copolymers having a most probable molecular-weight distribution and branch contents that range from 1.2 to 1.7 mol%. This figure shows an invariance in α_b with molecular weight. Similar results have also been found for a series of hydrogenated polybutadienes (randomly ethyl-branched ethylene copolymers) of slightly higher branch content (50). This result is quite different from that found for homopolymers. As was indicated in Figure 7, the interfacial content of linear polyethylene increases substantially with molecular weight. Although α_b is invariant with molecular weight, a continuous decrease in the core crystallinity occurs with increasing chain length. A very important point is that the changes in phase structure of linear polyethylene and its copolymers can be controlled and therefore can be used advantageously in the study of macroscopic properties of crystalline polymers.

The interfacial content (α_b) is a very strong function of the co-unit content, as is indicated in Figure 10 (47). The data compiled in this figure scatter somewhat because α_b involves the difference between two larger quantities. Much smoother data for α_b also were presented (48). However, for present purposes the data of Figure 10 suffice. Some very definite trends can be observed in Figure 10. The diazo copolymers have somewhat higher interfacial contents because of their much higher molecular weights (51). A very definite monotonic increase occurs in the interfacial content with increasing branching content. The value of α_b approaches 15–20% at the same time α_c is reduced to as low as 25–30%. Therefore, the interfacial region can represent an appreciable portion of the system. It deserves serious attention for higher-molecular-weight linear polymers and for copolymers. This quantity cannot be neglected in consideration of properties. Another important

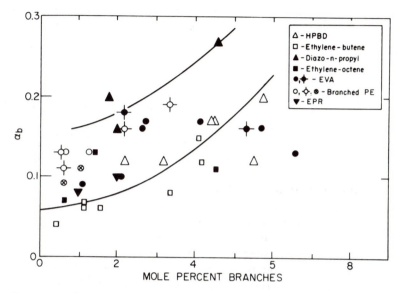

Figure 10. Plot of the fraction of the interfacial region, α_b, against the mole percent branch points for the indicated ethylene copolymer. (Reproduced from reference 48. Copyright 1989 American Chemical Society.)

phenomenon that still needs to be investigated is whether the details of the interfacial structure change as the chemical nature of the co-unit is varied. As yet, not enough data are available to make a definitive statement in regard to this important problem, which may also have serious consequences in terms of properties (48).

The preceding examples show that the important elements of phase structure can be varied over a very wide range by control of molecular weight and the structural regularity of a chain. These ideas can be applied to a study of properties by isolating the variables and seeing how individual elements of structure influence a particular property (1, 17, 46, 52–55). Of course, the other independent structural variables described earlier must also be taken into account. This approach to understanding properties can be applied to a variety of phenomena ranging from thermodynamic to mechanical. More recently, this approach was shown (54, 55) to lead to an understanding of large-range deformation properties as observed in experimental force–length relations of crystalline polymers. One example is found in the analysis of the yield stress that has been well established to be a linear function of crystallinity (54). However, careful examination in terms of the details of the phase structure reveals that when the yield stress is plotted against the density-determined level of crystallinity, the data do not extrapolate to the origin; that is, the yield stress does not become zero at zero crystallinity. However, when the same data are plotted against the core crystallinity, α_c, the result is a straight line that extrapolates very nicely to the origin. From this analysis, therefore, we can conclude that the core crystallinity is the major contributor to the yield stress and that the interfacial region does not have an important influence on this property. A more detailed analysis of the mechanical properties in terms of the phase structure and other independent variables was published elsewhere (55).

The analysis just outlined shows how the Raman internal modes and the enthalpy of fusion, as well as the density, give different but complementary results with respect to the phase structure and point out the structural features that play important roles in properties.

Interlamellar Structure. The basic crystallite of a homopolymer is of a lamellar habit and is observed over a wide molecular-weight range. As was pointed out earlier, the transformation from the melt is far from complete, and a substantial reduction in the level of crystallinity can be achieved depending on molecular weight and the structural regularity of the chains, as well as the crystallization conditions. Because this reduction in crystallinity can be substantial, the structure and amount of the interlamellar region is very important. We have already discussed the interfacial structure and are concerned at this point with noncrystalline chain units that connect the lamellae. Our analysis of the Raman internal modes indicated that this interlamellar region is liquidlike in character. Other physical measurements have substantiated this conclusion (44). All these conclusions have been essentially indirect. One method that directly examines the disordered liquidlike region is the D-LAM in the Raman spectra. This method has been pioneered by Snyder and co-workers (35, 36).

The D-LAM is a low-frequency Raman band that measures long-range conformational disorder. It has been observed in all noncrystalline polymers studied, including molten polyethylene (35, 36). It can be related to the average of the conformational disorder in a long-chain molecule (35). A basic question is whether the D-LAM band is found in semicrystalline polymers and, if so, what is its significance (37).

Figure 11 is a low-frequency Raman spectrum of molten polyethylene obtained at 150 °C. Here a well-defined band that is observed in the vicinity of 200 cm^{-1} is the D-LAM band for polyethylene. Figure 12 is the Raman spectrum at ambient temperature for a linear polyethylene with $(1 - \lambda)_{\Delta H}$ of 42%. Again, a broad, well-defined band in the vicinity of 200 cm^{-1} is observed. Two less intense bands near 350 and 410 cm^{-1} are also found in the semicrystalline polyethylenes, and, like the D-LAM band, they are associated with disordered chains. However, these bands are not of concern in the present context. A band identical to that of the molten polymer is observed in a semicrystalline system. Figure 13 also shows the changes in this band as a function of temperature. Here, the spectra were obtained at successively increasing temperatures from room temperature to 147 °C. As

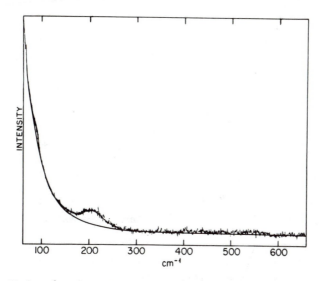

Figure 11. Low-frequency Raman spectrum of molten linear polyethylene at 150 °C. (Reproduced from reference 37. Copyright 1986 American Chemical Society.)

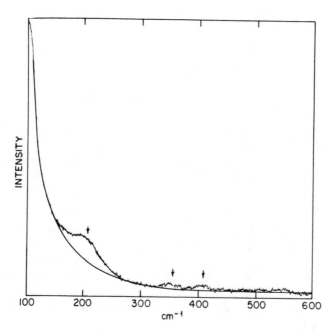

Figure 12. Low-frequency Raman spectrum of semicrystalline linear polyethylene $(1 - \lambda)_{\Delta H} = 0.42$. *(Reproduced from reference 37. Copyright 1986 American Chemical Society.)*

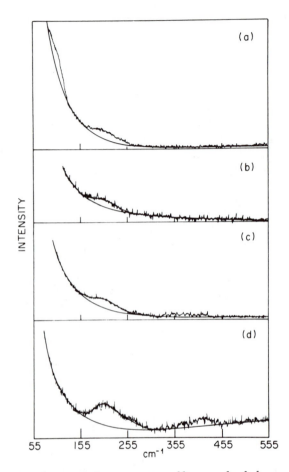

Figure 13. Low-frequency Raman spectra of linear polyethylene as a function of temperature. Initial sample $(1 - \lambda)_{\Delta H} = 0.42$. Spectra were obtained at (a) room temperature, (b) 57 °C, (c) 87 °C, and (d) 147 °C. (Reproduced from reference 37. Copyright 1986 American Chemical Society.)

the temperature of the sample is increased, the intensity of the D-LAM band also increases while the frequency of the band remains constant. In the melt at 147 °C, the D-LAM band is much more intense but is still found at the same frequency. There is no indication of any discontinuity in the position or shape of the D-LAM band in going from the crystalline state to the melt, a finding that is further evidence that in the crystalline polymer, the disordered regions have the same structural features as that of the pure liquid polymer.

It follows that the intensity of the D-LAM band for crystalline polyethylene should be closely related to the crystallinity of the sample. A comparison of these quantities is made in Figure 14 for samples having varying

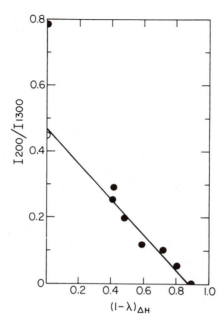

Figure 14. Relative intensity of the D-LAM band of linear polyethylene plotted against the degree of crystallinity $(1 - \lambda)_{\Delta H}$. *Solid circles are the experimental values. The open circle is the corrected value for the completely molten polymer. (Reproduced from reference 37. Copyright 1986 American Chemical Society.)*

levels of crystallinity at room temperature. In this figure the relative intensity of the D-LAM band is plotted against the calorimetrically determined crystallinity for a series of polyethylene samples whose crystallinities vary from about 90% to 0%. The solid circle at zero degree crystallinity is for the molten polymer at 150 °C. The data at ambient temperature are well represented by a straight line that extrapolates to a value for the completely amorphous polymer, which appears to be substantially lower than the value for the molten polymer at 150 °C. However, corrections for sample transparency and temperature must be applied to compare the result for the molten sample at 150 °C with the crystalline sample at ambient temperature. When these corrections are made (37), the open circle in Figure 14, corresponding to $(1 - \lambda)_{\Delta H} = 0$, falls on the straight line drawn through the experimental points.

From the plot in Figure 14, the general conclusion is that the fraction of the sample consisting of highly disordered chains, as measured by the D-LAM intensity, is linearly related to the degree of crystallinity. The D-LAM band could not be detected in the spectrum of a sample that was 90% crystalline and is consistent with the other data plotted. The absence of the D-LAM band for this sample indicates that at extremely high levels of crystallinity, the conformational structure of the disordered regions deviates from that of polyethylene in the liquid state. Under these conditions, the disordered sequence will be shorter, under strain, and influenced by the interfacial structure. Thus, the trans–gauche ratio in this region will be changed.

Summary

By means of a few limited and selected examples, we have tried to show that Raman spectroscopy and differential scanning calorimetry can be an effective combination to delineate many aspects of phase structure, as well as other features, of crystalline polymers composed of flexible chains. The understanding thus developed can then be applied to the study of macroscopic properties, as has also been indicated in a few examples. However, despite the effectiveness of these two experimental techniques in resolving many problems, a much broader repertoire of methods must be applied for an understanding of the wide variety of properties of interest.

Acknowledgments

Support of this work by the National Science Foundation Polymers Program Grant DMR 86–13007 is gratefully acknowledged.

References

1. Mandelkern, L. *Polym. J.* **1985**, *17*, 337.
2. Geil, P. *Polymer Single Crystals;* Wiley Interscience: New York, 1963.
3. Stuart, H. A. *Ann. N.Y. Acad. Sci.* **1959**, *83*, 3.
4. Lindenmeyer, P. H. *Science* **1956**, *147*, 1256.
5. Hoffman, J. D. *Soc. Plast. Eng. Trans.* **1964**, *4*, 315.
6. Hoffman, J. D.; Lauritzen, J. I. *J. Res. Natl. Bur. Stand. Sect. A* **1960**, *64*, 73; ibid. **1961**, *65*, 297.
7. Mansfield, M. L. *Macromolecules* **1983**, *16*, 914.
8. Flory, P. J.; Yoon, D. Y.; Dill, K. A. *Macromolecules* **1984**, *17*, 862.
9. Yoon, D. Y.; Flory, P. J. *Macromolecules* **1984**, *17*, 868.
10. Marqusee, J. A.; Dill, K. A. *Macromolecules* **1986**, *19*, 2420.
11. Kumar, S. K.; Yoon, D. Y. *Bull. Am. Phys. Soc.* **1988**, *33*, 249.
12. Voigt-Martin, I. G.; Fischer, E. W.; Mandelkern, L. *J. Polym. Sci. Polym. Phys. Ed.* **1980**, *18*, 2347.
13. Voigt-Martin, I. G.; Mandelkern, L. *J. Polym. Sci. Polym. Phys. Ed.* **1981**, *19*, 1769.
14. Stack, G. M.; Mandelkern, L.; Voigt-Martin, I. G. *Macromolecules* **1984**, *17*, 321.
15. Voigt-Martin, I. G.; Mandelkern, L. *J. Polym. Sci. Polym. Phys. Ed.* **1984**, *22*, 1901.
16. Voigt-Martin, I. G.; Alamo, R.; Mandelkern, L. *J. Polym. Sci. Polym. Phys. Ed.* **1986**, *24*, 1283.
17. Maxfield, J.; Mandelkern, L. *Macromolecules* **1977**, *10*, 1141.
18. Mandelkern, L.; Maxfield, J. *J. Polym. Sci. Polym. Phys. Ed.* **1979**, *17*, 1913.

19. Mandelkern, L.; Glotin, M.; Benson, R. S. *Macromolecules* **1981**, *14*, 22.
20. Glotin, M.; Mandelkern, L. *Macromolecules* **1981**, *14*, 1394.
21. Allen, R. C.; Mandelkern, L. *J. Polym. Sci. Polym. Phys. Ed.* **1982**, *20*, 1465.
22. Allen, R. C.; Mandelkern, L. *Polym. Bull.* **1987**, *17*, 473.
23. Allen, R. C., Ph.D. Dissertation, School of Materials Science and Engineering, Virginia Polytechnic Institute and State University, Blacksburg, VA, 1981.
24. Strobl, G. R.; Hagedorn, W. *J. Polym. Sci. Polym. Phys. Ed.* **1978**, *16*, 1181.
25. Glotin, M.; Mandelkern, L. *Coll. Polym. Sci.* **1982**, *260*, 182.
26. Mandelkern, L.; Peacock, A. J. *Polym. Bull.* **1986**, *16*, 529.
27. Mizushima, S.; Shimanouchi, T. *J. Am. Chem. Soc.* **1949**, *71*, 1320.
28. Schaufele, R. F.; Shimanouchi, T. *J. Chem. Phys.* **1967**, *47*, 3605.
29. Hartley, A.; Leung, Y. K.; Booth, C.; Shepherd, I. W. *Polymer* **1976**, *17*, 354.
30. Hendra, P. J.; Majid, H. A. *Polymer* **1977**, *18*, 573.
31. Strobl, G. R.; Eckel, R. *J. Polym. Sci. Polym. Phys. Ed.* **1976**, *14*, 913.
32. Fraser, G. V. *Polymer* **1978**, *19*, 857.
33. Glotin, M.; Mandelkern, L. *J. Polym. Sci. Polym. Phys. Ed.* **1983**, *21*, 29.
34. Glotin, M.; Mandelkern, L. *J. Polym. Sci. Polym. Lett. Ed.* **1983**, *21*, 807.
35. Snyder, R. G. *J. Chem. Phys.* **1982**, *76*, 3921.
36. Snyder, R. G.; Wunder, S. L. *Macromolecules* **1986**, *19*, 2404.
37. Snyder, R. G.; Schlotter, N. E.; Alamo, R.; Mandelkern, L. *Macromolecules* **1986**, *19*, 621.
38. Mattice, W. L.; Snyder, R. G.; Alamo, R.; Mandelkern, L. *Macromolecules* **1986**, *19*, 2404.
39. Alamo, R.; Mandelkern, L. *J. Polym. Sci. Polym. Phys. Ed.* **1986**, *24*, 2087.
40. Mandelkern, L.; Stack, G. M.; Mathieu, P. J. M. *Anal. Calorim.* **1984**, *5*, 223.
41. Flory, P. J. *J. Chem. Phys.* **1949**, *17*, 223.
42. Flory, P. J. *J. Am. Chem. Soc.* **1962**, *84*, 2857.
43. Mandelkern, L. *J. Phys. Chem.* **1971**, *75*, 3909.
44. Mandelkern, L. In *Characterization of Materials in Research, Ceramics, and Polymers*; Burke, J. J.; Weiss, V., Eds.; Syracuse University Press: Syracuse, NY, 1975; p 369.
45. Mandelkern, L.; Peacock, A. J. In *Studies in Physical and Theoretical Chemistry*; Lacher, R. C., Ed.; Elsevier Science Publishers: The Netherlands, 1988; Vol. 54, p 201.
46. Popli, R.; Glotin, M.; Mandelkern, L.; Benson, R. S. *J. Polym. Sci. Polym. Phys. Ed.* **1984**, *22*, 407.
47. Alamo, R.; Domszy, R.; Mandelkern, L. *J. Phys. Chem.* **1984**, *88*, 6587.
48. Alamo, R. G.; Mandelkern, L. *Macromolecules* **1989**, *22*, 1273.
49. Voigt-Martin, I. G.; Mandelkern, L. In *Handbook of Polymer Science and Technology*; Vol. 3; Cheremisinoff, N. P., Ed.; Marcel Dekker: New York, 1989, p 1.
50. Chan, E. K.; Alamo, R. G.; Mandelkern, L., unpublished data.
51. Richardson, M. J.; Flory, P. J.; Jackson, J. B. *Polymer* **1963**, *4*, 221.
52. Mandelkern, L.; Glotin, M.; Popli, R.; Benson, R. S. *J. Polym. Sci. Polym. Lett. Ed.* **1981**, *19*, 435.
53. Axelson, D. E.; Mandelkern, L.; Popli, R.; Mathieu, P. *J. Polym. Sci. Polym. Phys. Ed.* **1983**, *21*, 2319.
54. Popli, R.; Mandelkern, L. *J. Polym. Sci. Polym. Phys. Ed.* **1987**, *25*, 441.
55. Peacock, A. J.; Mandelkern, L. *J. Polym. Sci. B: Polym. Sci. Ed.* in press.

RECEIVED for review February 14, 1989. ACCEPTED revised manuscript November 7, 1989.

Free-Volume-Dependent Fluorescence Probes of Physical Aging in Polymers

Scott D. Schwab and Ram L. Levy

McDonnell Douglas Research Laboratories, St. Louis, MO 63166

Free-volume-dependent fluorescence (FVDF) probes were used to monitor physical aging in several types of polymers. Changes in fluorescence intensity (I_f) of the FVDF probes correlate with changes in microscopic free volume. The glass transition temperature (T_g) of probe-containing specimens was indicated by a slope change in a plot of I_f as a function of temperature. I_f increased monotonically with sub-T_g anneal time. The ability of these probes to monitor the thermoreversibility of the physical aging process was also demonstrated.

THE USE OF POLYMERS AS ENGINEERING MATERIALS for aerospace systems requires a thorough understanding of the nature and extent of the property changes taking place under service conditions. Physical aging of amorphous polymers is a universal, thermoreversible process that occurs below the material's glass transition temperature (T_g). Physical aging induces changes in many of the properties of polymeric glasses, and a growing number of experimental techniques is being used to monitor the kinetics and magnitude of these changes. Some techniques measure changes in macroscopic properties such as volume, enthalpy, and stress relaxations (1–3); stress–strain behavior (2); and ultimate tensile strength (2); other techniques respond directly to changes in microscopic parameters (i.e., local free volume) or molecular mobility such as dielectric (4), IR (5), NMR (6), and positron annihilation (7) spectroscopies as well as UV–visible spectroscopy for monitoring local free-volume-dependent photoisomerization of photochromic probes and labels (8–11). These methods, however, are not particularly suited to continuous, in-service monitoring of physical aging. This chapter

0065–2393/90/0227–0397$06.00/0

describes an application of free-volume-dependent fluorescence (FVDF) probes as a novel method for monitoring the sub-T_g free-volume relaxation that occurs during the physical aging of amorphous polymers.

Physical Aging of Polymeric Glasses

The volume (and enthalpy) relaxation in amorphous, glassy polymers below T_g is a fundamental property of the glassy state (1) and results from the nature of the glass transition. This phenomenon has long been studied in thermoplastics (1, 12), and in the last several years has been observed in some of the high-performance thermosets used in the aerospace industry (2, 13). A general diagram for how the relaxation process arises is given in Figure 1. As the polymer is cooled through T_g, segmental mobility becomes severely restricted, and excess, nonequilibrium volume is trapped in the matrix. Physical aging at a specified temperature, T_a, is defined as the slow loss of this excess free volume as the material approaches equilibrium. Because of the decrease in free volume, segmental mobility is further reduced, leading to embrittlement of the material. Thus, the physical and mechanical properties of the matrix are time-dependent and can change substantially from their initial values. Matrix properties strongly influence the overall properties of composite materials, and Kong and co-workers (2, 13) showed that the ultimate tensile strength, strain to break, toughness, and water sorption of carbon–epoxy composites decrease after physical aging.

Physical aging, as opposed to chemical aging (e.g., oxidative degradation), is thermoreversible. If the polymer is heated above T_g followed by a rapid quench to below T_g, the free volume lost during physical aging is restored.

The temperature range over which physical aging occurs is generally limited to between T_g and the highest temperature secondary transition (T_β in Figure 1). In most cases, the rate of physical aging increases as the temperature approaches T_g and then decreases at temperatures near T_g (1). Many polymeric materials used in aerospace systems operate in a service temperature envelope that includes at least part of their physical aging range. Therefore, the ability to monitor the aging process is important for understanding the long-term performance of these polymers.

Free-Volume-Dependent Fluorescence

The FVDF probe approach is based on the observation that the fluorescence quantum yields (Φ_f) of certain compounds are strongly dependent on the viscosity and free volume of the medium (14, 15). This dependence is the result of nonradiative deactivation of the excited state by intramolecular twisting or torsional motions that lead to a low value for Φ_f in low-viscosity, high-free-volume media. However, when the free volume of the medium

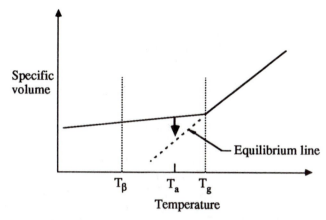

Figure 1. Specific volume as a function of temperature for an amorphous polymer in the region of the glass transition temperature (T_g). T_a is the aging temperature and T_β is the highest temperature secondary transition.

decreases, these torsional motions are more inhibited, leading to an increase in Φ_f and giving rise to FVDF behavior (*14–16*).

The utility of FVDF probes in polymer studies was first demonstrated by Oster and Nishijima (*14*). Loutfy exploited the FVDF approach to monitor the polymerization of vinyl monomers (*16*) and to determine polymer chain tacticity (*17*). Subsequently, we used the same FVDF probe to monitor epoxy cure kinetics (*18*). Loutfy (*19*) suggested the general utility of such "molecular rotor" fluorescence probes for monitoring polymer free volume. He showed that Φ_f of FVDF probes dispersed in a polymer matrix can be linked to the polymer free volume, V_f, according to

$$\phi_f = \frac{K_r}{K_n^0} \exp \beta \frac{V_0}{V_f} \qquad (1)$$

where K_r is the rate of radiative decay, K_n^0 is the intrinsic rate of molecular relaxation of the probe molecule, V_0 is the occupied (van der Waals) volume of the probe molecule, and β is a constant for the particular probe.

Experimental Details

Materials. The FVDF probes were [*p*-(*N,N*-dialkylamino)benzylidine malononitrile)] (DABM) and 4-(*N,N*-dimethylamino)-4'-nitrostilbene (DMANS, Aldrich Chemical Co.) (*see* structures). These probes were dissolved in diglycidyl ether of bisphenol A epoxy (DGEBA, Dow DER 332) at concentrations ranging from 5 to 40 ppm and then cured with the stoichiometric amount of phthalic anhydride (PA, Aldrich), 1,6-diaminohexane (DAH, Aldrich), *N,N*-dimethyl-1,6-diaminohexane (DDH, Aldrich), or a mixture of the amines. Epoxy–amine samples were cured at 50 °C in silicone rubber molds according to the method described by Fanter (*20*).

DABM

DMANS

To ensure that all chemical reaction had been completed, each sample was then postcured at 150 °C for 3 h. Epoxy–anhydride samples were cured at 135 °C for 24 h followed by a postcure at 170 °C for 6 h. DABM–poly(methyl methacrylate) (PMMA, Aldrich) samples were prepared by dissolving a mixture of the probe and polymer in methylene chloride and solvent casting onto a glass plate. The solvent was removed by allowing the sample to stand at room temperature for 48 h, followed by 24 h in a vacuum oven at 150 °C.

DABM could not be used in amine-cured systems because it reacts with amines to form a nonfluorescent (at visible wavelengths) compound. DMANS was found to be stable in amine-cured DGEBA provided it was stored in the dark. Prolonged exposure to light at wavelengths less than 450 nm caused the formation of an unidentified fluorophore whose maximum emission occurs at 415 nm. Because of the possibility of photodegradation, all samples containing DMANS were stored in the dark, and fluorescence measurements were made with the minimum possible excitation power.

Instrumentation. All fluorescence measurements were made with an SLM 4800 fluorescence spectrometer. The fluorescence intensity collected at a right angle to the excitation beam was focused onto a photomultiplier tube. Appropriate optical filters were used to isolate the fluorescence from the scattered excitation light. Physical aging was carried out in the spectrometer sample chamber, where the temperature was controlled to 0.1 °C with a water circulator. Spectra were corrected for the wavelength variation of the instrument response.

Differential scanning calorimetry (DSC) was used for determination of T_g and enthalpy relaxation and was carried out on a DuPont 9900 thermal analyzer. All scans were made at 10 °C/min in a flowing N_2 atmosphere. The T_gs of the four samples were, in degrees Celsius, PMMA, 105; DGEBA–PA, 90; DGEBA–DDH–DAH, 47; and DGEBA–DDH, 27.

Results and Discussion

The sub-T_g physical aging of epoxy specimens containing FVDF probes was monitored as a function of time by measurement of the fluorescence intensities and was verified by DSC. DSC monitoring of the enthalpy changes that accompany physical aging (*21, 22*) verifies that the observed changes in the fluorescence intensities are due to physical aging. A typical example of the changes that occur in the thermogram with sub-T_g annealing time in an amine-cured DGEBA polymer is shown in Figure 2. An endothermic peak that grows with annealing time corroborated the fluorescence data and indicated that physical aging, not additional chemical reaction, was responsible for the observed results. We found a linear relationship between the excess enthalpy (as compared to the quenched sample) and the logarithm of the annealing time. This relationship has been reported for many types of amorphous polymers (*2, 21, 22*).

The fluorescence intensities (I_f) of PMMA and DGEBA–PA containing

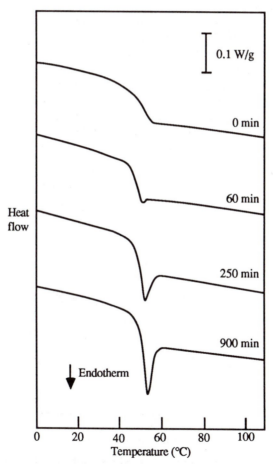

Figure 2. DSC scans of DGEBA cured with a mixture of DDH and DAH after the indicated physical aging times at 35 °C following a quench from 100 °C.

DABM increase with aging time at $T_g - 25\ °C$ after a quench from $T_g +$ 25 °C, as shown in Figure 3. The difference in slope between samples indicates a slower apparent aging rate for DGEBA–PA. The presence of cross-links in the epoxy network decreases the mobility of the chain segments above T_g and reduces the difference in free volume between the glassy and rubbery states. In a cross-linked system, as opposed to a linear thermoplastic such as PMMA, a sudden quench from above to below T_g produces a structure that has less nonequilibrium volume. Because the cross-linked structure starts the physical aging cycle closer to equilibrium, it approaches equilibrium at a slower rate. The results depicted in Figure 3 are consistent with the findings of Lee and McKenna (23), who determined from stress relaxation experiments that the physical aging rates of epoxy networks decreased with increasing cross-link density.

The study of physical aging in amine-cured epoxies was conducted with DMANS as the FVDF probe because DABM reacts with amines. The fluorescence spectrum of DMANS is highly sensitive to changes in the mobility and polarity of the local environment in liquids (24, 25) and thermoplastic polymers (26, 27). Substantial changes in both the intensity and wavelength of maximum emission (λ_{max}^{em}) of DMANS-containing DGEBA were observed as a function of cure time (Figure 4). The intensity increases in proportion to the free-volume decrease as bonds are formed in the curing reaction. The change in λ_{max}^{em} is the result of the probe's large dipole moment in the excited

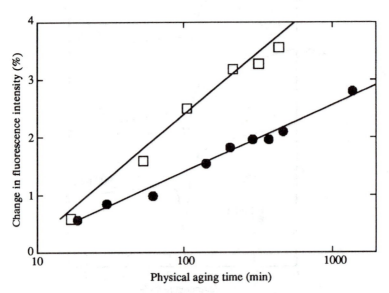

Figure 3. Fluorescence intensity of DABM in PMMA (□) and DGEBA–PA (●) after a quench from $T_g + 25\ °C$ to $T_g - 25\ °C$. Percent change is relative to intensity immediately after quench (time t = 0).

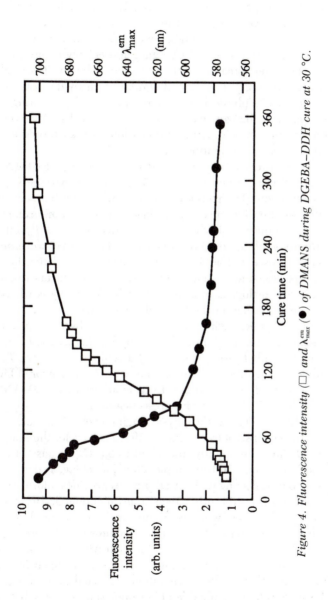

Figure 4. Fluorescence intensity (□) and λ_{max}^{em} (●) of DMANS during DGEBA–DDH cure at 30 °C.

state (28, 29). Before curing, when the matrix dipoles have sufficient mobility to stabilize the excited state during the fluorescence lifetime, the probe can relax to a lower energy level before emission. As the epoxy cures, however, the matrix dipole mobility decreases and the DMANS molecules are forced to emit fluorescence from progressively less relaxed states. Consequently, fluorescence emission occurs from higher energy levels and λ_{max}^{em} shifts toward the blue. Given that λ_{max}^{em} was still changing appreciably during the latter stages of cure, it may seem reasonable that λ_{max}^{em} should be sensitive to physical aging. However, after postcuring, λ_{max}^{em} of DMANS remained constant (within experimental error) with physical aging time. Apparently, postcuring induces chemical and physical changes that reduce mobility to the point where the dipole relaxation time of the matrix is much longer than the fluorescence lifetime, and λ_{max}^{em} reaches a limiting value.

The I_f as a function of temperature behavior of DMANS in DGEBA–DDH–DAH is given in Figure 5 and can be explained in terms of free-volume changes. The increase in free volume with temperature can be represented by the differences in the macroscopic volume expansivities (30). Therefore, the increase in the expansivity that occurs at T_g reflects an increase in the free-volume expansion as well. The slope is more negative above T_g because, as equation 1 indicates, an inverse relationship exists between I_f (proportional to Φ_f) and the free volume. The data presented in Figure 5, then, are analogous to a thermomechanical analysis where T_g is indicated by a slope change in the thermal expansivity.

In contrast to chemical aging, the effects induced by physical aging are thermoreversible and can be "erased" (1, 2) by heating above T_g and then rapid quenching to sub-T_g temperatures. Quenching from above T_g reinitiates the physical aging process. The thermoreversible behavior of DGEBA cured with a mixture of DDH and DAH was monitored with DMANS, and the results are shown in Figure 6. Fluorescence measurements were made at 35 °C (T_g – 12 °C). During the physical aging cycle, I_f increases in accordance with the collapse of free volume. After each cycle, the sample was heated to 80 °C and subsequently quenched to 35 °C. Because of the restoration of the free volume in the sample after quenching, I_f decreases.

Monitoring physical aging with FVDF probes provides information similar to that obtained from the other techniques mentioned. However, fluorescence-based techniques, when combined with fiber-optic fluorometry, are inherently more amenable to in-service monitoring. We are currently exploiting the FVDF phenomenon in conjunction with fiber-optic spectrofluorometry as a novel method for monitoring the composite curing process (31, 32). On the the basis of the observations reported here, a logical extension of this work is to use embedded fiber-optic sensors not only to follow the curing process, but also to monitor the extent of physical aging during service.

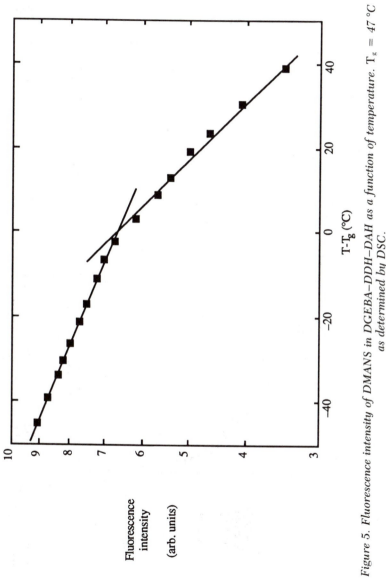

Figure 5. Fluorescence intensity of DMANS in DGEBA–DDH–DAH as a function of temperature. $T_g = 47 °C$ as determined by DSC.

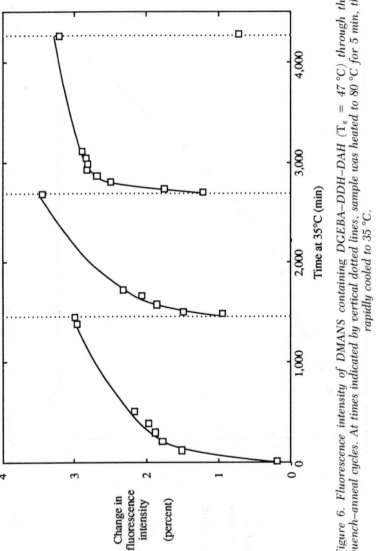

Figure 6. Fluorescence intensity of DMANS containing DGEBA–DDH–DAH ($T_g = 47\ °C$) *through three quench–anneal cycles. At times indicated by vertical dotted lines, sample was heated to 80 °C for 5 min, then rapidly cooled to 35 °C.*

Summary

We showed that FVDF probes can be used to monitor the free-volume relaxation associated with physical aging in both thermoplastic and thermoset polymers. Fluorescence intensity increased monotonically with physical aging time corresponding to the collapse of free volume in the probe's local environment. The restoration of free volume after a quench from above to below T_g could also be followed by changes in I_f. The demonstrated feasibility of using the FVDF probes to follow physical aging is important because this approach is particularly suitable for in-service application to carbon–epoxy composites through fiber-optic waveguides embedded in the structure.

Acknowledgments

This work was performed under the McDonnell Douglas Independent Research and Development program. We thank R. O. Loutfy for supplying the DABM fluorescence probe and T. C. Sandreczki and D. P. Ames for helpful discussions.

References

1. Struik, L. C. E. *Physical Aging in Amorphous Polymers and Other Materials;* Elsevier: Amsterdam, 1978.
2. Kong, E. S. W. *Adv. Polym. Sci.* **1986**, *80*, 120.
3. Cizmecioglu, M.; Fedors, R. F.; Hong, S. D.; Moacanin, *J. Polym. Eng. Sci.* **1981**, *21*, 940.
4. Gomez Ribelles, J. L.; Dias Calleja, R. *Polym. Bull. (Berlin)* **1985**, *14*, 45.
5. Joss, B. L.; Wool, R. P. *Polym. Mater. Sci. Eng.* **1985**, *53*, 307.
6. Kong, E. S. W.; Adamson, M.; Mueller, L. *Compos. Technol. Rev.* **1984**, *6*, 170.
7. Jean, Y. C.; Sandreczki, T. C.; Ames, D. P. *Polym. Mater. Sci. Eng.* **1985**, *53*, 185.
8. Sung, C. S. P.; Lamarre, L.; Chung, K. H. *Macromolecules* **1981**, *14*, 1839.
9. Victor, J. G.; Torkelson, J. M. *Macromolecules* **1987**, *20*, 2241.
10. Lamarre, L.; Sung, C. S. P. *Macromolecules* **1983**, *16*, 1729.
11. Yu, W. C.; Sung, C. S. P.; Robertson, R. E. *Macromolecules* **1988**, *21*, 355.
12. Kovacs, A. J. *Fortschr. Hochpolym. Forsch.* **1964**, *3*, 394.
13. Kong, E. S. W.; Lee, S. M.; Nelson, H. G. *Polym. Compos.* **1982**, *3*, 29.
14. Oster, G.; Nishijima, Y. *J. Am. Chem. Soc.* **1956**, *78*, 1581.
15. Forster, T.; Hoffman, G. *Z. Physik. Chem.* **1971**, *75*, 63.
16. Loutfy, R. O.; *J. Polym. Sci. Polym. Phys. Ed.* **1982**, *20*, 825.
17. Loutfy, R. O.; Teegarden, D. M. *Macromolecules* **1983**, *16*, 452.
18. Levy, R. L.; Ames, D. P. In *Adhesive Chemistry-Development and Trends;* Lee, L. Ed.; Plenum: New York, 1984; p 245.
19. Loutfy, R. O. *Pure Appl. Chem.* **1986**, *58*, 1239.
20. Fanter, D. L. *Rev. Sci. Instrum.* **1978**, *49*, 1005.
21. Petrie, S. E. B. *J. Polym. Sci. Part A-2* **1977**, *10*, 461.
22. Petrie, S. E. B. In *Physical Structure and the Amorphous State;* Allen, G.; Petrie, S. E. B., Eds.; Marcel Dekker: New York, 1977.

23. Lee, A.; McKenna, G. B. *Polymer* **1988**, *29*, 1812.
24. Lippert, E. *Angew. Chem.* **1961**, *73*, 605.
25. "25. " Lippert, E.; Luder, W.; Moll, F. *Spectrochim. Acta* **1959**, *10*, 858.
26. Eisenbach, C. D.; Sah, R. E.; Baur, G. *J. Appl. Polym. Sci.* **1983**, *28*, 1819.
27. Eisenbach, C. D.; Fischer, K. *Polym. Prepr.* **1988**, *29*, 501.
28. Czekalla, J.; Wick, G. *Z. Elektrochem.* **1961**, *65*, 727.
29. Liptay, W.; Czekalla, J. *Z. Elektrochem.* **1961**, *65*, 721.
30. Roberts, G. E.; White, E. F. T. In *The Physics of Glassy Polymers;* Haward, R. N., Ed.; Wiley: New York, 1973; p 153.
31. Levy, R. L.; Schwab, S. D. *Polym. Mater. Sci. Eng.* **1987**, *56*, 169.
32. Schwab, S. D.; Levy, R. L. *Polym. Mater. Sci. Eng.* **1988**, *59*, 591.

SUBMITTED November 22, 1988. RECEIVED for review February 14, 1989. ACCEPTED revised manuscript January 10, 1990.

24

Solid-State Nuclear Magnetic Resonance, Differential Scanning Calorimetric, and X-ray Diffraction Studies of Polymers

Alan E. Tonelli, Marian A. Gomez[1], Hajime Tanaka[2], Frederic C. Schilling, Madeleine H. Cozine[3], Andrew J. Lovinger, and Frank A. Bovey

AT&T Bell Laboratories, Murray Hill, NJ 07974

High-resolution NMR spectroscopy was coupled with differential scanning calorimetry and X-ray diffraction techniques to study the solid-state structures, conformations, dynamics, and phase transitions of several semicrystalline polymers. This combination of techniques was used to study the packing and dynamics of isotactic polypropylene chains in the α, β, and smectic crystalline polymorphs; the conformations and dynamics of poly(diethyloxetane) in form I and II crystals; the conformation of poly(butylene terephthalate) chains in the α and strain-induced β crystalline phases; the conformation and mobility of trans-1,4-polybutadiene chains in the high-temperature phase II crystals; the thermochromic phase transitions in several polydiacetylene single-crystal and melt-crystallized samples; and the thermotropic crystal to liquid-crystal transition in polyphosphazenes. Comparison of the structures, conformations, and dynamics of these polymer chains in their various solid phases provides a foundation upon which to build structure–property relationships.

[1]Current address: Instituto de Ciencia y Tecnologia de Polimeros, C.S.I.C., Madrid, Spain
[2]Current address: Department of Applied Physics, University of Tokyo, Tokyo, Japan
[3]Current address: Department of Chemistry, Yale University, New Haven, CT 06511

0065–2393/90/0227–0409$11.50/0

HIGH-RESOLUTION, SOLID-STATE NMR SPECTROSCOPY was coupled with differential scanning calorimetry (DSC) and X-ray diffraction techniques to study the solid-state structures, conformations, dynamics, and phase transitions of several crystalline polymers. The techniques of cross-polarization (CP), high-power proton dipolar decoupling (DD), and rapid magic angle sample spinning (MAS) were applied at various temperatures to achieve high-resolution ^{13}C and ^{31}P NMR spectra from solid samples of several crystalline polymers. This chapter presents a review of that work. Results were obtained for poly(diethyloxetane) (PDEO), *trans*-1,4-polybutadiene (TPBD), isotactic polypropylene (i-PP), poly(butylene terephthalate) (PBT), several polydiacetylenes (PDA), and poly(bis-4-ethylphenoxyphosphazene) (PBEPP). Each of these polymers can be crystallized into two or more polymorphs. Observation of the chemical shifts and spin-lattice relaxation times, T_1, for each of their chemically distinct nuclei permits an understanding of the conformations, packings, and mobilities of their chains in each of their solid phases monitored by DCS and X-ray diffraction measurements.

The ^{13}C chemical shifts of polymers observed in high-resolution ^{13}C NMR spectra of their solutions are sensitive to their microstructures, that is, stereoregularity, comonomer sequence, and defect structures (1). The microstructural sensitivity of polymer ^{13}C chemical shifts has its origin in the local polymer chain conformation (2, 3). Microstructural differences produce changes in the average local polymer chain conformation, which, in turn, are manifested as different ^{13}C chemical shifts for the carbon atoms in the vicinity of each unique microstructure.

The γ-gauche effect (2), as illustrated in Chart I, successfully accounts for the microstructurally dependent ^{13}C chemical shifts exhibited by polymers in their high-resolution solution spectra. In addition, several examples from the high-resolution spectra of polymers in the solid state (4–10) indicate that the γ-gauche effect also importantly influences the ^{13}C chemical shifts of solid polymers.

The amorphous carbons in semicrystalline polyethylene (PE) resonate 2–3 ppm upfield from the crystalline carbons (6, 11, 12). This observation is expected because the crystalline carbons reside in the all-trans, planar zigzag conformation (no γ-gauche shielding), but the C–C bonds in the amorphous portions of PE possess some gauche character, and therefore the amorphous carbons experience γ-gauche shielding (Chart I).

Bunn et al. (7) observed the methylene carbon resonance in crystalline syndiotactic polypropylene (s-PP) to be a doublet split by 8.7 ppm; for isotactic polypropylene (i-PP) the methylene carbon resonance is a singlet resonating midway between the s-PP methylene doublet. s-PP crystallizes (13) in the –TTGG– conformation, in which half the methylene carbons experience two γ-gauche effects and the remaining half experience no γ-gauche interactions. i-PP crystallizes (14) in the –TGTG– conformation, in which every methylene carbon experiences one γ-gauche shielding effect.

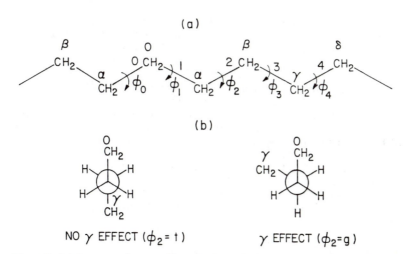

Chart I. (a) Portion of a paraffinic hydrocarbon chain in the all-T planar zigzag conformation. (b) Newman projections along bond 2 in (a) illustrating the γ-gauche shielding effect.

As is expected, the methylene carbon resonances in crystalline s-PP are split by 8–10 ppm, and the methylene carbons of crystalline i-PP resonate midway between them.

Poly(1-butene) crystallizes in three distinct helical conformations (15–22) characterized by 3/1, 11/3, and 4/1 monomer units/helical turn, and each backbone approximates the –TGTG– conformation. In passing from the 3/1 to the 4/1 crystal structure, the backbone rotation angles open up from perfectly staggered positions (60° dihedral angles) to significantly nonstaggered values (85°). This nonstaggered position results in a deshielding of ^{13}C resonances in the 4/1 form relative to the 3/1 form due to the reduction in the magnitude of γ-gauche interactions.

Aside from the three specific examples just discussed, several other reports (5, 23, 24) on the solid-state ^{13}C chemical shifts observed in crystalline polymers can be and have been analyzed in terms of γ-gauche shielding effects. In addition, the motional characteristics of solid polymers can be studied by high-resolution, solid-state ^{13}C NMR. Through observation of spectral relaxation parameters, such as the spin-lattice relaxation time (T_1), the motions of polymer chains in one or more crystalline phases can be compared to the mobility of the amorphous chains or to each other.

Experimental Methods

^{13}C and ^{31}P NMR spectra were recorded on a Varian XL-200 spectrometer at a static magnetic field of 4.7 T. Variable-temperature MAS was achieved with a Doty Scientific probe, which has a double air-bearing design. The standard Varian temperature controller was employed. Aluminum oxide rotors with Kel-F [poly(chlorotrifluoroethylene)] end caps were routinely spun at 2–4 kHz. A 45-kHz radio frequency field strength was used for ^{1}H DD, with a decoupling period of 200 ms. ^{13}C spectra were recorded under CP conditions with a contact time of 2 ms. No attempt was made to record absolute chemical shifts of the resonances observed for PDEO, TPBD, and i-PP. Instead, spectra were recorded consecutively on the same day without adjustment of the magnetic field and were then compared to obtain relative chemical shifts. ^{13}C chemical shifts observed for PBT, PDA, and PBEPP were compared to the resonance of poly(oxymethylene) (POM) placed in their rotors, which appears (25) at 89.1 ppm downfield from TMS under CP conditions. Phosphoric acid placed in the rotor served as a reference for the ^{31}P resonances observed in PBEPP. Spin-lattice relaxation times, T_1, were measured for each carbon under the CP condition by using the pulse sequence developed by Torchia (26) and without CP for carbon and phosphorus nuclei by the usual inversion-recovery method (27).

A Rigaku diffractometer at 1° (2θ) min^{-1} under Ni-filtered Cu K$_\alpha$ irradiation was used in the X-ray diffraction measurements. A Perkin-Elmer DSC-4 instrument operated at heating rates of 10–40 °C/min was used to obtain DSC scans for each polymer.

Poly(diethyloxetane) (PDEO) (28)

X-ray diffraction and calorimetric studies (29–31) of PDEO demonstrated the existence of two distinct crystalline polymorphs produced by altering the conditions for crystal growth. Form I melts at 73 °C and consists of PDEO chains in the all-trans T_4 conformation; form II melts at 57 °C, and its chains assume the T_2G_2 conformation. The principal purpose of the study reported here was the correlation of high-resolution solid-state ^{13}C NMR spectra with the two different crystalline chain conformations adopted by PDEO in the form I and II crystals.

$$CH_3-CH_2$$
$$|$$
$$(-CH_2-C-CH_2-O-)_n$$
$$|$$
$$CH_3-CH_2$$

PDEO

PDEO was obtained by cationic ring-opening polymerization of the corresponding monomer as described elsewhere (29). Two fractions of the bulk polymer with number-average molecular weights (M_n) 8 × 10^5 and 5 × 10^4, as measured by osmometry, were studied.

The samples were melted and crystallized inside the NMR rotors in thermostatic baths set at predetermined temperatures for periods of time sufficient to ensure complete crystallization as indicated by previous dilatometric studies (31). PDEO in form I was obtained by crystallizing from the melt at 60 °C for 2 days and in form II through crystallization at 0 °C for 15 h. Crystallization at 35 °C (for the low-molecular-weight sample) and at 20 °C (for the high-molecular-weight sample) yielded comparable amounts of both crystalline forms in the same sample.

Small portions of the PDEO samples were removed before and after performing the NMR experiments, and X-ray diffraction and DSC measurements were conducted to verify the crystalline form obtained. Figure 1 presents the CPMAS spectra of PDEO ($\overline{M}_n$ = 50,000) crystallized at three different temperatures from the melt to produce samples containing form I (Figure 1a), form II (Figure 1c), and form I and II (Figure 1b) crystals. The spectra of PDEO ($\overline{M}_n$ = 800,000) containing form I crystals, obtained with and without cross polarization, are compared in Figure 2, where crystalline and amorphous resonances are seen to be easily discriminated. Table I presents the relative ^{13}C chemical shifts observed for form I and II PDEO referenced to the corresponding resonances of amorphous PDEO.

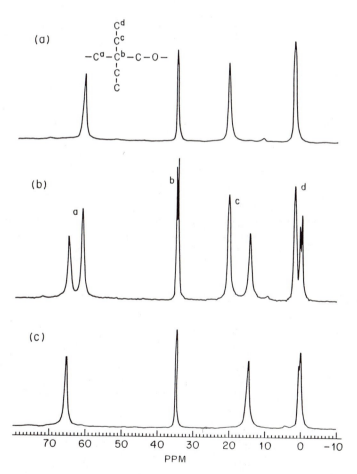

Figure 1. CPMAS spectra of PDEO ($\overline{M}_n$ = 50,000) crystallized at three different temperatures: (a) form I (60 °C); (b) forms I and II (35 °C); and (c) form II (0 °C). All three spectra were recorded at room temperature with no reference employed.

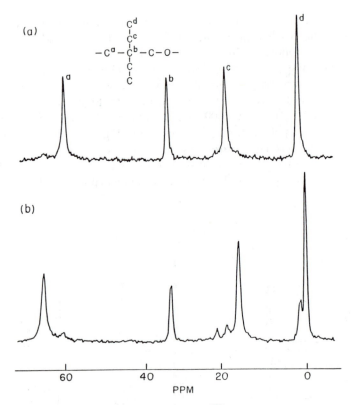

Figure 2. ¹³C NMR spectra at 60 °C of PDEO ($\overline{M}_n = 800,000$) *containing form I crystals obtained (a) with cross polarization (crystalline resonances) and (b) without cross polarization (amorphous resonances), with a 1–2-s pulse-repetition delay. No reference was employed.*

Spin-lattice relaxation times, T_1, observed for each carbon in form I and II PDEO crystals are presented in Table II.

A pertinent example is a solid-state ¹³C NMR study of polyoxetane (PTO) and poly(3,3-dimethyloxetane) (PDMO) by Perez and VanderHart (24). Both PTO and PDMO can be crystallized in at least two different crystalline forms with different chain conformations in each polymorph. Perez and VanderHart produced and reported spectra of the two crystalline modifications with T_2G_2 and T_3G conformations for each polymer. Single resonances are observed for all the carbons in the T_2G_2 crystalline forms of both polymers, but in the T_3G crystalline forms, two resonances are observed for the CH_2–O or α-methylene and methyl carbons (PDMO only).

In the T_3G conformation, one α-methylene carbon experiences one γ-gauche interaction, while the other α-methylene carbon has no γ-gauche interactions. Both α-methylene carbons in the T_2G_2 crystalline forms experience a single γ-gauche interaction. On the basis of γ-gauche shielding

Table I. ^{13}C NMR Chemical Shifts (ppm) for PDEO in the
Solid State

$$^dCH_3$$
$$|$$
$$^cCH_2$$
$$|$$
$$-(^aCH_2\text{-}C^b\text{-}CH_2\text{-}O)-$$
$$|$$
$$CH_2$$
$$|$$
$$CH_3$$

PDEO	a	b	c	d
Form I				
(T_4, T_m = 73 °C)	−5.1	1.1	3.1	1.7
Form II				
(T_2G_2, T_m = 57 °C)	−1.2	0.6	−2.8	−0.3
Amorphous	0.0	0.0	0.0	0.0

NOTE: ^{13}C NMR chemical shifts were observed at 25 °C in PDEO ($\overline{M}_n$ = 50,000) containing both form I and II crystals (*see* Figure 1) and referenced to the resonances observed for the amorphous PDEO carbons of each type.

effects, we expect both α-methylene carbons in the T_2G_2 forms to resonate near the most upfield α-methylene carbon in the T_3G forms, and the other T_3G α-methylene carbon should be downfield from these resonances by one γ-gauche interaction with oxygen. The α-methylene carbons are observed to split by 2.2 (PTO) and 1.3 (PDMO) ppm in the T_3G crystalline forms, but the two α-methylene carbons in the T_2G_2 forms appear 1.8 (PTO) and 1.9 (PDMO) ppm further upfield from the most shielded T_3G α-methylene carbon resonance.

On the basis of a comparison (32) of the ^{13}C NMR spectra of *n*-alkanes and *n*-alkyl ethers, the shielding produced by an oxygen atom in a γ-gauche arrangement with the α-methylene carbons in PTO and PDMO would be expected to be at least −5 ppm. The fact that the α-methylene carbons are split by only 1.3–2.2 ppm in the T_3G crystalline forms, coupled with the observation that both T_2G_2 α-methylenes resonate an additional 1.8–1.9 ppm upfield from the most shielded T_3G α-methylene carbon, despite also being in a single γ-gauche arrangement, indicates that γ-gauche shielding

Table II. Spin-Lattice Relaxation Times, T_1, for Carbons
in Form I and II PDEO

Carbon	Form I	Form II
>C<	62	15
α-CH$_2$	40	14
CH$_2$ (side chain)	22	1
CH$_3$	2	1

interactions do not play the principal role in determining the solid-state ^{13}C NMR chemical shifts observed in PTO and PDMO.

This conclusion is further strengthened by the ^{13}C chemical shifts observed for the methyl carbons in PDMO, all of which resonate within a 1.1-ppm range in both crystalline forms, even though all methyls have one γ-gauche interaction with oxygen except one of the T_3G methyls, which is γ-gauche to two oxygens. In addition, the β-methylene (PTO) and quaternary (PDMO) carbons, which have no γ-gauche interactions in either crystalline form, are separated by 1.0 (PTO) and 1.8 (PDMO) ppm between the T_2G_2 and T_3G forms. Clearly some source other than γ-gauche interactions, such as crystalline packing effects, must be sought to explain the ^{13}C chemical shifts observed by Perez and VanderHart (24) in crystalline PTO and PDMO samples.

$$-C\frac{G}{G}C\frac{\overset{\displaystyle C}{\underset{\displaystyle C}{|}}}{}\frac{G}{T}C\frac{T}{T}O\frac{T}{T}$$

PDMO

Poly(3,3-diethyloxetane) (PDEO) crystallizes (29, 30) into two polymorphs, I and II, with T_4 and T_2G_2 conformations, respectively. Chart II presents Newman projections about the four bonds attached to the quaternary carbon. The two backbone bonds (1 and 2) are either both T or G in the T_4 and T_2G_2 conformations, respectively. In the form I T_4 crystals, the side-chain bonds are both T; their conformations are not known in the form II T_2G_2 crystals. The C–O bonds are T in both crystalline forms.

Table III presents a summary of the numbers and kinds of γ-gauche interactions occurring in both PDEO crystalline forms. Based on these γ-gauche arrangements, we expect the following behavior for the ^{13}C chemical shifts in PDEO forms I and II:

1. CH_3 should resonate at the same field in I and II,

2. $>C<$ should resonate at the same field in I and II,

3. α-CH_2 II should resonate upfield from α-CH_2 I by one $\gamma(O)$ or by $\gamma(O) - \gamma(CH_3)$, depending on whether bonds 3 and 4 are T or G in form II, and

4. CH_2 (side chain) I should resonate upfield from CH_2 (side chain) II by one $\gamma(O)$ or by $\gamma(O) - \gamma(CH_3)$, depending on whether bonds 3 and 4 are T or G in form II (see Chart II).

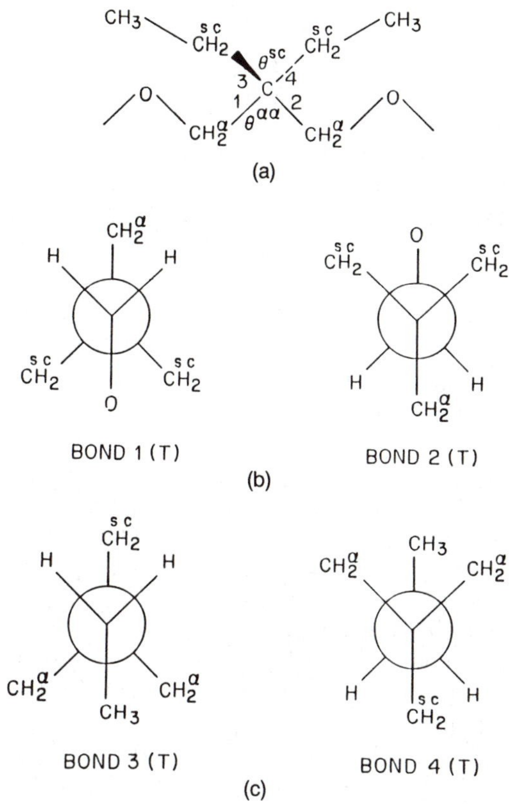

Chart II. *(a) PDEO chain structure. (b) Newman projections along the C–C backbone bonds (1,2) in PDEO. (c) Newman projections along the >C< CH$_2$ side-chain bonds (3,4) in PDEO.*

Table III. Number of γ-gauche Interactions in Form I and II PDEO

Carbon	Form I (T_4)	Form II (T_2G_2)
>C<	0	0
CH$_3$	2 (CH$_2$)	2 (CH$_2$)
α-CH$_2$	2 (CH$_3$)	1 (O) +
		if bonds 3 and 4 = T, 2 (CH$_3$)
		if bonds 3 and 4 = G, 1 (CH$_3$)
CH$_2$ (side chain)	2 (O)	1 (O) +
		if bonds 3 and 4 = T, 0 (CH$_3$)
		if bonds 3 and 4 = G, 1 (CH$_3$)

A comparison of the three CPMAS spectra of PDEO ($\overline{M}_n$ = 50,000) in Figure 1 readily shows that the ^{13}C chemical shifts expected for >C< and CH_3 are similar to those observed. However, instead of α-CH_2 II being upfield from α-CH_2 I, it resonates 3.9 ppm downfield. Similarly, form II CH_2 (side chain) comes upfield from form I CH_2 (side chain) by 5.9 ppm instead of being downfield as expected. As observed for PDMO by Perez and VanderHart (24), the ^{13}C chemical shifts observed in forms I and II PDEO are not predominantly influenced by γ-gauche shielding effects.

Different packing of the PDEO chains in forms I and II is a possible source of the difference observed in their ^{13}C chemical shifts. The >C< and CH_3 carbons have identical γ-gauche interactions in both PDEO crystalline forms, yet there are 0.5 (>C<) and 2.1 (CH_3) ppm chemical shift differences between the two polymorphs. On the basis of the ^{13}C chemical shift packing effects seen in n-alkane (33), PE (34), and i-PP (35) crystals, the differences observed between form I and II >C< and CH_3 chemical shifts do not seem atypical.

As another example, the ^{13}C chemical shifts of PTO in the T_2G_2 crystalline form may be compared with those observed for its cyclic tetramer, c-$(TO)_4$, which also adopts (36) the T_2G_2 conformation in the crystalline state (28). The α-CH_2 ^{13}C chemical shifts differ by 0.3–0.5 ppm, and the β-CH_2 shifts by 1.4 ppm, between crystalline PTO (T_2G_2) and c-$(TO)_4$. These shift differences are similar to the packing effects observed on the >C< and CH_3 carbons in PDEO. Thus, packing effects in PTO, PDMO, and PDEO can be as large as 2 ppm.

However, packing effects of this magnitude are not nearly sufficient to explain the observed differences in the ^{13}C chemical shifts of the α-CH_2 and CH_2 (side chain) carbons between forms I and II PDEO, which are >4–6 ppm.

In addition to interchain packing effects, some differences in intramolecular chain geometries must exist between forms I and II PDEO. Possibly the valence angles differ significantly between the T_4 conformation in form I crystals and the T_2G_2 conformation in the form II crystals. Such valence angle differences would be expected (38) to produce large ^{13}C chemical shift effects; however, it is a bit more difficult to understand why the >C< and CH_3 carbons would not also be affected.

Suppose that the C^α–C–C^α backbone valence angle ($\theta^{\alpha\alpha}$) is sensitive to the rotational states of the C–C bonds (1 and 2 in Chart II) and adopts different values for the T and G rotational states. If $\theta^{\alpha\alpha}(G)$ > or < $\theta^{\alpha\alpha}(T)$, then the valence angle C^{SC}–C–C^{SC} between ethyl side chains (θ^{SC}) should follow (39), and $\theta^{SC}(T)$ > or < $\theta^{SC}(G)$. The ^{13}C chemical shift of >C< should be independent of complimentary changes in $\theta^{\alpha\alpha}$ and θ^{SC}, because all four quaternary carbon substituents are methylene carbons. In addition, the methyl carbons are probably sufficiently removed from $\theta^{\alpha\alpha}$ and θ^{SC} to have their ^{13}C chemical shifts unaffected by changes in these valence angles.

Spin-lattice relaxation times, T_1, observed for each carbon in forms I and II PDEO crystals are compared in Table II. The most dramatic differences in T_1s are those observed for the side-chain methylene carbons, that is, $T_1 = 22$ s (I) and 1 s (II). Apparently the side chains in the form II crystals possess considerable mobility, most probably as a result of rotations about side-chain bonds 3 and 4 (*see* Chart II). This finding implies that the ethyl side chains in form II crystals are conformationally disordered and are likely to be interconverting rapidly between T and G conformations.

In summary, the large chemical shift differences observed for the methylene carbon resonances in the solid-state ^{13}C NMR spectra of forms I and II PDEO are likely the result of intramolecular geometrical differences between the chains residing in these two polymorphs. Valence angle distortion at the quaternary carbon, which depends on the conformational states of the bonds attached to this carbon, rather than conformationally sensitive γ-gauche effects or differences in interchain packing, seem the most likely source of the large chemical shift differences observed in the solid state for the forms I and II methylene carbons in PDEO.

In addition, spin-lattice relaxation times observed for both crystalline forms of PDEO indicate that the side chains in the form II crystals are motionally labile and disordered. Rapid interconversion between T and G side-chain conformations seems a likely source of the short T_1s observed for the side-chain methylene carbons in form II PDEO.

Poly(butylene terephthalate) (PBT) (41)

It was first observed (*42–44*) over a decade ago that the uniaxial extension of PBT fibers is accompanied by a crystal–crystal transition. In the relaxed or α-form crystals, the molecular chain or fiber repeat is 10% shorter than observed in the stretched or β-form crystals. The transition between the α and β forms of PBT produced by mechanical deformation (uniaxial stretching) is reversible (*44–46*), and only the α form is stable in the relaxed, unstretched state at ambient temperature.

X-ray structural studies have been reported (*45, 47–51*) for both PBT crystalline forms. Infrared and Raman spectroscopy (*52*) suggest a nearly trans–trans–trans sequence (ϕ_a, ϕ_b, ϕ_c) for the glycol residue in the extended structure (Chart III). However, the crystal structures proposed by Yokouchi et al. (*45*) and Hall, Stambaugh, and co-workers (*48–51*) depart significantly from the extended, all-trans glycol structure. All crystal structures proposed for the relaxed, contracted α form approximate a gauche-trans-gauche conformation for the glycol residue, although there are differences in detail among them.

As noted by Davidson et al. (*53*), the low scattering power of hydrogen atoms makes X-ray diffraction unsuitable for defining the conformation of

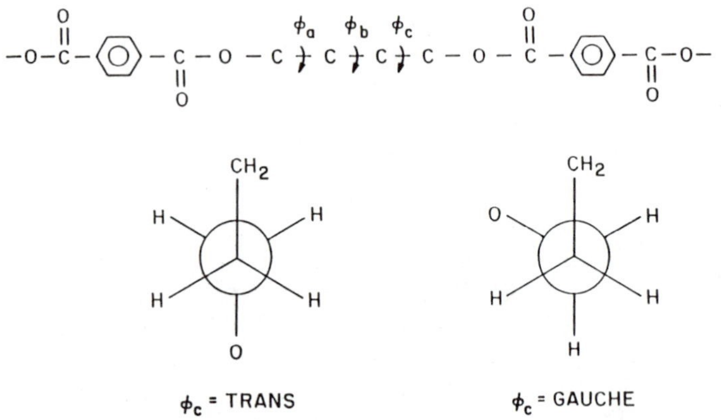

Chart III. PBT chain with torsions about C–C bonds indicated and Newman projections of the trans and gauche conformers about the terminal C–C bonds.

the glycol residues in PBT. Instead, these same authors applied broad-line ^{1}H NMR measurements to oriented PBT in both the relaxed α and strained β forms and determined the second moments of the proton line shapes as a function of specimen orientation. They found the ^{1}H NMR results to be consistent with a nearly fully extended (trans–trans–trans) conformation for the β stretched form and to agree quantitatively with the X-ray structure proposed by Hall and Pass (48). However, their NMR results for the relaxed α form were not consistent with any of the proposed crystal structures and suggest instead that the conformation and orientation of the central methylene pairs in the glycol residues are not substantially altered in the strain-induced transformation from the α to β form.

More recently, Grenier-Loustalot and Bocelli (54) studied the structures of four PBT model compounds by X-ray diffraction and high-resolution ^{13}C NMR spectroscopy in the solid state. Single-crystal X-ray diffraction revealed that two of the PBT model compounds crystallized with *trans–trans–trans*-glycol residues; one had a *trans–trans–gauche*-glycol conformation, and the remaining compound had its glycol residue in the gauche–trans–gauche conformation.

In the high-resolution solid-state ^{13}C NMR spectra of the PBT model compounds, they observed the central methylene carbons that are gauche to their ester oxygens to resonate 3.0–3.7 ppm upfield from those central methylene carbons adopting the trans arrangement (Chart III). This finding is consistent with the often observed shielding of carbon nuclei whose γ substituents are in a gauche arrangement, that is, the γ-gauche effect (2, 3, 55). The central methylene carbons in α form PBT were observed to resonate midway between the corresponding methylene resonances in the model compounds.

Because of the broadness of the central methylene resonance in PBT

($>$3 ppm), which is likely a consequence of contributions from carbon nuclei in both the crystalline and amorphous regions of the sample (as discussed later), Grenier-Loustalot and Bocelli (54) were unable to draw conclusions regarding the conformation of the glycol residue in α-PBT. A similar study was attempted by Havens and Koenig (56), but it too was plagued by broad resonances and, in addition, by an erroneous conformational assignment to one of their PBT model compounds, as pointed out by Grenier-Loustalot and Bocelli (54).

Most recently, Perry et al. (57) employed high-resolution solid-state ^{13}C NMR techniques to study the crystalline conformations and dynamics of PBT chains in both the α and β crystalline forms. Their spectra also exhibited broad resonances, especially for the central methylene carbons (4–5 ppm). However, they concluded that, as the amount of trans content in the glycol residue increases, the interior methylene resonance shifts to a higher field, so the β-form resonance moves upfield from the α-form resonance. This conclusion is in direct opposition to the model compound study of Grenier-Loustalot and Bocelli (54) and to the expected order of chemical shifts based on the conformationally sensitive γ-gauche effect (2, 3, 55).

In an attempt to determine the conformations of PBT chains in their α- and β-form crystals, we conducted variable-temperature, high-resolution solid-state ^{13}C NMR studies. Above $\sim$100 °C, the spectra are significantly better resolved, with resonance line widths not exceeding 1–2 ppm. This narrowing of resonances is apparently a consequence (58) of removing contributions made by the amorphous carbons, which no longer cross-polarize efficiently at temperatures well above the glass transition of PBT (59). Comparison of the high-temperature, high-resolution solid-state ^{13}C NMR spectra recorded for α- and β-PBT with those of the PBT model compounds reported by Grenier-Loustalot and Bocelli (54) leads to several conclusions concerning the conformations of PBT chains in the relaxed α and stretched β crystals.

PBT pellets (Aldrich 19,094–2) were cryogenically ground to a fine powder, which was annealed at 150 °C for 3 days to produce the α form (45, 60). Melt pressing of the pellets at 250 °C produced thin films that were quenched into liquid nitrogen. Strips were cut from the film and placed in an Instron tensile testing machine. The strips were drawn to 300% elongation, held under tension, and annealed at 150 °C for several hours to produce (45) the β form sample.

X-ray diffraction photographs recorded for both samples in the film form confirmed that we did indeed produce both α- and β-PBT samples. The PBT samples were spun in an aluminum oxide rotor with Kel-F [poly-(chlorotrifluoroethylene)] end caps. PBT in the α form was placed in the rotor as a powder, while a strip of the β-form sample was wound under tension onto a spindle to form a spool, and the ends of the strip were glued to prevent relaxation to the α form. The spool wound with the β-form strip was then inserted into the rotor, and special end caps (Doty Scientific) were

used to secure the ends of the spindle so that the β-form PBT spool would
rotate at the same speed as the rotor.

A comparison of the CPMAS/DD spectra of α- and β-PBT recorded at
105 °C is made in Figure 3. Table IV contains the observed solid-state ^{13}C
NMR chemical shifts. ^{13}C spin-lattice relaxation times, T_1, measured in the
solid state at 105 °C, are given in Table V.

The comparison of CPMAS spectra recorded at 105 °C and presented
in Figure 3 and the corresponding chemical shifts listed in Table IV show
that, aside from the protonated aromatic carbons (PAR), the carbon nuclei
in α- and β-PBT resonate at nearly identical frequencies. This observation
is at variance with the results of Perry et al. (57), who found the CPMAS/
DD spectra of α- and β-PBT at 20 °C to be closely similar except for the

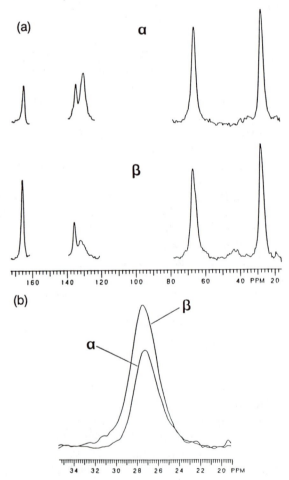

Figure 3. CPMAS/DD spectra measured at 105 °C for α- and β-PBT: full
spectra (a) and expansion of the central methylene carbon regions (b).

Table IV. ^{13}C Chemical Shifts of α- and β-PBT

Carbon	α	β	Amorphous[a]
CH$_2$	27.2	27.6	26.6
OCH$_2$	66.2	66.7	65.9
PAR	130.8	131.7	130.2
NPAR[b]	135.2	135.2	134.9
C=O	165.6	165.5	166.0

NOTE: All ^{13}C shifts are given in parts per million versus tetramethylsilane (TMS); spectra were measured at 105 °C and referenced to the POM resonance at 89.1 ppm from TMS (25), although we observed a 0.3-ppm downfield shift with respect to ambient temperature.
[a]Measured from spectra obtained without CP, but with MAS/DD, using a 3-s pulse-repetition delay.
[b]NPAR stands for nonprotonated aromatic.

resonances of the interior methylene carbons. The source of this disparity is revealed by comparing the spectra in Figure 3 with those presented by Perry et al. (57). Perry's spectra are characterized by broad resonances, 2–3 times as broad as the resonances seen in Figure 3, presumably a consequence of the different local environments, both conformational and packing, experienced by the carbon nuclei of the sluggish, amorphous PBT chains, which result in a dispersion of chemical shifts.

Recording the CPMAS/DD spectra of PBT at elevated temperatures (105 °C) results in enhanced resolution, because the amorphous carbons, which constituted 30–50% of the samples, are sufficiently above their glass transition temperature (T_g = 50–55 °C) (59) to be mobile enough not to cross-polarize efficiently (58). Comparison of the spectra recorded with and without CP (pulse-repetition delay of 3 s) makes apparent that the crystalline and amorphous carbons resonate at similar frequencies, and the amorphous carbon resonances appear upfield as expected. Thus at temperatures sufficiently close to T_g, where the amorphous PBT chains are relatively rigid and immobile, CPMAS/DD spectra would be expected (58) and in fact are observed to be significantly broadened by the overlap of crystalline and amorphous resonances.

The near coincidence of methylene carbon chemical shifts observed for α- and β-PBT strongly suggests that in both the relaxed (α) and strained (β)

Table V. Spin-Lattice Relaxation Times, T_1, for α- and β-PBT

Carbon	α[a]	β[a]	Amorphous[b]
CH$_2$	0.13	0.20	0.16
OCH$_2$	0.32	0.27	0.24
PAR	5.7	7.0	0.31
NPAR	18.9	12.7	—
C=O	18.0	15.0	—

NOTE: All T_1 values are in seconds and were measured at 105 °C.
[a]Obtained under CP conditions with the Torchia (26) pulse sequence.
[b]Obtained without CP by the inversion-recovery method (27)

crystals, the glycol residues of the PBT chains are adopting very similar conformations. Just what conformation is adopted by the glycol residues in crystalline PBT?

Chart IV presents schematic structures of the four PBT model compounds studied by Grenier-Loustalot and Bocelli (54) and of PBT. A comparison of the central methylene carbon chemical shifts clearly indicates that the glycol residues in both α- and β-PBT crystals are in the nearly extended trans–trans–trans conformation found by Grenier-Loustalot and Bocelli (54) for the PBT model compounds 3 and 4.

If the glycol residues of both α- and β-PBT are nearly fully extended, then what conformational differences can account for the 10% increase in the fiber repeat of the β-form crystals that are formed upon extension of α-PBT? The ester bonds [C(=O)O] in PBT are likely trans planar, as they are in the four PBT model compounds. In each crystalline model compound the

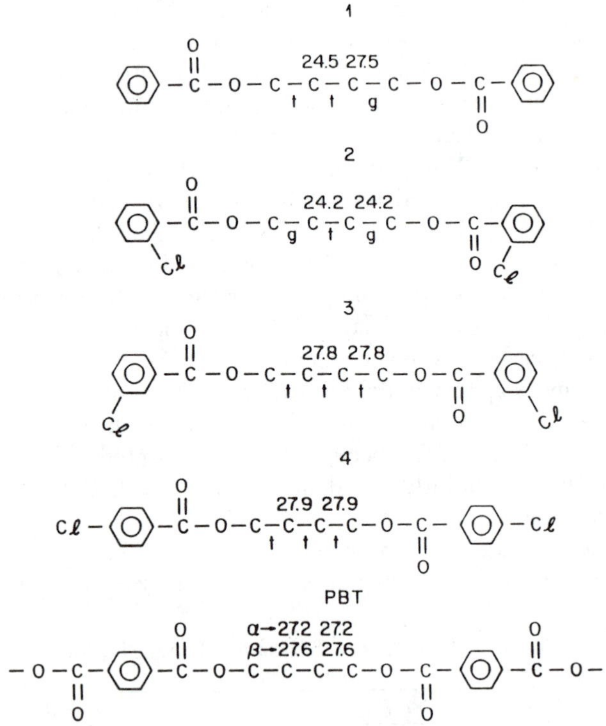

1

$$\text{⬡}-\overset{\overset{O}{\|}}{C}-O-C\underset{t}{-}C\underset{t}{-}C\underset{g}{-}C-O-\overset{\underset{O}{\|}}{C}-\text{⬡} \qquad \text{24.5 27.5}$$

2

$$\text{⬡}-\overset{\overset{O}{\|}}{C}-O-C\underset{g}{-}C\underset{t}{-}C\underset{g}{-}C-O-\overset{\underset{O}{\|}}{C}-\text{⬡} \qquad \text{24.2 24.2}$$
(Cl substituents)

3

$$\text{⬡}-\overset{\overset{O}{\|}}{C}-O-C\underset{t}{-}C\underset{t}{-}C\underset{t}{-}C-O-\overset{\underset{O}{\|}}{C}-\text{⬡} \qquad \text{27.8 27.8}$$
(Cl substituents)

4

$$\text{Cl}-\text{⬡}-\overset{\overset{O}{\|}}{C}-O-C\underset{t}{-}C\underset{t}{-}C\underset{t}{-}C-O-\overset{\underset{O}{\|}}{C}-\text{⬡}-\text{Cl} \qquad \text{27.9 27.9}$$

PBT

$$-O-\overset{\overset{O}{\|}}{C}-\text{⬡}-\overset{\overset{O}{\|}}{C}-O-C-C-C-C-O-\overset{\underset{O}{\|}}{C}-\text{⬡}-\overset{\underset{O}{\|}}{C}-O- \qquad \begin{array}{l}\alpha\rightarrow 27.2\ 27.2 \\ \beta\rightarrow 27.6\ 27.6\end{array}$$

Chart IV. The four PBT model compounds studied by Grenier-Loustalot and Bocelli (54). The conformation of each glycol residue, as determined by X-ray diffraction, is indicated (t is trans; g is gauche), and the chemical shifts (parts per million versus tetramethylsilane) for the central methylene carbons observed by CPMAS/DD ^{13}C NMR spectroscopy are also listed. The structure of PBT is presented, and the chemical shifts observed in this work for the central methylene carbons in the α- and β-form crystals are indicated.

CH_2–O bonds are also in the trans conformation. Because the chemical shifts of the central methylene carbons should be sensitive (2, 3, 55) to the conformations of these bonds (*see* Charts III and IV), as well as to the conformations of the terminal CH_2–CH_2 bonds, the CH_2–O bonds in α- and β-PBT are also most likely trans. If they were gauche in either or both polymorphs, then different chemical shifts would be expected for the central methylene carbons in α-and β-PBT, or chemical shifts reflecting this shielding (2) would be produced by γ-gauche carbonyl carbons. Instead, the chemical shifts of the central methylene carbons in both α- and β-PBT are very similar to the chemical shifts observed for the same carbons in model compounds 3 and 4, where both the CH_2–O and terminal CH_2–CH_2 bonds are trans.

The only conformational degree of freedom that remains to distinguish the α and β forms of PBT is the relative orientation of the carbonyl groups in the terephthaloyl residues, which are determined by the torsional angles about the sp^2–sp^2, carbonyl to aromatic C–C bonds (*see* Charts III and IV). In terms of γ-gauche shielding effects (2, 3, 55), the chemical shifts of the protonated aromatic carbons (PAR) are expected to reflect the conformations of these bonds.

In fact the 0.9-ppm chemical shift difference observed between the PARs of α- and β-PBT is by far the largest difference between their high-resolution, solid-state ^{13}C NMR spectra (*see* Figure 3). Thus our results support the conclusion reached by Davidson et al. (53) via analysis of broad-line ^{1}H NMR data; changes in the conformation of the terephthaloyl residue, but not in the glycol residue, must accompany the solid-state transformation of PBT from the α to the β form.

As suggested by the comparison of solid-state ^{13}C NMR chemical shifts observed for α- and β-PBT and several of their model compounds, all of the bonds in crystalline PBT are nearly trans except the bonds connecting the ester groups to the aromatic rings. The bonds in amorphous PBT chains would be expected to be a mixture of trans and gauche conformations, the latter producing upfield chemical shifts via the γ-gauche effect. It follows that the amorphous carbon nuclei should resonate upfield from the corresponding carbons in α- and β-PBT. The solid-state ^{13}C NMR chemical shifts presented in Table IV for the amorphous and crystalline carbons in PBT confirm this expectation.

Spin-lattice relaxation times, T_1, for both crystalline forms and amorphous PBT, as presented in Table V, serve to indicate the motional characteristics of solid PBT chains. The most striking observation is the near coincidence of the T_1s measured for the methylene carbons of the glycol residues in the α and β crystallites of PBT with those observed for the amorphous methylene carbons. By contrast, the T_1s measured for the crystalline PAR carbons are more than an order of magnitude longer than those observed for the amorphous PAR carbons. Apparently, the methylene carbons of the glycol residue are undergoing significant motion, independent of whether they are included in the α and β crystallites or not.

In agreement with Perry et al. (57), whose measurements were performed more than 80 °C below ours (105 °C), we do not find any significant differences between the spin-lattice relaxation times of the crystalline methylene carbons in α- and β-PBT. Similarly, the T_1s of the crystalline carbon nuclei belonging to the terephthaloyl residues in α- and β-PBT are also not markedly different. The α and β phases apparently do not constrain the motions of their constituent PBT chains in any significantly different manner (*also see* Jelinski et al. (61) and Garbow and Schaefer (62) for further discussion of solid-state PBT motion).

High-resolution, solid-state ^{13}C NMR studies of α- and β-PBT revealed several important features concerning the conformations and motions of PBT chains in both crystalline phases. The glycol residues are in the nearly extended (trans-trans-trans) conformations in both crystalline forms, and different orientations of the ester groups and phenyl rings probably account for the 10% difference in the fiber repeats of α- and β-PBT. In both crystals the methylene carbons are sampling rapid motions, which are significantly faster than the motions experienced by the carbons of the terephthaloyl residues.

Isotactic Polypropylene (i-PP) (63)

Isotactic polypropylene is a stereoregular vinyl polymer that normally develops significant crystallinity below 200 °C. The thermodynamically stable crystalline form, or α-form, consists of i-PP chains in the 3_1 helical conformation (...*tgtgtg*...) packed in a monoclinic unit cell (64, 65). Left- and right-handed helices are in close proximity. The metastable β-form crystals of i-PP contain hexagonally packed 3_1 helical chains arranged in groups of the same helical handedness (left or right) resulting in the distant packing of left- and right-handed chains (65, 66).

The smectic form of i-PP is (64, 67–73) only partially ordered compared to the α- and β-crystalline forms, although the i-PP chains in the smectic form remain in the 3_1 helical conformation (64, 68). Smectic i-PP is therefore primarily disordered in the intermolecular packing of its chains.

The purpose of the investigation reported here was to learn more about the structures of the crystalline regions in the α-, β-, and smectic forms of i-PP using high-resolution ^{13}C NMR spectroscopy as a structural probe.

Isotactic PP in the smectic form was made by cryogenic grinding of a Hercules Profax-6523 i-PP sample (74) as described by Lovinger et al. (60). The α-form i-PP was obtained from the smectic sample by annealing for 1 h at 160 °C. The β-form sample was made (75) by unidirectional crystallization at a growth rate of 10 μm/min with a temperature gradient of 300 °C/cm.

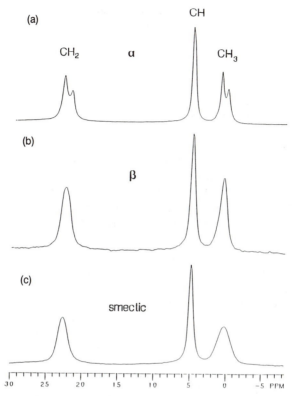

Figure 4. CPMAS/DD spectra of i-PP in (a) α form, (b) β form, and (c) smectic form. Spectra were recorded at ambient temperature with no reference employed.

X-ray diffractograms of all three i-PP samples were recorded before and after the ^{13}C NMR experiments to ensure that the high-speed (3 kHz) magic angle spinning of the samples did not induce any solid–solid transitions.

Figure 4 presents the CPMAS/DD spectra of the three forms of i-PP recorded at ambient temperature. Table VI contains the observed solid-state ^{13}C NMR chemical shifts. [13]C spin-lattice relaxation times, T_1, measured in the solid state at ambient temperature are given in Table VII. In addition, the T_1 values measured (76) for i-PP in solution are also presented in Table VII for comparison. A comparison of the CPMAS ^{13}C NMR spectra in Figure 4 and the relative chemical shifts presented in Table VI leads to several observations. First, both the methylene and methyl carbon resonances in α-form i-PP are split by 1 ppm, as first reported by Bunn et al. (77). The ratio of intensities (peak heights) of the downfield to the upfield component is 2:1 for both carbon types. Bunn et al. (77) interpreted this splitting as due to the inequivalent sites, A and B, produced by pairing of helices of opposite handedness (64) (*see* Figure 5a), which are also present in the ratio

Table VI. ^{13}C NMR Chemical Shifts for i-PP in the
Solid State

i-PP Form	CH_2	CH	CH_3
α	0, 1.07	0.20	0, 0.88
β	0.37	0.02	0.27
Smectic	0.47	0	0.08

NOTE: All ^{13}C NMR chemical shifts are in parts per million and were observed at ambient temperature and referenced to the most upfield resonance of each carbon type.

A:B = 2:1. The A sites correspond to a separation of 5.28 Å between helical axes, and for the B sites the helices are 6.14 Å apart.

In the spectrum of β-form i-PP (Figure 4b), on the other hand, each carbon exhibits a single resonance. The β-form methylene and methyl resonances are close to the upfield member of each pair of the same resonances in the α form, which were attributed to the B sites (see Table VI). Figure 5b indicates the interchain packing proposed (65) for β-form i-PP. Unlike the α-form packing (64), 3_1 i-PP helices of the same handedness are packed together in groups in the β-form crystals. The interhelical separation (68) is 6.36 Å, very similar to the smallest interhelical separation involving the B sites of α-form crystals. Thus, the near-coincidence of the ^{13}C chemical shifts in β-form i-PP with those corresponding to the B sites in the α form can be understood on the basis of similar interhelical separations in both packing modifications.

Bunn et al. (77) also reported the ^{13}C NMR CPMAS spectrum of β-form i-PP. They found, in contrast to our results, that the chemical shifts of the A site methylene and methyl carbon resonances in α-form i-PP were closer than the B site resonances to those observed in their β-form i-PP. We were unable to obtain β-form i-PP following their preparation method, a finding confirmed by X-ray diffraction. The difficulties in obtaining pure β-form i-PP exclusively via thermal treatment are noted in the literature (75, 78–81). Instead, special nucleating agents (80, 81) or unidirectional crystallization (75) are employed. Consequently, Bunn et al. (77) may not have studied i-PP in the pure β-form.

Table VII. ^{13}C T_1 Relaxation Times for Solid i-PP at
Ambient Temperature

i-PP Form	CH	CH_2	CH_3
α	37[a]	52[a]	0.32[a], 0.80[a], 0.48[b]
β	29[a]	34[a]	0.75[a], 0.44[b]
Smectic	22[a]	33[a]	0.75[a], 0.51[b]
Solution[c]	0.40[b]	0.20[b]	0.75[b]

NOTE: All T_1 values are given in seconds.
[a]Measured with CP using the pulse method of Torchia (26).
[b]Measured by inversion-recovery (27) without CP.
[c]Measured in solution at 46 °C by Randall (76) using the inversion-recovery method (27).

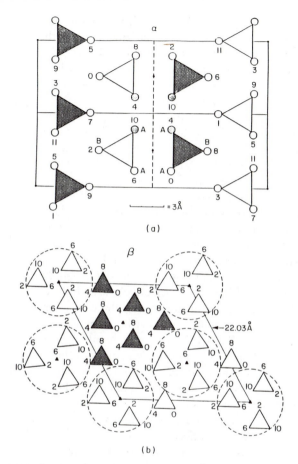

Figure 5. Crystal structures of (a) α form (64) and (b) β form (65) of i-PP. Full (right-handed) and open (left-handed) triangles indicate 3₁ helical i-PP chains of different handedness. A and B label the inequivalent sites discussed in the text, and are applicable to all three carbon types, because the CH–CH₂ bond is nearly parallel with the c-axis. Numerals at the triangle vertices indicate heights of methyl groups above a plane perpendicular to the c-axis in twelfths of c. The circles at the triangle vertices in part a correspond to methyl carbons, and the pairs of circles at numbers 10 and 4 correspond to the enmeshed A-site methyls.

^{13}C NMR chemical shifts recorded for the smectic form of i-PP are nearly coincident with those found for the β form. This finding suggests that the local packing of 3_1 i-PP helices in the smectic form closely resembles that found in the β form.

Having concluded that the local packing of chains is similar in the β and smectic forms of i-PP on the basis of their observed ^{13}C chemical shifts, let us look at the spin-lattice relaxation time (T_1) behavior of the carbon nuclei in the three polymorphs of i-PP. The T_1 relaxation times of the crystalline carbons were obtained while cross-polarizing by using the Torchia (26) pulse sequence. Smectic-form i-PP T_1 values (see Table VII) are very similar to those measured for the β-form crystals, but the α-form T_1 values appear unique. The T_1 values reported previously for α-form i-PP by Fleming et al. (82) are in good agreement with the values shown in Table VII.

The T_1 values obtained by the inversion-recovery method (27) are dominated by the relaxation of the amorphous carbons and are obtained without CP. As expected from the spectra obtained without CP [not shown (63)], only the T_1 of the methyl carbons are obtained by this method, and, within experimental error, they are the same for all three i-PP polymorphs. In addition, the T_1 values measured for the methyl carbons in the crystals and in solution are similar to the amorphous methyl spin-lattice relaxation times. Clearly the spin-lattice relaxation times of methyl carbons are dominated by their internal rotations and not by the segmental motions of the i-PP chains.

The two methyl resonances observed in the CPMAS spectrum of α-form i-PP (see Figure 4a) relax at different rates; $T_1 = 0.32$ and 0.80 s for the downfield and upfield peaks, respectively. Having identified these resonances with the A and B packing sites in the α-form crystalline lattice (see Figure 5a), it is worth mentioning that both the β- and smectic-form crystalline methyls have $T_1 = 0.75$ s in agreement with the T_1 of the α-form, B-site methyl carbons. This observation supports the conclusion, obtained previously from a comparison of ^{13}C chemical shifts, that the interhelical separation of chains is similar for the B sites in α-form crystals and in the β- and smectic-form i-PP crystals.

The α-form methyl carbons associated with the A and B packing sites exhibit T_1 relaxation times different by almost a factor of 2. The B-site methyls are rotating twice as fast as the enmeshed methyl carbons at the A sites because they are apparently on the fast side of the T_1 minimum (82).

The results of our solid-state ^{13}C NMR studies of the three polymorphs of i-PP are consistent with several known structural features of the α- and β-form crystals and permit inferences about the local chain-packing structure in the smectic form. Both the ^{13}C chemical shifts and spin-lattice relaxation times observed for the smectic-form carbons indicate that the packing of their 3_1 helices is similar (at least on a very local scale) to the packing of i-PP chains in the β-form crystals.

This conclusion is consistent with the proposal of Gailey and Ralston (69), who suggested that smectic-form i-PP is composed of small (50–100 Å) hexagonal or β-form crystals. Suggestions that the smectic form is composed of monoclinic, or α-form, microcrystals made by Bodor et al. (70), or a smectic-form packing characterized locally by a core of α structure surrounded by chains in a pseudohexagonal arrangement made by Corradini et al. (71), are not consistent with the results of our ^{13}C NMR study. Also the suggestion made by Miller (72) and Zannetti et al. (73) that α-form paracrystallinity (distortions of the monoclinic lattice with loss of long-range order) characterizes the structure of smectic-form i-PP is not supported by our results.

trans-1,4-Polybutadiene (TPBD) (*83*)

trans-1,4-Polybutadiene exists in two crystalline polymorphs (*84*). At room temperature, the chain conformation of form I is as follows (*84, 85*):

$$E \quad s^\pm \quad t \quad s^\mp \quad E$$

$$\bullet\bullet\bullet-CH_2-CH_2-CH=CH-CH_2-CH_2-CH=CH-CH_2-CH_2-\bullet\bullet\bullet$$

where the double bond is of course trans ("E") and $s^\pm$ (or $s^\mp$) designates approximate skew conformations:

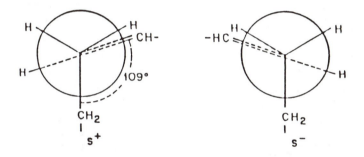

In the exact skew conformation, the $C=C$ and $C-H$ bonds are eclipsed, and the dihedral angle is 120°. The chain packing is a hexagonal array. Above approximately 75 °C, the stable form, called form II, is of lower density but with the chains still parallel to each other and still in a hexagonal array (*86–89*). They are believed to be in a disordered state, as judged by the blurring of all nonequatorial reflections in the X-ray diffraction pattern, and the marked decrease of the second moment of the wide-line proton NMR spectrum (*90*) indicates the onset of molecular motion. However, the details of the form II chain conformation and the nature of the motion are not well established. Suehiro and Takayanagi (*86*) proposed that the chain has a single definite structure, similar to that of form I except that the skew angle is decreased from 109° to 80°. They further proposed that the motion consists of large torsional oscillations about the carbon–carbon single bonds. Evans and Woodward (*88*) employed this conformation to calculate the heat capacity of form II and reported good agreement with experiment below and above the form I → form II transition; they did not consider chain motion or attempt to calculate the heat capacity during the transition. Iwayanagi and Miura (*90*) proposed that, instead of undergoing large torsional oscillations, the chains are rotating about their long axes. Grebowicz and co-workers (91, 92) made thermodynamic calculations similar to those of Evans and Woodward (88) but assumed a conformationally disordered state for form II. DeRosa et al. (89) also proposed a disordered conformation—based on packing energy calculations—consisting of a 50:50 equilibrating mixture of a and b:

$$
\begin{array}{cccccc}
\text{a: E} & \text{s}^{\pm}(90°) & \text{t} & \text{s}^{\mp}(90°) & \text{E} \\
\text{b: E} & \text{s}^{\pm}(90°) & \text{t} & \text{cis} & \text{E}
\end{array}
$$

This conformation corresponds to a 25% probability of cis for $CH–CH_2$ bonds.

We previously reported (93) the solid-state ^{13}C NMR spectra of TPBD at room temperature. The results demonstrate that independent carbon nuclei can be observed from the crystalline and the mobile fold-surface regions of TPBD single crystals (94, 95). The olefinic and methylene carbons in the folds appear from their chemical shifts to have essentially the same average conformation as the 1,4-trans sequences in amorphous bulk polybutadiene. In addition, the ^{13}C spin-lattice relaxation times (T_1) of the folds are observed to be the same as for amorphous polymer (96), an observation indicating that chain motions in the two phases are similar.

Figure 6 shows CPMAS/DD spectra as a function of temperature. The spectrum at 23 °C, obtained by using a 1.0-ms contact time, shows single olefinic and methylene resonances for the crystalline stems of form I. As the temperature is increased, new resonances appear at higher field positions for both carbons, reflecting the onset of the solid–solid phase transition. At temperatures where both form I and II are present, a contact time of 2.0 ms was chosen to permit observation of both forms. However, because of substantial differences in chain mobility (vide infra), the intensities of these resonances do not quantitatively reflect the ratio of these phases. Despite this fact, we can estimate the midpoint of the transition to be ~60 °C; it is essentially complete at 65 °C. The midpoint observed in the initial heating of solution-crystallized TPBD is ~50 °C. The higher transition point observed following cooling and subsequent heating is probably the result of crystalline annealing during the first heating.

At 23 °C the chemical shifts of the olefinic and methylene carbons of form I are 130.7 and 35.2 ppm, respectively (97). Those of form II are more shielded by 1.2 and 1.8 ppm, differences very close to those reported for fold-surface (i.e., amorphous) carbons versus crystalline stem carbons for form I (93). Despite the close similarity in chemical shift between the crystalline stem carbons of form II and amorphous carbons, individual resonances can be observed in non-cross-polarized spectra. These phases can be clearly differentiated in the course of an inversion-recovery (27) T_1 measurement.

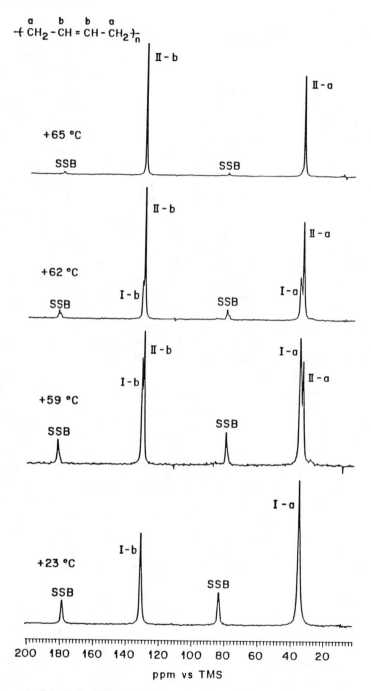

Figure 6. *^{13}C NMR CPMAS/DD spectra, 50.31 MHz, of 1,4-trans-polybuta-diene (I is form I; II is form II). SSB indicates spinning side bands.*

Figure 7 shows the T_1 spectra for the methylene carbons observed at 70 °C and therefore representing only form II. The fully relaxed spectrum ($\tau = 25$ s) shows partial resolution of the two phases. Near the null, both positive and inverted resonances are observed, clearly indicative of two phases relaxing at different rates. As a result of the partial overlap of the methylene resonances in Figure 7, accurate T_1s cannot be determined from this inversion-recovery data. The T_1 data measured by the cross-polarization method (CPT_1) (26) are shown in Table VIII. The form II stem carbons exhibit values of 10.5 and 12.2 s for CH_2 and CH, respectively. The value for the methylene carbons in the fold is estimated from the null point in Figure 7 to be ~0.7 s. These carbons are more shielded than the stem carbons by 0.6 ppm. For form I at 23 °C (Table VIII), the T_1 values are 0.33 and 0.65 s for the surface-fold CH_2 and CH, respectively, and the crystalline stem carbons exhibit much longer values having two components: 55 and 130 s for CH_2 and 53 and 123 s for CH. The shorter values may correspond to monomer units near the crystal surface (98).

The markedly greater shielding of the form II stem carbons is difficult to understand on the basis of the conformation proposed by Suehiro and Takayanagi (86), nor can the profound difference in carbon T_1 values for form I and II stems be explained merely by torsional oscillations. Both observations seem consistent, however, with the disordered conformation suggested by DeRosa et al. (89), possibly combined with chain rotation, as proposed by Iwayanagi and Miura (90).

Additional insight into the nature of the chain motion in TPBD can be obtained from observing the nonspinning ^{13}C spectra as a function of temperature. The CP/DD spectrum at 23 °C in Figure 8a, corresponding to form I, shows that the olefinic carbon has an axially asymmetric chemical shift anisotropy, as previously reported (93). The value of $\sigma_{11}-\sigma_{33}$ is ~178 ppm. The spectrum of form II recorded at 83 °C without CP (100) (Figure 8b) shows a dramatic narrowing in the powder pattern as a result of chain motion. The form of the pattern at 83 °C indicates a very anisotropic motion because the shift tensor does not simply average to the isotropic value, σ_i; instead the pattern of both the olefinic and methylene carbons shows an unsymmetrical change in addition to a growth in intensity at the isotropic positions. Of course, without CP a fraction of the intensity observed at σ_i must be attributed to the amorphous carbons.

Additional evidence for the presence of conformational disorder is found in the differential scanning calorimetry (DSC) for TPBD, recorded on a Perkin-Elmer DSC-4 with a heating rate of 10 °C/min. A large endotherm at 67 °C is associated with the form I → form II transition, and a much smaller endotherm at 133.2 °C is associated with melting. The entropy of the solid–solid transition (ΔS_{tr}) in TPBD has been shown experimentally to be 1.6–2.0 times the entropy of melting (ΔS_m), depending on sample preparation (101).

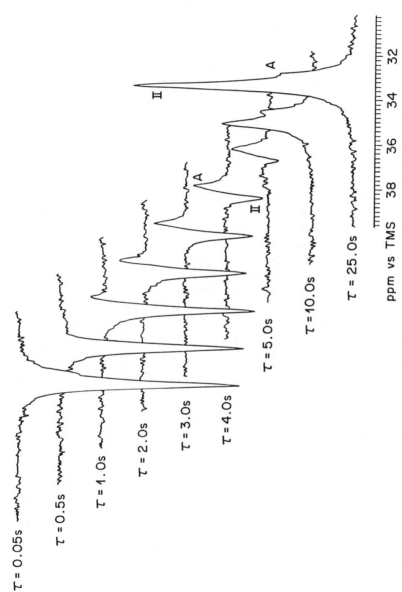

Figure 7. ^{13}C *NMR inversion-recovery spectra, 50.31 MHz, of the methylene carbons in 1,4-trans-polybutadiene at 70 °C (II is form II; A is amorphous). τ is inversion recovery delay time.*

Table VIII. ^{13}C Cross-Polarization T_1 Values for
1,4-*trans*-Polybutadiene

Temp., °C	Form	Stem		Fold	
		CH_2	$CH=$	CH_2	$CH=$
23.0	I	55, 130	53, 123	0.33[a]	0.65[a]
50.5	I	28, 69	40, 75	—	—
60.0	I	23, 56	28, 66	—	—
60.0	II	8.5	9.1	—	—
70.0	II	10.5	12.2	~0.7[b]	—

NOTE: All T_1 values are given in seconds.
[a]Inversion-recovery measurement.
[b]Estimate from inversion-recovery null point (Figure 7).

Earlier work (93, 99) showed that the treatment of TPBD crystals with m-chloroperbenzoic acid results in the epoxidation of the surface folds only and not the crystalline stems. The solid-state ^{13}C NMR spectra showed that the epoxidation results in the immobilization of the surface folds and that the oxirane rings have probably raised the T_g of the fold surface above room temperature. Our present work supports this conclusion by the observation of a ^{13}C T_1 of ~5 s for the oxirane carbons as compared to 0.6 s for the CH carbons in the folds of nonepoxidized TPBD.

We also examined the epoxidized polymer (28% total epoxidation) at various temperatures to observe the effect of this treatment upon the solid–solid transition. The sample's behavior changes little as compared to that of untreated TPBD. The ^{13}C CPMAS/DD spectra shown in Figure 9 indicate that the midpoint of the transition is 47 °C, similar to the midpoint observed in the initial heating of the nonepoxidized TPBD material. Subsequent temperature cycling of the epoxidized sample does not change the midpoint of the transition because the immobilized surface prevents thickening of the crystals by annealing. In addition, examination of the X-ray diffractograms of TPBD and epoxidized TPBD shows identical crystalline structures for the two materials as indicated by the position of the main reflection at $2\theta = 22.5°$. The fact that the presence of immobilized oxirane rings on the surface of the crystal does not prevent or even perturb the solid–solid phase transition indicates that the folds are not involved in the phase transition of TPBD in any significant manner. Also, there is probably little motion along the direction of the crystalline stem in form II. Such motion would require movement of the oxirane folds, which appear to be immobile at 47 °C, as evidenced by the broadened resonances (Figure 9) for the oxirane CH and CH_2 carbons.

High-energy irradiation of TPBD crystals initiates an expansion of the crystalline lattice at room temperature in a manner similar to that observed at the form I → form II transition. Such irradiation will probably cause cross-linking, which in turn will inhibit motion within the crystalline regions of this analogue of form II.

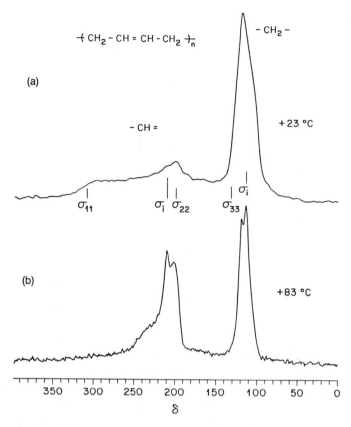

Figure 8. ^{13}C *NMR nonspinning DD spectra, 50.31 MHz, of 1,4-trans-poly-butadiene: (a) form I with CP, with a 5-s pulse-repetition delay, and (b) form II without CP, with a 60-s pulse-repetition delay.*

Polydiacetylenes (PDA) (102–105)

PDAs prepared (*106*) by the topochemical polymerization of single-crystal

$$= C - C \equiv C - C =$$

with R substituents on the first and third carbons

diacetlyene (DA) monomers are an interesting class of polymers, much studied (*107, 108*) because of their availability as large single crystals and for their unusual optical properties. Poly(ETCD), whose substituent R is $(CH_2)_4$–OCONH–CH_2–CH_3, is typical of the PDAs that exhibit a thermochromic transition, as evidenced by a change in color from blue to red at ~115 °C in the heating process. (*Also see* Figure 10.)

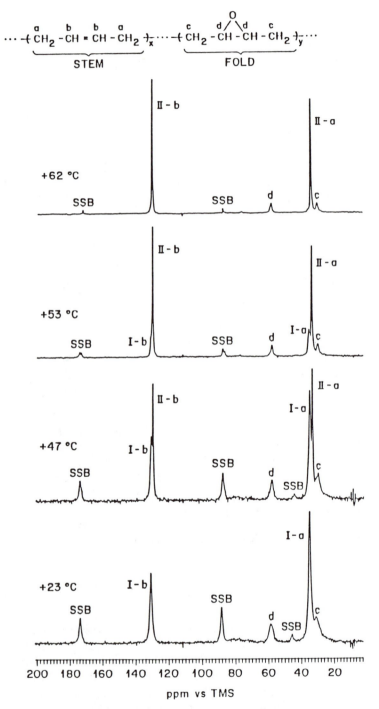

Figure 9. ^{13}C NMR CPMAS/DD spectra, 50.31 MHz, of crystal surface epoxidized 1,4-trans-polybutadiene (I is form I; II is form II).

Figure 11 shows the temperature dependence of ^{13}C NMR spectra in the heating process. The coexistence of both phases can clearly be seen at 115 °C in the heating process. The ^{13}C chemical shifts and T_1s measured for poly(ETCD) are displayed for the blue ($T < 115$ °C) and red ($T > 115$ °C) crystalline phases in Table IX. We observed no resonances characteristic of the butatrienic form of backbone conjugation in either phase. The constancy

$$
\begin{array}{ccc}
\text{R} & & \text{R} \\
| & & | \\
-\text{C}=\text{C}=\text{C}=\text{C}- &
\end{array}
$$

of the $C=O$ chemical shift implies that the hydrogen-bonded network of side chains (*see* poly(ETCD) structure) is retained in both phases. On the basis of the differences in ^{13}C chemical shifts of the $-C\equiv$ and $\beta,\gamma-CH_2$ carbons in the blue and red phases, the thermochromic blue-to-red phase transition observed in solid poly(ETCD) is most likely accompanied by a

The intramolecular structure of poly(ETCD) with hydrogen bonding. Protons are not drawn, and — represents the hydrogen bond.

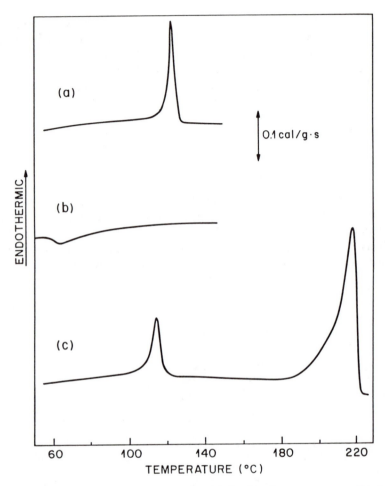

Figure 10. DSC scans at 10 °C/min for (a) the first heating process, (b) the following cooling process, and (c) the second heating process to above the melting point in poly(ETCD).

planar-to-nonplanar conformational change in the backbone and an extension of the side chains to a nearly all-trans conformation.

The chemical shift of the backbone $-C\equiv$ carbon is especially sensitive to the transition changing from 107 to 103 ppm on transition from the blue to the red phase. The observed chemical shift change of the resonance cannot be explained by defects in the backbone structure, because only a single resonance is observed for $-C\equiv$ in both phases. The backbone must be uniformly distorted. We have found (*104*) that, irrespective of side chains, all blue-phase PDAs have their $-C\equiv$ resonances at ~107 ppm, and their

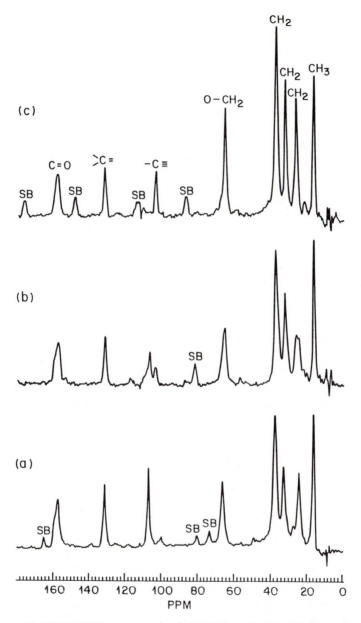

Figure 11. CPMAS/DD spectra of poly(ETCD) at 23 (a), 115 (b), and 127 (c) °C. Spectra were referenced to the resonance of POM (89.1 ppm from tetramethylsilane (TMS) (25)). The peaks labeled as SB correspond to the spinning side bands.

Table IX. ^{13}C Chemical Shifts and Spin-Lattice Relaxation Times, T_1, for Poly(ETCD)

$$\overset{\displaystyle R}{\underset{\displaystyle R}{\overset{\displaystyle |}{\underset{\displaystyle |}{= C - C \equiv C - C =}}}}$$

$$R = \underset{\alpha}{CH_2} - \underset{\beta}{CH_2} - \underset{\gamma}{CH_2} - \underset{\delta}{CH_2} - \underset{\epsilon}{OCONH - CH_2 - CH_3}$$

| | ^{13}C (ppm vs. TMS) | | T_1 (s) | |
| | Blue Phase | Red Phase | Blue Phase | Red Phase |
Carbon	(Low T)	(High T)	(Low T)	(High T)
C=O	157.5	158.3	116	47
>C=	131.6	132.0	153	25
-C=	107.4	103.6	172	30
δ-CH$_2$	66.6	65.5	11, 110	6
α-CH$_2$	37.3	37.8	9, 103	8
ϵ-CH$_2$	32.9	32.6	9, 108	6
β,γ-CH$_2$	24.5	26.4	2, 97	4
CH$_3$	16.2	16.7	2, 14	6

$-C\equiv$ carbons resonate at ~103 ppm for red-phase PDAs. The backbone structure of the blue phase was found to be more planar than that of the red phase on the basis of the overall $-C\equiv$ shift positions. The transition between blue and red phases is likely achieved by small rotations of opposite sign about the single C–C backbone bonds. The chemical shifts of the β,γ-

CH$_2$ carbons are 2 ppm downfield in the red phase compared to their position in the blue phase. This result strongly suggests that the alkyl side-chain bonds have more trans or planar character in the red phase. Comparison of the chemical shifts of the β,γ carbons in poly(ETCD) with those of model systems whose solid-state conformations are known (see Chart IV, for example), makes it appear that the alkyl portions of the side chains in poly(ETCD) have the gt$\bar{g}$ conformation in the blue phase and the t′tt′ conformation in the red phase for the bonds of the tetramethylene fragment, where g,$\bar{g}$ are gauche rotations of opposite sign and t′ represents an imperfect or nonplanar trans (t) conformation. This change in side-chain conformation is consistent with the expansion of the crystalline lattice in the side-chain direction observed by X-ray diffraction to accompany the transition from the blue to the red phase (*105, 109*). In addition, the conformational transition gt$\bar{\text{g}}$ ⇄ t'tt' suggested for the alkyl portion of the side chains is of the Helfand type (*110*) thought to be the most facile for alkane chains in condensed media.

When poly(ETCD) is recrystallized, upon cooling from its melt, the –C≡ resonance moves to 102 ppm and the β,γ-CH$_2$ resonances to 27.5 ppm (*105*). Thus, the backbone appears even more nonplanar than in the red phase, and the side chains appear more extended (ttt) when crystallized from the melt (Figure 12).

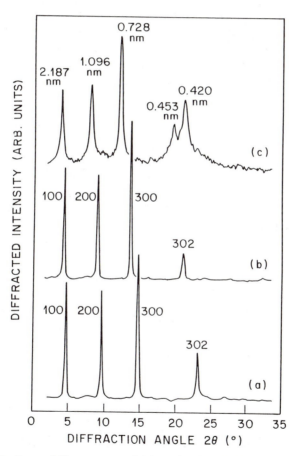

Figure 12. X-ray diffractograms of (a) single-crystal blue phase, (b) single-crystal red phase, and (c) once-melted poly(ETCD). The 100 peak corresponds to the spacing between methyl carbons belonging to side chains attached to adjacent –C = C– carbons.

Poly(bis-4-ethylphenoxyphosphazene) (PBEPP) (111, 112)

Polyphosphazenes are an interesting class of inorganic polymers that can

O–R
|
(–N=P–)
|
O–R

exhibit (113) several phase transitions dependent on their thermal histories. The most unique among these transitions is a thermotropic crystal–liquid-crystal transition, T(1), preceding the final melting. Figures 13 and 14 present DSC scans and X-ray diffractograms of PBEPP (R = $C_6H_4CH_2CH_3$) that reflect this transition. In Figure 15, the ^{31}P MAS/DD spectra of PBEPP recorded at several temperatures are compared. The sudden decrease in line width observed above 100 °C (T(1)) reflects the crystal–liquid-crystal

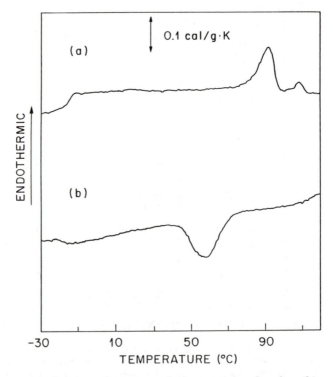

Figure 13. DSC scans of PBEPP in the heating (a) and cooling (b) process.

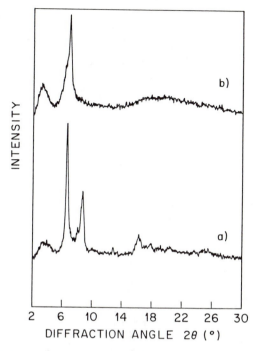

Figure 14. X-ray diffractograms of PBEPP at 25 (a) and 120 (b) °C.

phase transition and indicates considerable backbone motion in the liquid-crystalline phase. This result is consistent with the disappearance of all intrachain diffraction peaks (*see* Figure 14) in the liquid-crystalline state (above 100 °C).

The CPMAS/DD ^{13}C NMR spectra of PBEPP are presented in Figure 16 both below and above the T(1) crystal–liquid-crystal transition. The spin-lattice relaxation times, T_1, presented in Table X, indicate that the side chains are also mobile in the liquid-crystalline phase. In addition, the short T_1s observed only for the protonated aromatic carbons in the crystal indicate mobile phenyl rings rotating rapidly about their 1,4-axes even in this phase. However, crystalline phenyl-ring rotation is not rapid enough to average the chemical shifts of the methyl and several aromatic carbon resonances that are split into multiplets (Figure 16a).

Conclusions

We hope the several examples discussed here will serve to demonstrate the utility of high-resolution NMR spectroscopy to the study of solid polymers. Particularly when coupled with DSC and X-ray diffraction techniques, solid-

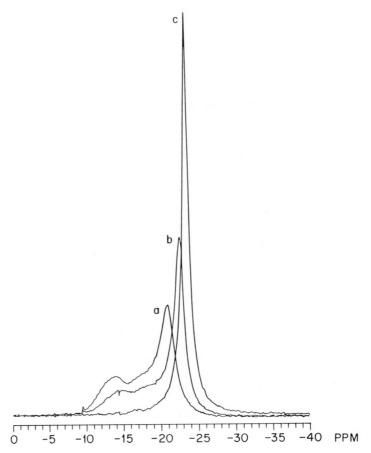

Figure 15. *³¹P MAS/DD spectra of PBEPP at 25 (a), 80 (b), and 120 (c) °C.*

**Table X. ¹³C Spin-Lattice Relaxation Times, T_1, for
PBEPP**

$$-O-\overset{a}{\underset{b\quad c}{\langle\bigcirc\rangle}}\overset{}{\underset{1\quad 2}{d}}-CH_2-CH_3$$

Carbon	T = 25 °C	T = 100 °C
C-a	17	4
C-d	15	3
C-b	1.5	0.6
C-c	1.5	0.5
CH₂	10	0.8
CH₃	2	2

NOTE: All T_1 values are given in seconds.

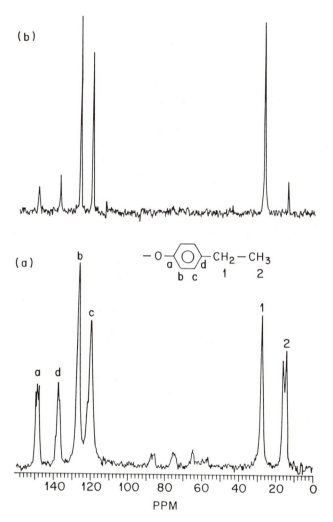

Figure 16. ^{13}C spectra below and above T(1): CPMAS/DD spectra at 24 (a) and 120 (b) °C.

state NMR spectroscopy can be a valuable tool for studying the structures, conformations, dynamics, and phase transitions of solid polymers, and will serve as a foundation upon which to build structure–property relations for polymers.

Acknowledgments

We are grateful to M. Thakur for providing all PDA samples and to S. V. Chichester-Hicks and R. C. Haddon for synthesizing the sample of PBEPP.

References

1. Bovey, F. A. *High-Resolution NMR of Macromolecules;* Academic: New York, 1972.
2. Tonelli, A. E.; Schilling, F. C. *Acc. Chem. Res.* **1981**, *14*, 233.
3. Bovey, F. A. *Chain Structure and Conformation of Macromolecules;* Academic: New York, 1982.
4. VanderHart, D. L.; Earl, W. L.; Garroway, A. N. *J. Magn. Reson.* **1975**, *44*, 361.
5. Moller, M. *Adv. Polym. Sci.* **1985**, *66*, 59.
6. Earl, W. L.; VanderHart, D. L. *Macromolecules* **1979**, *12*, 762.
7. Bunn, A.; Cudby, M. E. A.; Harris, R. K.; Packer, K. J.; Say, B. J. *J. Chem. Soc. Chem. Commun.* **1981**, 15.
8. Belfiore, L. A.; Schilling, F. C.; Tonelli, A. E.; Lovinger, A. E.; Bovey, F. A. *Macromolecules* **1984**, *17*, 2561.
9. Sacchi, M. C.; Locatelli, P.; Zetta, L.; Zambelli, A. *Macromolecules* **1983**, *17*, 483.
10. Terao, T.; Maeda, S.; Saika, A. *Macromolecules* **1983**, *16*, 1535
11. Kitamaru, R.; Horii, F.; Murayama, M. *Polym. Bull.* **1982**, *7*, 583.
12. Ando, I.; Yamanobe, T.; Sorita, T.; Komoto, T.; Sato, H.; Deguchi, K.; Imanari, M. *Macromolecules* **1981**, *17*, 1955.
13. Corradini, P.; Natta, G.; Ganis, P.; Temussi, P. A. *J. Polym. Sci. Part C* **1967**, *16*, 2477.
14. Turner-Jones, A.; Aizlewood, J. M.; Beckett, D. R. *Makromol. Chem.* **1961**, *75*, 134.
15. Natta, G.; Corradini, P.; Bassi, I. W. *Nuovo Cimento Suppl.* **1960**, *15*, 52.
16. Miyashita, T.; Yokouchi, M.; Chatani, Y.; Tadokoro, H. Annual Meeting of the Society of Polymer Science of Japan; Tokyo, 1974, preprint p 453. Quoted in Tadokoro, H. *Structure of Crystalline Polymers;* Wiley-Interscience: New York, 1979; p 405.
17. Turner-Jones, A. *J. Polym. Sci. Part B* **1963**, *1*, 455.
18. Petraccone, V.; Priozzi, B.; Frasci, A.; Corradini, P. *Eur. Polym. J.* **1976**, *12*, 323.
19. Zannetti, R.; Manaresi, P.; Buzzori, G. C. *Chim. Ind. (Milan)* **1961**, *43*, 735.
20. Danusso, F.; Gianotti, G. *Makromol. Chem.* **1963**, *61*, 139.
21. Geacintov, C.; Schottand, R.; Miles, R. B. *J. Polym. Sci. Polym. Lett. Ed.* **1963**, *1*, 587.
22. Miller, R. C.; Holland, V. F. *J. Polym. Sci. Polym. Lett. Ed.* **1961**, *2*, 519.
23. Schilling, F. C.; Bovey, F. A.; Tonelli, A. E.; Tseng, S.; Woodward, A. E. *Macromolecules* **1981**, *17*, 728.
24. Perez, E.; VanderHart, D. L. *Polymer* **1987**, *28*, 733.
25. Earl, W. L.; VanderHart, D. L. *J. Magn. Reson.* **1982**, *48*, 35.
26. Torchia, D. A. *J. Magn. Reson.* **1978**, *30*, 613.
27. Farrar, T. C.; Becker, E. D. *Pulse Fourier Transform NMR;* Academic Press: New York, 1971.
28. Gomez, M. A.; Cozine, M. H.; Schilling, F. C.; Tonelli, A. E.; Bello, A.; Fatou, J. G. *Macromolecules* **1987**, *20*, 1761.
29. Perez, E.; Gomez, M. A.; Bello, A.; Fatou, I. G. *Colloid Polym. Sci.* **1983**, *261*, 571.
30. Gomez, M. A.; Atkins, E. D. T.; Upstill, C.; Bello, A.; Fatou, J. G. *Polymer*, **1988**, *29*, 224.
31. Gomez, M. A.; Fatou, J. G.; Bello, A. *Eur. Polym. J.* **1986**, *22*, 43.

32. Stothers, I. B. *Carbon-13 NMR Spectroscopy;* Academic: New York, 1972; Chapters 3 and 5.
33. VanderHart, D. L. *J. Magn. Reson.* **1981**, *44*, 117.
34. VanderHart, D. L.; Khoury, F. *Polymer* **1983**, *25*, 1589.
35. Bunn, A.; Cudby, M. E. A.; Harris, R. K.; Packer, K. J.; Say, E. J. *Polymer* **1982**, *23*, 694.
36. c-(TO)$_4$ actually adopts the (T$_2$G$_2$T$_2\overline{G}_2$)$_2$ conformation in the crystal (37) because of its cyclic structure. However, both the α and β CH$_2$ carbons experience the same kinds and numbers of γ-gauche interactions in both PTO and c-(TO)$_4$ crystals.
37. Groth, P. *Acta Chem. Scand.* **1971**, *25*, 725.
38. Reference 32, Chapter 4.
39. This is a consequence (40) of the conservation of total s and p character of the sp^3 molecular orbitals around >C<.
40. Huheey, J. E. *Inorganic Chemistry: Principles of Structure and Reactivity;* Harper and Row: New York, 1972; Chapter 4.
41. Gomez, M. A.; Cozine, M. H.; Tonelli, A. E. *Macromolecules* **1988**, *21*, 388.
42. Boye, C. A., Jr.; Overton, J. R. *Bull. Am. Phys. Soc. Sci.* 2 **1974**, *19*, 352.
43. Jakeways, R.; Ward, I. M.; Wilding, M. A.; Hall, I. H.; Desborough, I. J.; Pass, M. G. *J. Polym. Sci. Polym. Phys. Ed.* **1975**, *17*, 799.
44. Jakeways, R.; Smith, T.; Ward, I. M.; Wilding, M. A. *J. Polym. Sci. Polym. Lett. Ed.* **1976**, *14*, 41.
45. Yokouchi, M.; Sakakibara, Y.; Chatani, Y.; Tadokoro, H.; Tanaka, T.; Yoda, K. *Macromolecules* **1976**, *9*, 266.
46. Brereton, M. G.; Davies, C. R.; Jakeways, R.; Smith, T.; Ward, I. M. *Polymer* **1978**, *19*, 17.
47. Mencik, Z. *J. Polym. Sci. Polym. Phys. Ed.* **1975**, *13*, 2173.
48. Hall, I. H.; Pass, M. G. *Polymer* **1977**, *17*, 807.
49. Desborough, I. J.; Hall, I. H. *Polymer* **1977**, *18*, 825.
50. Stambaugh, B. D.; Koenig, J. L.; Lando, J. B. *J. Polym. Sci. Polym. Lett. Ed.* **1979**, *15*, 299; *J. Polym. Sci. Polym. Phys. Ed.* **1979**, *171*, 1053.
51. Hall, I. H. In *Fiber Diffraction Methods;* French, A. D.; Gardner, K. H., Eds.; ACS Symposium Series No. 141; American Chemical Society: Washington, DC, 1980, p 335.
52. Ward, I. M.; Wilding, M. A. *Polymer* **1977**, *18*, 327.
53. Davidson, I. S.; Manuel, A. I.; Ward, I. M. *Polymer* **1983**, *24*, 30.
54. Grenier-Loustalot, M.-F.; Bocelli, G. *Eur. Polym. J.* **1983**, *20*, 957.
55. Axelson, D. E. In *High Resolution NMR Spectroscopy of Synthetic Polymers in Bulk;* Komoroski, R. E., Ed.; VCH: Deerfield Beach, FL, 1986, Chapter 6.
56. Havens, J. R.; Koenig, J. L. *Polym. Commun.* **1983**, *24*, 194.
57. Perry, B. C.; Koenig, J. L.; Lando, J. B. *Macromolecules* **1987**, *20*, 422.
58. Lyerla, J. R.; Komoroski, R. A.; Axelson, D. E. In ref. 55, Chapters 2–5.
59. The T_g of PBT has been reported in the literature as varying from 22 to 80 °C (*see*, for example: Lewis, O. G. *Physical Constants of Linear Homopolymers;* Springer-Verlag: New York, 1968; pp 150–151; and Lee, W. A.; Knight, G. J. In *Polymer Handbook;* Brandrup, J.; Immergut, E. H., Eds.; Interscience: New York, 1966; p III-79). DSC measurements performed on our α-PBT sample and a quenched amorphous PBT sample yielded T_g 50–55 °C.
60. Lovinger, A. J.; Belfiore, L. A.; Bowmer, T. N. *J. Polym. Sci. Polym. Phys. Ed.* **1985**, *23*, 1449.
61. Jelinski, L. W.; Dumais, J. J.; Watnick, P. I.; Engel, A. K.; Sefcik, M. D. *Macromolecules* **1983**, *16*, 409.
62. Garbow, J. R.; Schaefer, J. *Macromolecules* **1987**, *20*, 819.

63. Gomez, M. A.; Tanaka, H.; Tonelli, A. E. *Polymer* **1987**, *28*, 2227.
64. Natta, G.; Corradini, P. *Nuovo Cimento Suppl.* **1960**, *15*, 40.
65. Turner-Jones, A.; Aizlewood, J. M.; Beckett, D. R. *Makromol. Chem.* **1964**, *75*, 134.
66. Turner-Jones, A.; Cobbold, A. *J. Polymer Lett.* **1968**, *6*, 539.
67. Natta, G.; Peraldo, M.; Corradini, P. *Rend. Accad. Naz. Lincei* **1959**, *26*, 14.
68. Wyckoff, H. *J. Polym. Sci.* **1962**, *62*, 83.
69. Gailey, J. A.; Ralston, P. H. *Soc. Plast. Eng. Trans.* **1964**, *4*, 29.
70. Bodor, G.; Grell, M.; Kallo, A. *Faserforsch. Textiltech.* **1964**, *15*, 527.
71. Corradini, P.; Petraccone, V.; De Rosa, C.; Guerra, G. *Macromolecules* **1986**, *19*, 2689.
72. Miller, R. L. *Polymer* **1960**, *1*, 135.
73. Zannetti, R.; Celotti, G.; Armigliato, A. *Eur. Polym. J.* **1970**, *6*, 879.
74. ^{13}C NMR of an i-PP solution (Profax) indicated that it contained 94% isotactic (mm) triads.
75. Lovinger, A. J.; Ching, J. O.; Gryte, C. C. *J. Polym. Sci. Polym. Phys. Ed.* **1977**, *15*, 641.
76. Randall, J. C. *J. Polym. Sci. Polym. Phys. Ed.* **1976**, *14*, 1693.
77. Bunn, A.; Cudby, M. E. A.; Harris, R. K.; Packer, K. J.; Say, E. *J. Polymer* **1982**, *23*, 694.
78. Varga, J.; Toth, F. *Makromol. Chem. Macromol. Symp.* **1986**, *5*, 213.
79. Shi, G.; Huang, E.; Cao, Y.; He, Z.; Han, Z. *Makromol. Chem.* **1986**, *187*, 643.
80. Leugering, H. J. *Makromol. Chem.* **1967**, *109*, 204.
81. Duswalt, A. *Am. Chem. Soc. Div. Org. Coat.* **1970**, *30*, 93.
82. Fleming, W. W.; Fyfe, C. A.; Kendrick, R. D.; Lyerla, J. R.; Vanni, H.; Yannoni, C. S. In *Polymer Characterization by ESR and NMR*; Woodward, A. E.; Bovey, F. A., Eds.; ACS Symposium Series No. 142, American Chemical Society: Washington, DC, 1980.
83. Schilling, F. C.; Gomez, M. A.; Tonelli, A. E.; Bovey, F. A.; Woodward, A. E. *Macromolecules* **1987**, *20*, 2954.
84. Natta, G.; Corradini, P.; Porri, L. *Atti Accad. Naz. Lincei Cl. Sci. Fis. Mat. Nat. Rend.* **1956**, *20*, 728.
85. Natta, G.; Corradini, P. *Nuovo Cimento Suppl.* **1960**, *1*, 9.
86. Suehiro, J.; Takayanagi, M. *J. Macromol. Sci. Phys.* **1970**, *B4*, 39.
87. Stellman, J. M.; Woodward, A. E.; Stellman, S. D. *Macromolecules* **1973**, *6*, 330.
88. Evans, H.; Woodward, A. E. *Macromolecules* **1978**, *11*, 685.
89. DeRosa, C.; Napolitano, R.; Pirozzi, B. *Polymer* **1986**, *26*, 2039.
90. Iwayanagi, S.; Miura, I. *Rep. Progr. Polym. Phys.* **1965**, *8*, 1965.
91. Grebowicz, J.; Lau, S.-F.; Wunderlich, B. *J. Polym. Sci. Polym. Symp.* **1984**, *71*, 19.
92. Grebowicz, J.; Aycock, W.; Wunderlich, E. *Polymer* **1986**, *27*, 575.
93. Schilling, F. C.; Bovey, F. A.; Tonelli, A. E.; Tseng, S.; Woodward, A. E. *Macromolecules* **1984**, *17*, 728.
94. Canale, A.; Hewett, W. A.; Shryne, T. M.; Youngman, E. A. *Chem. Ind.* **1962**, 1054. Rinehart, R. E. *Polym. Prepr. (Am. Chem. Soc. Div. Polym. Chem.)* **1966**, *7*, 556.
95. Wang, P.; Woodward, A. E. *Macromolecules* **1987**, *20*, 1818, 1823.
96. Jelinski, L. W.; Dumais, J. J.; Watnick, P. I.; Bass, S. V.; Shepherd, L. I. *Polym. Sci. Polym. Chem. Ed.* **1982**, *20*, 3285.
97. All chemical shifts in this work are referenced to the amorphous methylene carbon at 32.79 ppm versus TMS (*93*).

98. Perez, E.; VanderHart, D. L. *J. Polym. Sci. Polym. Phys. Ed.* **1987**, *25*, 1637.
99. Schilling, F. C.; Bovey, F. A.; Tseng, S.; Woodward, A. E. *Macromolecules* **1983**, *16*, 808.
100. The nonspinning spectrum recorded with CP at 83 °C shows a complete loss of intensity in the region of σ_i. This distortion of the powder pattern for the mobile chains of form II is thought to result from a failure to cross-polarize those nuclei whose C–H vectors are oriented at or close to the magic angle of 54.7°. We thank Dr. L. W. Jelinski for bringing this point to our attention.
101. Ng, S. B.; Stellman, J. M.; Woodward, A. E. *J. Macromol. Sci. Phys.* **1973**, *7*, 539.
102. Tanaka, H.; Thakur, M.; Gomez, M. A.; Tonelli, A. E. *Macromolecules* **1987**, *20*, 3084.
103. Tanaka, H.; Gomez, M. A.; Tonelli, A. E.; Thakur, M. *Macromolecules* **1989**, *22*, 1208.
104. Tanaka, H.; Thakur, M.; Gomez, M. A.; Tonelli, A. E. unpublished data.
105. Tanaka, H.; Gomez, M. A.; Tonelli, A. E.; Lovinger, A. J.; Davis, D. D.; Thakur, M. *Macromolecules* **1989**, *22*, 2427.
106. Wegner, G. *Discuss. Faraday Soc.* **1980**, *68*, 494.
107. Bloor, D.; Chance, R. R. *Polydiacetylenes: NATO ASI Series E, Applied Science;* Martinus Nijhoff Publisher (Kluwer Academic: Norwell, MA), 1985.
108. Chance, R. R. *Encyclopedia of Polymer Science and Engineering;* Wiley: New York, 1986; Vol. 4, p 767.
109. Downey, M. J.; Hamill, G. P.; Rubner, M.; Sandman, D. J.; Velazquez, C. S. *Makromol. Chem.* **1988**, *189*, 1199.
110. Helfand, E. *J. Chem. Phys.* **1971**, *54*, 4651.
111. Tanaka, H.; Gomez, M. A.; Tonelli, A. E.; Chichester-Hicks, S. V.; Haddon, R. C. *Macromolecules* **1988**, *21*, 2301.
112. Tanaka, H.; Gomez, M. A.; Tonelli, A. E.; Chichester-Hicks, S. V.; Haddon, R. C. *Macromolecules* **1989**, *22*, 1031.
113. *Inorganic and Organometallic Polymers;* ACS Symposium Series No. 360; American Chemical Society: Washington, DC, 1988, Chapters 19–25, pp 250–312.

RECEIVED for review February 14, 1989. ACCEPTED revised manuscript November 30, 1989.

MORPHOLOGY

Small-Angle Neutron Scattering

Recent Applications to Multicomponent Polymer Systems

L. H. Sperling, J. N. Yoo, S. I. Yang, and A. Klein

Materials Research Center, Center for Polymer Science and Engineering, Materials Science and Engineering Department, Department of Chemical Engineering, Lehigh University, Bethlehem, PA 18015

Small-angle neutron scattering, SANS, is a new instrumental method of determining polymer chain radii of gyration, domain dimensions in phase-separated materials, and diffusion constants. Instruments currently in service include those at Grenoble, Jülich, Oak Ridge, National Institute of Standards and Technology, and Los Alamos. The basic theory is reviewed, and examples of results are illustrated.

THREE MAJOR SCATTERING METHODS exist for the determination of the morphology in multicomponent polymer materials: light-scattering, small-angle X-ray scattering (SAXS), and small-angle neutron scattering (SANS) (*1–9*). Each is more useful through particular ranges of domain sizes, contrast requirements, and information desired. Several terms common to light scattering and neutron scattering are summarized in Table I (*1*). To achieve contrast, for example, a good light-scattering experiment requires a reasonable difference in refractive index between the two polymers, but a reasonable difference in electron density is required for SAXS experiments. For SANS experiments, good contrast can be obtained by deuterating one of the phases or adding deuterated polymer to one of the phases.

The general problem of the size and shape of domain in multicomponent polymer materials stands at the very heart of the field. The morphology is influenced by the synthetic method and by processing. Once formed, it

0065–2393/90/0227–0455$06.00/0

Table I. Scattering Term Equivalents

Light Scattering	Neutron Scattering
R, Rayleigh ratio	$d\Sigma/d\Omega$, cross section
n, refractive index	$\Sigma b_i/V$, scattering length density
λ, wavelength ($\sim$5000 Å)	λ, wavelength ($\sim$5 Å)
$(dn/dc)^2$, determines scattering intensity	$(a_H - a_D)^2$, determines scattering intensity

SOURCE: Reproduced with permission from reference 1. Copyright 1984.

determines the usefulness of the final product. At one time, almost the only instrumental techniques available for the examination of morphology were electron microscopy and dynamic mechanical spectroscopy. Although enormous amounts of new fundamental information were obtained, many more types of samples could not easily be examined by these routes.

Multicomponent polymer materials, such as interpenetrating polymer networks, (IPNs), latex dispersions, and block copolymers, can be characterized via small-angle neutron scattering. For phase-segregated systems, the experiments reveal the correlation length, specific surface areas, and domain diameters. For example, domains of the order of several hundred angstroms were found for IPN systems, roughly in agreement with transmission electron microscopic (TEM) studies. For block copolymers, the domains are larger via SANS. The value of the difference is that SANS can be used in some cases where TEM cannot be conveniently used, and in other cases SANS provides entirely new information. For example, the kinetics of phase separation can also be analyzed, because phase separation via spinodal decomposition is far different from that of nucleation and growth. Depending on the ratio of end-to-end distance to particle diameter, latices may undergo a type of core-shell segregation. Recent experiments show that maximum segregation occurs when the chains are about half as big as the latex particle. Each of these areas will be briefly reviewed in this chapter.

Unlike ordinary light and X-rays, neutrons are actually particles of matter. The relationship between the particle mass characteristics, velocity through space, and corresponding wave-like behavior have been known since the works of Louis deBroglie in 1924 (10). The wavelength of thermal neutrons is about 1 Å. The longer the wavelength, and the smaller the angle, the larger the objects that can be studied. Quantitatively, this relationship is expressed in terms of the wave vector $\mathbf{K}$

$$\mathbf{K} = \frac{4\pi}{\lambda}\sin\theta \tag{1}$$

where the quantity λ represents the wavelength of the neutron radiation, and 2θ is the angle of scatter.

Table II. Selected SANS Instrumentation

Location	S-D (m)	Max. λ (Å)	Min. **K** (Å⁻¹)
Grenoble, France	40	15	0.0008
Jülich, Federal Republic of Germany	20	—[a]	0.001
Oak Ridge, TN	19	4.75	0.002
NIST, Gaithersburg, MD	3.5	18	0.005
NIST (under construction)	20	18	0.002
Los Alamos, NM	4	16.5	0.004

[a]— indicates not available.

The maximum particle size (or morphology) that can be characterized depends on the minimum value of **K** obtainable by a given instrument. In general, the maximum particle size that can be studied is the inverse of the minimum value of **K**. Several internationally important instruments are described in Table II. The sample-to-detector distance, S-D, controls the minimum angle obtainable. The maximum sizes now characterizable are about 1000 Å with appropriate allowance for resolution (3).

At the time of this writing, the facility at Oak Ridge National Laboratories is restarting after being closed for inspection and repairs. The facility at the National Institute of Standards and Technology (NIST) is closed for 3–6 months to install beam guides, and major new facilities are being built. A schematic of the low-Q (low-angle) diffractometer at Los Alamos is illustrated in Figure 1 (4).

One way to obtain smaller values of **K** is to design instruments that use colder and colder neutrons in order to increase their wavelength. The practical limit seems to be the temperature of liquid helium. The wavelength distribution obtained at NIST as a function of temperature is shown in Figure 2 (11–13).

The alternative is to design equipment with smaller and smaller attainable angles. This condition requires longer and longer distances between the sample and the detector. Eventually, such techniques become self-defeating because of the inverse square relationship between distance and beam intensity. Another way to reach low **K** is via the double-crystal diffractometer (DCD) technique. The only available instrument in the United States is at Oak Ridge, where data have been taken down to $\mathbf{K} \simeq 2 \times 10^{-4}$ Å⁻¹. The current state-of-the-art DCD has been built by Schwahn and co-workers at Jülich (12, 13). The evolution of SANS instrumentation is given in Table III.

Some instrumentation now available includes

1. The Los Alamos Neutron Scattering Center, Los Alamos National Laboratory

   - low-Q diffractometer (Q = **K**)
   - surface profile analysis reflectometer

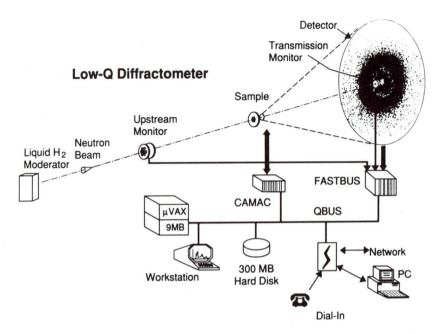

Figure 1. Schematic diagram of the low-Q diffractometer (LQD). A well-collimated pulse of neutrons with a broad wavelength range strikes the sample. Those neutrons scattered at forward angles between 0.3° and 3.5° are detected. The angular position in the detector and the elapsed time since the neutron pulse (time of flight) for each event are tallied in memory in the FASTBUS for subsequent examination and analysis by computer. (Reproduced with permission from reference 4.)

2. NIST reactor, National Institute of Standards and Technology, small-angle scattering facility

3. Argonne National Laboratory intense pulsed neutron source (IPNS)

4. Brookhaven National Laboratory

Oak Ridge, Los Alamos, and NIST are available to the public, and proposals are required.

This review will focus on small-angle neutron scattering.

Theory

One of the more important approaches to the application of SANS is derived from the correlation function, $\gamma(r)$, of Peter Debye (*14*). The scattering intensity, I, is related to the correlation function as follows:

$$I = AC \int_0^\infty \gamma(r)\, r^2\, \frac{\sin(\mathbf{K}r)}{\mathbf{K}} r\, \mathrm{d}r \qquad (2)$$

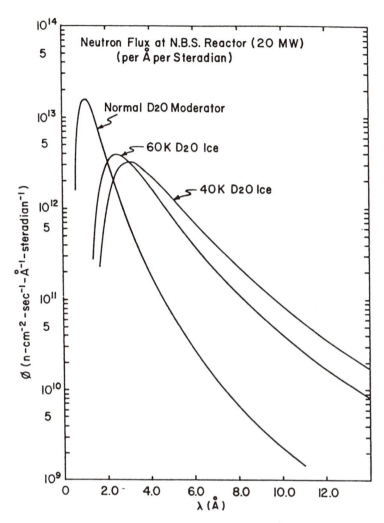

Figure 2. Neutron wavelength dependence on the temperature of the moderator. (Reproduced from reference 11.)

Table III. Evolution of SANS Instrumentation

Method	Location	Comments
Long flight path	(a) ILL, Grenoble (b) Oak Ridge (c) Jülich	Inverse square distance law means long experiment times
Long wavelength	NIST	Neutrons cooled via liquid He or H_2
Time-of-flight (TOF)[a]	Los Alamos	"White" neutrons, liquid H_2, pulsed source

[a]Pulsed neutrons separated via wavelength via TOF, equivalent of FTIR to IR.

where A is proportional to the total volume of the sample divided by the square of the difference between the coherent neutron scattering length of the mers, C represents the mean square deviation of the fluctuations in scattering length density, and

$$\gamma(r) = \exp\left[\frac{-r}{a}\right] \tag{3}$$

and a represents the correlation length. As shown in Figure 3, the quantity a can be interpreted in terms of the correlation length, the average distance across a domain,

$$l_1 = \frac{a}{\phi_2} \tag{4}$$

$$l_2 = \frac{a}{\phi_1} \tag{5}$$

or in terms of the specific surface area, S_{sp}

$$S_{sp} = 4\phi_2 \frac{1 - \phi_1}{a} \tag{6}$$

where ϕ_1 and ϕ_2 represent the volume fractions.

A maximum in the scattering due to spinodal decomposition can be interpreted in terms of "wavelength", Λ, a characteristic distance across a domain

$$\Lambda = \frac{2\pi}{K_m} \tag{7}$$

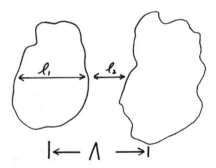

Figure 3. Schematic illustration of correlation lengths (l), specific surface areas (S_{sp}), and wavelength (Λ), as determined by SANS.

where K_m is the wave vector at which the angular maximum is observed. Other mathematical formulations exist to treat spherical domains, etc. Of course, for molecular solutions, radii of gyration of random coils can be calculated.

Selected Results to Date

Block Copolymers. Block copolymers such as polybutadiene-*block*-polystyrene are known to phase separate. The triblock copolymer forms thermoplastic elastomers because of phase separation, certainly not in spite of it. Previously, TEM on samples stained with osmium tetroxide revealed that the domains may be spherical, cylindrical, or lamellar shaped. Important parameters to be determined for spherical domains are diameters and the shape of the polymer chains inside of the domains.

The most important SANS studies on block copolymers were carried out by Cohen and co-workers (*15–17*). Domain diameters radii were calculated from the maxima as a function of angle. The angular peaks of spherical particles bear a relationship to the rainbow, where diffraction maxima occur from uniform water droplets. Domain radii are compared to TEM results in Table IV. The SANS studies result in significantly larger radii than TEM.

By partially deuterating the polybutadiene phase, chain dimensions can be determined. The highest molecular weight has relaxed chain dimensions that exceed the diameter of the domains. As might be expected, the chains are elastically compressed.

Latex Dispersions. Similar results were obtained by Linne et al. (*18*) on 6×10^6 g/mol polystyrene in latices of 380-Å diameter. In this case, the polymer chains are restricted by a water phase rather than an immiscible polymer phase.

A main finding of the investigations on latex dispersions using SANS was a significant segregation of the first polymerized material from that polymerized later (*see* Figure 4) (*19*). Segregation goes through a maximum when the chain dimensions are about half the latex diameter. The cause of the segregation is theorized to be related to the thermodynamic difficulties

Table IV. Mean Radii of Polybutadiene Spheres from Electron Microscopy and SANS

Sample	R_{EM} (Å)	R_{SANS} (Å)	R_{EM}/R_{SANS}
SB-1	90 ± 7	117 ± 3	0.77
SB$_d$-1	94 ± 6	124 ± 2	0.76
SB$_d$-7	136 ± 9	197 ± 3	0.69
(S)SB$_d$-3	172 ± 14	221 ± 3	0.78
SB-7	187 ± 17	222 ± 3	0.84

SOURCE: Reproduced with permission from reference 16. Copyright 1982.

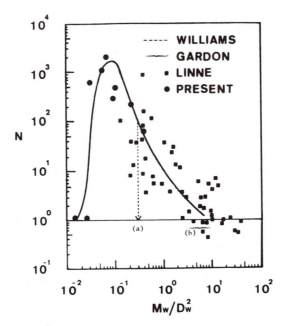

Figure 4. Supermolecular structure in polystyrene latices. The quantity N
represents the number of chains segregated near the center of the latex particle.
(Reproduced with permission from reference 19. Copyright 1989 Wiley.)

of having elbows or chain ends extending into the aqueous phase. This
extension into the aqueous phase results in a repulsive wall, repelling the
chains from the surface (20). The problem is reduced at low molecular
weights, and entropic restrictions overcome segregation at high molecular
weights.

This problem has also been treated theoretically. Joanny et al. (21) and
deGennes (22) showed that the concentration profile of confined polymer
solutions between flat walls or in cylindrical tubes leads to a depletion layer
near the wall. This work was continued for spheres by Jones and Richmond
(23). The concentration of polymer was shown to vanish at the surface of an
ideal wall and increase to the bulk concentration toward the interior, depending on the polymer molecular weight and solution variables.

Film Formation from Latex Particles.   The process of film formation from an emulsion latex can be divided into three stages: evaporation
of water, coalescence and deformation of latex particles, and interdiffusion
of polymer chains between adjacent particles. Extensive studies have been
devoted to the first and second stages of film formation, but the interdiffusion
problem has received little attention because of instrumental limitation. The
recent advent of SANS has provided a method to determine the radius of

gyration in films from which the interdiffusion and film strength build-up behavior can be calculated.

Using SANS technique, Linne et al. (*24*) investigated the early stage of polymer chain interdiffusion. Polystyrene latex that was synthesized contained 50% deuterated polystyrene chains with a molecular weight of 5.85×10^6 g/mol and particle size of 380 Å. After removing the aqueous phase and surfactants, the dried polystyrene particles were subjected to molding under vacuum, just enough to form fully dense films. Molded samples were annealed at 160 °C for various times and then evaluated by SANS and the indentation toughness test.

Initially, the polystyrene chains were constrained in the latex particles with a compressed dimension 4 times smaller than that in the relaxed state as a result of high molecular weight and small particle size. Figure 5 (*24*) shows the hindered initial interdiffusion of constrained chains, which is most likely caused by residual interfacial effects such as the presence of sulfate chain end groups. The diffusion mode of the molecules in the film was reported (*24*) to be transient because of the nonequilibrium state of chain conformation, and the diffusion process was expected to change to classical translational diffusion as the chains relaxed. The segmental diffusion rate was calculated to be faster than the translational diffusion rate. Indentation toughness test results suggested that the diffusion across the particle boundaries to a level of 50–60 Å, as evidenced by an increase in the radius of gyration, was sufficient to develop fully healed films of polystyrene. The mechanism of the film formation during annealing is modeled in Figure 6 (*24*).

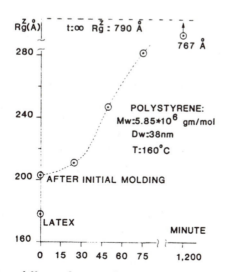

Figure 5. Initial interdiffusion during polystyrene film formation. (Reproduced with permission from reference 24. Copyright 1988 Marcel Dekker.)

TIME   : 0   SHORT   LONG

TOUGHNESS : LOW   HIGH   HIGH

TEMPERATURE OF ANNEALING HIGHER THAN Tg

Figure 6. Schematic of molecular diffusion processes toward the film formation from latices. (Reproduced with permission from reference 24. Copyright 1988 Marcel Dekker.)

Film formation from latex particles and crack healing in bulk polymers have an analogy in terms of chain interdiffusion. Kim and Wool (25, 26) extensively studied the healing problems of the polymer interface and developed a minor chain motion model based on deGennes' reptation model. Kim and Wool concluded that the mechanical strength in crack healing has one-fourth power dependence on time.

Yoo et al. (27) examined the relationship between the depth of interdiffusion during film formation from latex particles and the mechanical strength build-up feature of the film. High- and low-molecular-weight latices that were synthesized contained 100% deuterated polystyrene. Each latex was mixed with 16 times its weight of a latex of identical protonated polystyrene. Annealed bulk films at 140 °C were prepared as mentioned previously. SANS measurements were made to evaluate the size of the deuterated polystyrene-rich particles, while tensile strength tests were carried out on all protonated samples.

The results of current analysis on the SANS data are shown in Table V (27). For high-molecular-weight samples, the initial diffusion was hindered, a finding confirming the result of Linne et al., whereas no appreciable initial hindrance in diffusion was observed in low-molecular-weight samples. In particular, a Zimm plot method could be applied to the data for the low-molecular-weight samples annealed after more than 4 h, a result indicating that the deuterated polystyrene-rich particles, through chain diffusion, started to contact neighboring particles of the same kind. As seen in Table V, the molecular weight determined from the Zimm plot is approaching the real molecular weight of the sample as the annealing time increases. In both series, the apparent particle radii data immediately after molding show an abnormality that is attributed to the initial anisotropy of particles resulting from the high-pressure molding conditions. The anisotropy is believed to disappear through physical relaxation on annealing, because the high pressure is not applied in the annealing process.

Table V. Apparent Radii of the Deuterated Polystyrene-Rich Particles Measured by SANS

Annealing Time	HMW Series (M_w = 2,200,000) Radius (Å)	LMW Series (M_w = 380,000) Radius (Å)	$M_w{}^a$
Original[b]	244.6	246.8	—[c]
0	270.6	294.8	—
5 min	240.7	258.2	—
15 min	241.5	267.6	—
30 min	240.4	278.4	—
1 h	240.6	304.7	—
2 h	248.1	340.8	—
4 h	254.5	358.0	—
8 h	261.7	—	703,000
19 h	266.1	—	645,000
48 h	278.2	—	572,000

SOURCE: Reproduced from reference 27. Copyright 1990 American Chemical Society.
[a]Determined by Zimm plot.
[b]Radius of the 100% deuterated polystyrene particle before molding.
[c]— indicates does not apply.

Figure 7 (27) shows a preliminary result of tensile strength build-up characteristics for the low-molecular-weight polystyrene latex particles. Interestingly, tensile strength seems to fit one-fourth power dependence on time, suggested by Wool et al. for unidirectional diffusion, even though radial diffusion is involved in this system.

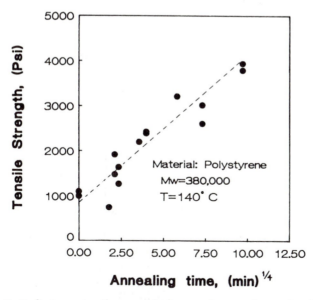

Figure 7. Preliminary tensile strength data confirming the result of Wool et al. on 490-Å diameter PS latex.

Considering, however, the scatter in the tensile strength data, it is premature to draw a firm conclusion about the relationship between the depth of interdiffusion and the strength build-up characteristics. Supplementary experiments are in progress.

SANS on IPNs and Blends. From an instrumental point of view, samples of about 1 cm × 1 cm × 1 mm or a little larger are required. SANS studies by our group at Lehigh University have emphasized *cross*-polybutadiene-*inter-cross*-polystyrene IPNs, both fully polymerized (28), and as polymerization is proceeding (29–31). A collage from the latter, Figure 8

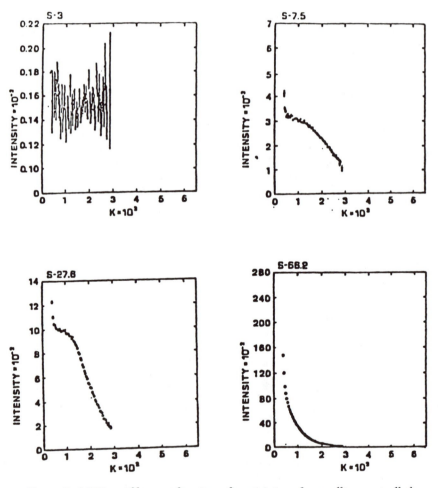

Figure 8. SANS profile as a function of conversions for swelling-controlled cross-*polybutadiene*-inter-cross-*polystyrene IPNs. Designations S-3, S-7.5, S-27.5, and S-66.2 indicate percent conversion of the polystyrene. (Reproduced from reference 30. Copyright 1988 American Chemical Society.)*

(30) illustrates both a maximum in the scattering intensity and later, a smooth decrease with angle. Interpretation of these data according to equations 2–5 yields the results in Table VI.

Most importantly, the data from these experiments yield an interpretation strongly suggesting the importance of spinodal decomposition (32) as a major mechanism of phase separation. The most important supporting result is a negative diffusion coefficient (32).

Although both light-scattering and SANS studies have yielded negative diffusion coefficients and domain sizes, light-scattering domain sizes are consistently larger (*see* Table VII) (33). Whether this result means that there are actually two sizes of different orders in the system or that an artifact is arising from the very different wavelengths and types of experiments is not yet known.

Miscibility and Phase Diagrams. Questions of major importance in polymer science and engineering today relate to the miscibility of two polymers. No longer is it sufficient to say that the polymers are "compatible" or "incompatible", words that also mean acceptable and unacceptable prop-

Table VI. Phase Dimensions of Swelling-Controlled *cross*-Polybutadiene-*inter-cross*-polystyrene IPNs

Sample I.D.	Correlation Length a (Å)	Special Interfacial Surface Area S_{sp} (m²/g)	Transverse Length l_{PB} (Å)	Transverse Length l_{PS}(Å)	Wavelength Λ (Å)
S-3	n.a.[a]	—[b]	—	—	—
S-7.5	35.7	(84)[c]	476	39	600
S-14.8	49.2	(103)	332	58	600
S-18.2	42.2	(141)	232	52	600
S-27.6	61	(131)	221	84	600
S-40.7	82.2	117	202	139	—
S-51.7	87.4	114	169	181	—
S-66.2	99.1	—	90	150	293

SOURCE: Reproduced from reference 31. Copyright 1987 American Chemical Society.
[a] n.a. means not available.
[b] — indicates not calculated.
[c] From the angular portion above the SANS intensity maxima are given in parenthesis. Values are approximate.

Table VII. Wavelength Characteristics of *cross*-Polybutadiene-*inter-cross*-polystyrene IPNs

Method	Diffusion Coefficient D (cm²/s)	Angular Position of Maximum K_m (cm⁻¹)	Wavelength Λ (Å)
Light scattering	-1.8×10^{-13}	3.9×10^4	16,000
Neutron scattering	-1.1×10^{-16}	1.2×10^6	540

SOURCE: Data from reference 33.

erties industrially. Although it is a big advance to be able to speak of one glass transition temperature and transparent samples as indicating miscibility, that too has shortcomings.

What is lacking is related to thermodynamics and phase diagrams (32). In the last 15 years, enormous progress has been made in this area, and now very many polymer blend pairs are known to exhibit lower critical solution temperatures (LCST), above which they phase separate. Polymer blends are becoming less miscible as the molecular weight of the components is increased, but relatively little is known about the effect of cross-linking. This problem was explored via SANS by Bauer et al. (34) for semi-II IPNs based on poly(vinyl methyl ether) and cross-linked deuterated polystyrene.

One way to determine the spinodal temperature of a composition is to plot the inverse zero-angle scattering intensity versus inverse absolute temperature. As shown in Figure 9 (34), Sample 1, which contains no divinylbenzene (DVB), gave a linear plot, as predicted by mean field theory. This plot yielded an intercept of the spinodal temperature, 151 °C. Sample 4, which had the lowest degree of cross-linking, also gave a linear plot in this range with a spinodal temperature of 146 °C. Sample 3, which had a higher degree of cross-linking, yielded a pronounced curvature. This result suggests that, as the cross-linking increases, the sample gets closer to the spinodal temperature, and with the highest degree of cross-linking, phase separation has already taken place at room temperature. Bauer et al. (34) commented that, if a sample can be made in the one-phase region, the presence of cross-links would deter phase separation. On the other hand, if phase separation takes place at the temperature of cross-linking, especially before an infinite network is formed, then the presence of the cross-linking destabilizes the single-phase region to the un-cross-linked blend.

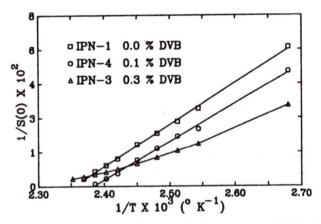

Figure 9. Determination of spinodal temperatures on semi-II IPNs based on poly(vinyl methyl ether) and polystyrene. (Reproduced from reference 34. Copyright 1987 American Chemical Society.)

To summarize, SANS is a powerful tool for investigating the morphology of multicomponent and heterogeneous polymer materials.

Acknowledgments

The authors are pleased to thank the National Science Foundation for support through grant numbers DMR–8405053 and CBT–8512923.

References

1. Sperling, L. H. *Polym. Eng. Sci.* **1984**, *24*, 1.
2. Wignall, G. D. In *Encyclopedia of Polymer Science and Engineering*; Kroschwitz, J., Ed.; Wiley: New York, 1987; pp 10, 112.
3. Wignall, G. D.; Christen, D. K.; Ramakrishnan, V. *J. Appl. Crystallogr.* **1988**, *21*, 438.
4. Seeger, P. A. In *LANSCE*; Hyer, D., Ed.; Los Alamos National Laboratory: Los Alamos, NM; Spring, 1989; p 9.
5. Hjelm, R. P., Jr. *J. Appl. Crystallogr.* **1988**, *21*, 618.
6. Wignall, G. D.; Bates, F. S. *J. Appl. Crystallogr.* **1987**, *20*, 28.
7. Hjelm, R. P.; Seeger, P. A. In *Advanced Neutron Sources*; Hyer, D., Ed.; Institute of Physics Conference Series No. 97, Institute of Physics Publications, Ltd.: Bristol, England, 1989.
8. Higgins, J. S.; Stein, R. S. *J. Appl. Crystallogr.* **1978**, *11*, 346.
9. Maconnachie, A.; Richards, R. W. *Polymer* **1978**, *19*, 739.
10. *McGraw-Hill Encyclopedia of Science and Technology*; McGraw-Hill: New York, 1977; Vol. 14, p 481.
11. *NBS Cold Neutron Research Facility*; National Bureau of Standards: Washington, DC, 1988.
12. Schwahn, D.; Milsovsky, A.; Rauch, H.; Seidl, E.; Zugarek, G. *Nucl. Inst. Methods Phys. Res.* **1985**, *A239*, 229.
13. Schwahn, D.; Yee-Madeira, H. *Colloid Polym. Sci.* **1987**, *265*, 867.
14. Debye, P.; Anderson, H. R.; Brumberger, H. *J. Appl. Phys.* **1957**, *28*, 679.
15. Bates, F. S.; Cohen, R. E.; Berney, C. V. *Macromolecules* **1982**, *15*, 589.
16. Berney, C. V.; Cohen, R. E.; Bates, F. S. *Polymer* **1982**, *23*, 1222.
17. Bates, F. S.; Berney, C. V.; Cohen, R. E.; Wignall, G. D. *Polymer* **1983**, *24*, 519.
18. Linne, M. A.; Klein, A.; Sperling, L. H.; Wignall, G. D. *J. Macromol. Sci. Phys.* **1988**, *B27*, 181.
19. Yang, S.-I.; Klein, A.; Sperling, L. H., *J. Polym. Sci. Polym. Phys. Ed.* **1989**, *27*, 1658.
20. deGennes, P. G. *Scaling Concepts in Polymer Physics*; Cornell University Press: Ithaca, NY, 1979.
21. Joanny, J. F.; Leibler, L.; deGennes, P. G. *J. Polym. Sci. Polym. Phys. Ed.* **1979**, *17*, 1073.
22. deGennes, P. G. *Adv. Colloid Interface Sci.* **1987**, *27*, 189.
23. Jones, J. S.; Richmond, P. *Faraday Trans. II* **1977**, *73*, 1062.
24. Linne, M. A.; Klein, A.; Miller, G. A.; Sperling, L. H.; Wignall, G. D. *J. Macromol. Sci.* **1988**, *1327*, 217.
25. Kim, Y. H.; Wool, R. P. *Macromolecules* **1983**, *16*, 1115.
26. Wool, R. P., private communication, October 1988.

27. Yoo, J. N.; Sperling, L. H.; Klein, A. *Macromolecules* accepted for publication.
28. Fernandez, A. M.; Sperling, L. H.; Wignall, G. D. In *Multicomponent Polymer Materials;* Advances in Chemistry Series No. 211; Paul, D. R.; Sperling, L. H., Eds.; American Chemical Society: Washington, DC, 1986.
29. An, J. H.; Fernandez, A. M.; Sperling, L. H. *Macromolecules* **1987**, *20*, 191.
30. An, J. H.; Sperling, L. H. In *Cross-Linked Polymers: Chemistry, Properties, and Applications;* ACS Symposium Series No. 367; Dickie, R. A.; Labana, S. S.; Bauer, R. S., Eds.; American Chemical Society: Washington, DC, 1988.
31. An, J. H.; Fernandez, A. M.; Wignall, G. D.; Sperling, L. H. *Polym. Mat. Sci. Eng.* **1987**, *56*, 541.
32. Olabisi, O.; Robeson, L. M.; Shaw, M. T. In *Polymer–Polymer Miscibility;* Academic Press: New York, 1979.
33. Sperling, L. H.; Heck, C. S.; An, J. H. In *Multiphase Polymers: Blends and Ionomers;* ACS Symposium Series No. 395; Utracki, L. A.; Weiss, R. A., Eds.; American Chemical Society: Washington, DC, 1989.
34. Bauer, B. J.; Briber, R. M.; Han, C. C. *Polym. Prepr.* **1987**, *28(2)*, 169.

RECEIVED for review February 14, 1989. ACCEPTED revised manuscript August 30, 1989.

Gels and Foams from Ultrahigh-Molecular-Weight Polyethylene

Lucy M. Hair and Stephan A. Letts

Lawrence Livermore National Laboratories, Livermore, CA 94550

Crystallization–gelation of ultrahigh-molecular-weight polyethylene (UHMW PE) was used to make stiff gels that were supercritically dried to make low-density, small-cell-size foams. The effects of solvent and cooling conditions on gelation and morphology were investigated. X-ray diffractometry showed that the size of the crystalline lamellae in the finished foam decreased with increased cooling rate for foams made from UHMW PE in tetralin, but not in dodecane or decalin. This difference may be attributable to the greater expansion of the polyethylene chain in tetralin than in dodecane, as revealed by viscometry. However, the superstructure of the foam, which includes the pore sizes and homogeneity, was found to be affected by solvent as well as by cooling conditions.

$\mathbf{G}$ELS ARE FORMED BY CRYSTALLIZATION of polyethylene from solution (*1, 2*). For a time, the belief persisted that shearing of the solution was required for gelation of polyethylene (*3, 4*), but Edwards and Mandelkern (*5*) and Smith and Lemstra (*6*) showed that shearing is not required, and in fact, gels will form on cooling from quiescent solutions. In addition, crystals formed during gelation appear to be the same type as those formed from dilute solution, that is, lamellar rather than fringed micelle (*5*).

The variables involved in forming the crystalline cross-linked structures (several chains physically locked into the crystal) can be understood by looking at the process. As the solution is cooled, chains form crystalline lamellae that serve as reversible cross-links if chain entanglement in the solution was sufficient. Chain entanglement is sufficient if for each lamella, at least three chain ends leave and participate in the formation of another

0065–2393/90/0227–0471$06.00/0

lamella. Several variables seem important in this process. The degree of entanglement of the chains prior to gelation will obviously affect the formation of the cross-linked system. A high-molecular-weight polyethylene with a narrow molecular-weight distribution would be expected to give the most entanglement. This hypothesis has been well proved by Domszy et al. (7), who showed that gel formation is proportional to molecular weight and concentration. Also, the expansion of the polyethylene chain in solution will differ depending on the solvent used. Thus, the aspect of gelation related to chain entanglement should be a function of the solvent.

The cooling rate, crystallization temperature, and solvent are important in the formation of crystalline nuclei. The formation of crystalline nuclei requires energy because new surfaces are being created, so a certain amount of subcooling is required, even though the crystalline form of polyethylene is more thermodynamically stable once formed (8, 9). Thus, the interfacial surface energy at the crystal face–solvent boundary strongly influences the rate of crystal formation and the crystallization temperature. Geil (8) pointed out that, for a given molecular weight and concentration of polyethylene in a particular solvent, the lamellar thickness is solely a function of the cooling conditions for dilute solutions. However, the lamellar thickness is changed by different solvents (10).

Given this background, we investigated the effects of the solvent, cooling rate, and crystallization temperature on polyethylene gel formation for ultrahigh-molecular-weight polyethylene (UHMW PE) over a narrow concentration range. To study the crystalline structure and more details of the morphology, the gels were dried to foams via a very gentle technique, supercritical drying. Cloud point measurements, viscometry, differential scanning calorimetry (DSC), optical microscopy (OM), scanning electron microscopy (SEM), and X-ray diffraction measurements (XRD) were used to investigate the effects of the processing parameters.

Experimental Details

Preparation of UHMW PE Gels and Foams. An entangled solution was made by injecting 3–30% (w/v) PE powder, 0.4% (w/v) antioxidant, and the solvent into a 10-mL ampule. Hercules PE 1900, a powder of about 30-μm diameter, with a molecular weight of 3×10^6 daltons, was generally used in the experiments. The antioxidant, 4,4′-methylenebis(2,6-di-*tert*-butylphenol), retards chain scission that would reduce the degree of entanglement. The filled ampule was then sealed to prevent evaporation of the solvent and reduce oxidation of the PE. The contents were shaken to disperse the PE powder and immediately placed in a 150 °C oil bath, where dissolution and/or melting occurred within 10 min. One to three days was given for complete disentanglement and randomization of the polymer chains. The solution was now cooled to crystallize and phase separate, during which the gel shrinks slightly.

The now-solid, white gel was slid into a Soxhlet extractor for 7 days of exchange with isopropyl alcohol. In the next step, isopropyl alcohol was replaced with liquid carbon dioxide for critical point drying. The gel was placed in a pressure vessel, and the vessel was filled with liquid CO_2 at 14 °C and 900 psi. The CO_2 was replaced three to four times each day for 1 week. Then, the CO_2 level was dropped to half, the temperature was raised to 40 °C, and the pressure was raised to between 1200 and 1600 psi. (The critical temperature (T_c) is 31 °C, and the critical pressure (P_c) is 1100 psi). After 10 min, the pressure valve was cracked open, and the CO_2 was allowed to vent down to atmospheric pressure over 3 h without ever passing a liquid meniscus through the fragile foam material. The finished foam was cylindrical, white, and deformable. It shrank to 70–80% of its original volume when tetralin was the solvent.

Apparatus for Characterization. An Ubbelohde tube with Schott Gerate AVS 300 automatic dilution viscometer was used to determine intrinsic viscosities of the polyethylene samples. Gel points were determined via a Nametre oscillating viscometer. A Leitz Orthoplan-Pol polarizing microscope with Vario-Orthomat 2 camera attachment was used for optical microscopy. A Perkin-Elmer DSC-4 was used for the differential scanning calorimetry. The X-ray diffractometer consisted of a Norelco X-ray source with copper anode and nickel filter coupled to a Phillipps scanning ganiometer. Scanning electron micrographs were taken via the Hitachi S-800, generally at 10 kV.

Results and Discussion

Solvent Effects on Solution and Gelation. We screened 14 solvents for solubility and behavior during gelation. Cloud point and dissolution temperature measurements of the eight most promising solvents were made to determine the degree of subcooling. These were determined visually in the oil bath while the solutions were cooled at about 10 °C/h. The cloud point data are shown in Figure 1 as a function of solution concentration. The cloud points for n-alkanes are typically 10–15 °C higher than those for aromatic and hydroaromatic compounds. However, the degrees of subcooling (ΔT_m), that is, the initial dissolution temperature minus the cloud point, are virtually the same for both tetralin and dodecane at 2 wt % (where ΔT_m is 38 °C for tetralin and 41 °C for dodecane) and at 9 wt % (where ΔT_m is 31 and 21 °C), respectively. The lowered values of dissolution temperature and cloud point for PE in tetralin are probably related to its greater effectiveness as a solvent. But because the degree of subcooling is not changed by different solvents, this measure gives no indication that crystalline structure should be affected by the solvent.

We next investigated the effect of solvent on chain dimension via viscometric techniques. Tetralin and n-dodecane were chosen for further comparison of the aromatic compounds with the n-alkanes. Dilute solution capillary flow viscometry was used to measure the limiting viscosity number, $[\eta]$. In a good solvent, the chain is swollen with solvent and expanded, a condition that increases the viscosity. Einstein's viscosity equation (11)

$$\eta = \eta_0(1 + 2.5\phi)$$

$$(1)$$

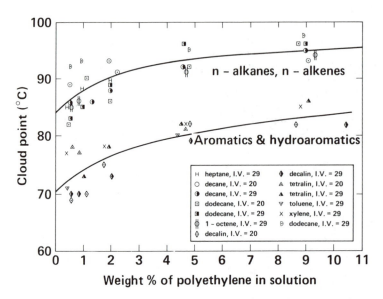

Figure 1. The effect of solvent and concentration on cloud point of solutions of UHMW PE. I.V. stands for intrinsic viscosity of the polyethylene. I.V.s 29 and 20 correspond to molecular weights of approximately 6×10^6 and 3×10^6 daltons, respectively.

relates the solution viscosity, η, to the solvent viscosity, η_0, and the volume fraction of the particles, ϕ. This equation can be rewritten in terms of limiting viscosity number, partial molar volume of the polymer, v_2, and the molecular weight, M (*11*).

$$[\eta] = \frac{2.5v_2}{M} \tag{2}$$

For the UHMW PE used here, the molecular weight had been determined to be about 3×10^6 daltons by applying the Mark–Houwink (*12*) equation to intrinsic viscosity measurements. The limiting viscosity numbers were 20 dL/g in tetralin and 5 dL/g in dodecane, corresponding to partial molar volumes of 3.98×10^{-15} and 9.99×10^{-15} cm^3/molecule, respectively, as calculated from equation 2. If we assume the particles to be spherical, the radii are 9.83×10^{-6} cm^3 in tetralin and 6.19×10^{-6} cm^3 in dodecane, results indicating one-third higher expansion in tetralin.

The average radius of a polymer molecule can be calculated with the assumption that the polymer takes on a random flight configuration in solution. Then the mean square radius is

$$r^2 = nl^2 \tag{3}$$

where n is the number of repeating units, and l is the length of each bond (11). For PE chains with molecular weight of 3×10^6 daltons, n is 2.1×10^5, l is 1.54×10^{-8} cm, and the mean radius is 7.06×10^{-6} cm, which is between the experimental values for tetralin and dodecane.

Actual molecules do not assume truly random flight configurations in solution because bond angles and steric effects restrict flexibility and tend to increase the size of the molecule in solution. The mean square radius of a chain with a bond angle θ is

$$r^2 = \frac{nl^2(1 - \cos \theta)}{(1 + \cos \theta)} \tag{4}$$

For PE in which the polymer is linked by C–C single bonds, θ is the tetrahedral angle, 109.5°. The mean radius was found to be 9.98×10^{-6} cm, almost exactly that found experimentally from analysis of the limiting viscosity data for UHMW PE in tetralin. This result suggests that PE is collapsed somewhat in dodecane relative to the unperturbed state.

Next, the concentrations and temperatures at which entanglement of the UHMW PE occurs were investigated with vibrational viscometry. The probe, a 0.5-in. diameter cylinder, oscillated at approximately 600 Hz with an amplitude of 25 μm. The extremely low amplitude of oscillation avoided any stirring motion that may have induced polymer chain alignment. Figure 2 shows the rise in viscosity as the solutions were cooled from 150 °C to their crystallization temperatures. Viscosity increases sharply for solution concentrations high enough to produce chain entanglement and is determined by the amount of chain entanglement. The concentration at which the viscosity shows a sharp rise is defined as the gel concentration, c_g. If gelation is the condition when the polymer chains just touch, the chain radii can be calculated. With closest packing, the total polymer particle volume is then

$$V_p = 0.74 V_t \tag{5}$$

where V_t is the total volume of the solution. The total polymer particle volume is also given by the product of the volume of an individual particle with the number of particles,

$$V_p = \left[\frac{4\pi r^3}{3} \right] \left[\frac{c_g V_t N_0}{M} \right] \tag{6}$$

where N_0 is Avogadro's number. Equations 5 and 6 can be combined to give

$$r = \left[\frac{0.177 M}{c_g N_0} \right]^{1/3} \tag{7}$$

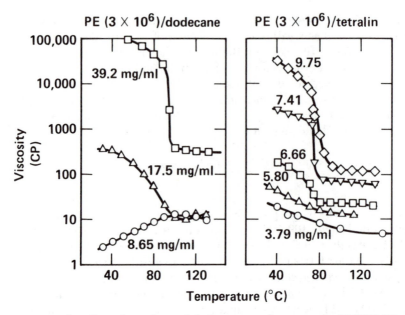

Figure 2. The effect of tetralin and dodecane as solvents on the UHMW PE gel concentration and temperature (gel point defined here as place where viscosity increases sharply) demonstrated via results of oscillating viscometry.

The critical gel concentrations were found to be approximately 0.74% (w/v) at 75 °C for tetralin and 3.0% (w/v) at 90 °C for dodecane. From equation 7, the polymer particle radius was found to be 4.88×10^{-6} cm in tetralin and 3.08×10^{-6} cm in dodecane, each approximately half the dimension found from analysis of dilute solution viscometry. The ratio of the radius in tetralin to that in dodecane is about 1.6, exactly that found from the dilution viscometry experiments. The higher solution concentrations may change the polymer dimensions, or the polymer chains may already be overlapping at our definition of critical gelation. Nevertheless, analyses and experiments show that entanglement should and does occur at a lower concentration in tetralin.

Conditions during Crystallization. The strength and structure of the PE gel depend directly on the crystalline structure induced during cooling to gelation. Previous work and theory (8) on crystallization from dilute solution indicated that a high degree of subcooling would result in a smaller crystal size. We therefore studied the effects on gel structure of cooling rate and crystallization temperature via DSC, OM, SEM, and XRD.

Heating and cooling cycles for a 5% (w/v) PE–tetralin gel were performed directly on the DSC. The sample was cooled at 0.1, 1, 10, 30, 60, 100, and 320 °C/min and then heated at 10 °C/min. The results for the heating cycle, shown in Figure 3, were reproducible. A higher onset and melting point were obtained when the sample was cooled more slowly. A very fast cool (100 or 320 °C/min) resulted in double peaks, but a cool of 10 °C/min resulted in a left shoulder. This finding suggests that more rapid cooling may result in smaller morphology, or perhaps a dimorphism, where different size crystals are formed.

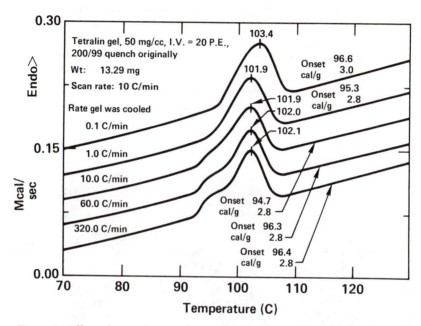

Figure 3. *Effect of quench rate of UHMW PE gels on crystal size, as seen by differential scanning calorimetry.*

However, the DSC scans of the gels permeated with isopropyl alcohol in the first solvent-exchange step and of the foams showed about 75% crystallinity and no substantial differences attributable to cooling rate effects. Probably, tetralin acts to stabilize the amorphous regions around the smaller crystals, and these amorphous regions crystallize to form larger crystals during exchange, considering that the PE–isopropyl alcohol gels do not show any differences.

If the rate of cooling is fast and the crystallization temperature sufficiently low, the crystal size and the degree of crystallinity in the gel are significantly reduced. Analysis of another series of 5% (w/v) PE–tetralin gel samples via DSC revealed that the heat of melting was 2.2 cal/g for a gel quenched to −5 °C, as compared to 5.7 cal/g for samples quenched to 45 and 99 °C. In addition, the gel quenched to −5 °C was noticeably less rigid than the others.

Optical micrographs of the gels and scanning electron micrographs of foams made from those gels support the DSC results. Gels made from 5% (w/v) PE in tetralin, dodecane, and decalin were cooled by quenching to –5 °C and by slower cooling rates of 5–8 °C/h and 2–3 °C/h. These results are shown in Figures 4–6, respectively. For all three solvents, the structure is most homogeneous for the samples quenched to –5 °C. With the PE–tetralin gels, optical micrographs of those more slowly cooled show spherulitic structures of about 200 μm, but the quenched gel has a fine, homogeneous structure of 20-μm sliverlike crystals. These changes are reflected in the foams by an increase in homogeneity as seen in the SEMs. The gels from dodecane were influenced the least by the cooling rates, and those from decalin gave the most dramatic results as well as the most obvious relationship between the OMs and the SEMs, as seen in Figure 6. At the slowest cooling rate, OMs show large spherulites of approximately 100 μm, and the SEMs show "sea sponges" near the same size. These structures

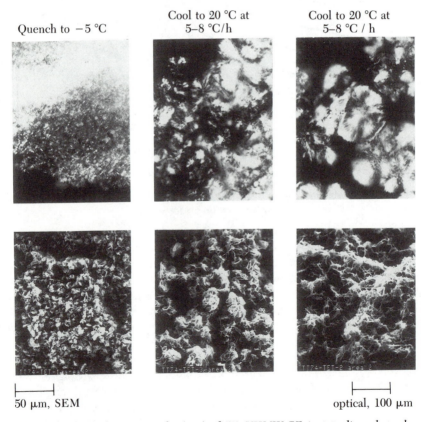

Figure 4. Optical micrographs (top) of 5% UHMW PE in tetralin gels and scanning electron micrographs (bottom) of the supercritically dried foams. Increased cooling rate decreases the spherulitic size and increases homogeneity.

Quench to −5 °C   Cool to 20 °C at   Cool to 20 °C at
                              5–8 °C/h   2–3 °C/h

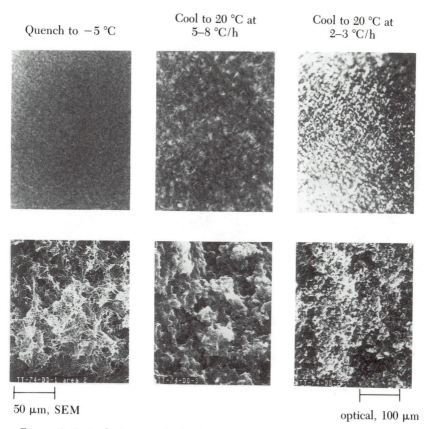

50 μm, SEM

optical, 100 μm

Figure 5. Optical micrographs (top) of 5% UHMW PE in dodecane gels and scanning electron micrographs (bottom) of the supercritically dried foams. Spherulitic size decreases slightly, and homogeneity increases.

clearly decrease in size as the cooling rate is increased, so that for the quenched sample, they are from 10 to 20 μm in diameter.

The thickness of the folded polymer chain lamella that make up the walls of the cellular foam can be measured by X-ray diffraction techniques. For a foam whose walls are relatively perfect crystalline lamella, the crystalline thickness can be calculated from the Scherrer equation (*13*),

$$L_{hkl} = \frac{K\lambda}{\beta_0 \cos \theta} \tag{8}$$

In the Scherrer equation, L_{hkl} is the crystallite dimension perpendicular to the *hkl* plane, K is a constant commonly set to unity, β_0 is the full width at half maximum intensity of the reflection occurring at a diffractometer angle of θ corresponding to the *hkl* reflection, and λ is the radiation wavelength.

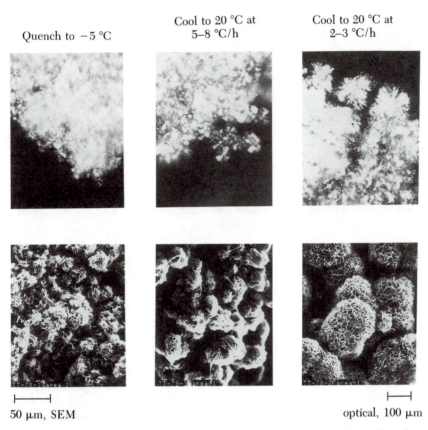

Quench to −5 °C   Cool to 20 °C at   Cool to 20 °C at
                            5–8 °C/h   2–3 °C/h

50 μm, SEM   optical, 100 μm

*Figure 6. Optical micrographs (top) of 5% UHMW PE in decalin gels and
scanning electron micrographs (bottom) of the supercritically dried foams.
Increased cooling rate significantly decreases spherulitic size and increases
homogeneity.*

Accurate analysis of crystal dimensions by the Scherrer equation is complicated by the broadening caused by lattice distortions. However, comparisons
between similar materials provide useful qualitative information on crystal
thickness. The thickness of polyethylene lamella is a function of the magnitude of the undercooling during crystallization.

We first compared two samples of polyethylene with molecular weight
3.5×10^6 daltons: one was melted and slowly cooled as a film; the other
was a foam prepared as a gel and then supercritically dried. Analysis of the
diffractometer scans from the two samples gave a crystal thickness of 17 nm
for the melt and 8 nm for the foam. Because of lattice distortion, both values
may be low by a factor of 2. A cell-wall thickness of about 35 nm was
previously calculated (14) from a foam structural model; therefore, the cell
walls must be composed of two to four layers of polyethylene lamella.

Next, the gels prepared from tetralin, dodecane, and decalin under different cooling conditions as discussed were dried to foams. A thin slice of foam was mounted and scanned by the X-ray diffractometer. Analysis of XRD scans shows that crystal sizes of the 5% PE–tetralin foam, as determined from the 110 and 200 reflections, decrease with decreased crystallization temperature, as shown in Table I. However, no correlation is evident between crystallization conditions and crystal sizes for the foams made from dodecane or decalin.

Table I. Effect of Cooling Rate on Crystal Size As Seen by 110 and 200 Reflections

Cooling Rate (°C/h)	Tetralin		Dodecane		Decalin	
	110	200	110	200	110	200
2–3	14	13	10	10	11	11
5–8	10	8	14	13	12	9
–5 °C quench	7	6	14	12	13	11

NOTE: All values are given in nanometers.

Obviously, the combination of the SEM, OM, and XRD results suggests that the crystal lamellae forming the structure are affected by cooling conditions in the gelation process only for tetralin. But the superstructure that uses the crystal lamellae to form the gels is affected by solvent and cooling conditions. A further example of this effect is the fact that the percent shrinkage on going from gel to foam is different for both solvent and cooling rate. On the average, the percent shrinkage was 26, 37, and 47 for tetralin, decalin, and dodecane, respectively. And, for the 2–3 °C/h, 5–8 °C/h, and –5 °C quench samples, the percent shrinkage was 31, 36, and 43, respectively. This result indicates that the gels made in tetralin and cooled at the slowest rate, 2–3 °C/h, are the most cohesive and strong.

Summary

We studied the effects of solvent and cooling conditions on the structure and properties of crystallization gels and foams made from those gels by supercritical drying. Dilute and concentrated viscometric studies showed that PE is more expanded in tetralin than in dodecane, a necessary condition for entanglement and gelation. The effects of cooling rate on gel morphology were studied via DSC and OM. Increased cooling rate resulted in a more homogeneous and overall finer structure. Studies of the foams via SEM and XRD verified these results. Also, increased cooling rate and decreased crystallization temperature decreased the crystal size in foams made from PE–tetralin gels. However, some structural change may occur during the solvent exchange or drying process, considering that DSC of the foam samples shows no significant effect of cooling rate.

Acknowledgments

We acknowledge the fine work of Michael Saculla in XRD and Paul McCarthy in SEM. This work was performed under the auspices of the U.S. Department of Energy by the Lawrence Livermore National Laboratory under contract No. W–7405–ENG–48.

References

1. Flory, P. J. *Discuss. Faraday Soc.* **1974**, *57*, 7.
2. Mandelkern, L. *Crystallization of Polymers;* McGraw-Hill: New York, 1964, pp 113, 303.
3. Barham, P. J.; Hill, M. J.; Keller, A. *Colloid Polym. Sci.* **1980**, *258*, 899.
4. Narh, K. A.; Barham, P. J.; Keller, A. *Macromolecules* **1982**, *15*, 464.
5. Edwards, C. O.; Mandelkern, L. *J. Polym. Sci. Polym. Lett. Ed.* **1982**, *20*, 355.
6. Smith, P.; Lemstra, P. J. *Br. Polym. J.* **1980**, *12*, 212.
7. Domszy, R. C.; Alamo, R.; Edwards, C. O.; Mandelkern, L. *Macromolecules* **1986**, *19*, 310.
8. Geil, P. H. *Polymer Single Crystals;* Robert E. Krieger Publishing Company: Huntington, NY, 1973.
9. Adamson, A. W. *Physical Chemistry of Surfaces;* Interscience: New York, 1967.
10. Keller, A. *J. Polym. Sci. Symp.* **1975**, *51*, 7.
11. Hiemenz, P. C. *Polymer Chemistry;* Marcel Dekker: New York, 1984.
12. *Polymer Handbook;* Brandrup, J.; Immergut, E. H., Eds.; Wiley: New York, 1975.
13. Alexander, L. E. *X-Ray Diffraction Methods in Polymer Science;* Wiley-Interscience: New York, 1969.
14. Letts, S. A.; Lucht (Hair), L. M. "Polymer Foam Developments," Section 3–4 in *1985 Laser Program Annual Report;* Document 50021–85, National Technical Information Service, U.S. Department of Commerce: Springfield, VA, 1985.

RECEIVED for review February 14, 1989. ACCEPTED revised manuscript October 18, 1989.

INDEXES

AUTHOR INDEX

AFFILIATION INDEX

SUBJECT INDEX

Copy editing and production: Janet S. Dodd
Indexing: Karen McCeney Belton
Acquisition: Robin M. Giroux

Typeset by Techna Type Inc., York, PA
Books printed and bound by Maple Press, York, PA

The paper in this book meets the minimum requirements of the American National Standard for Information Sciences—Permanence of Paper for Printed Library Materials, ANSI Z39.48—1984 ∞

Related Titles

Silicon-Based Polymer Science: A Comprehensive Resource
Edited by John M. Zeigler and F. W. Gordon Fearon
Advances in Chemistry Series 224; 801 pages; ISBN 0–8412–1546–4

Polymers in Microlithography: Materials and Processes
Edited by Elsa Reichmanis, Scott A. MacDonald, and Takao Iwayanagi
ACS Symposium Series 412; 449 pages; ISBN 0–8412–1701–7

Radiation Curing of Polymeric Materials
Edited by Charles E. Hoyle and James F. Kinstle
ACS Symposium Series 417; 567 pages; ISBN 0–8412–1730–0

Barrier Polymers and Structures
Edited by William J. Koros
ACS Symposium Series 423; 406 pages; ISBN 0–8412–1762–9

Sound and Vibration Damping with Polymers
Edited by Robert D. Corsaro and L. H. Sperling
ACS Symposium Series 424; 469 pages; ISBN 0–8412–1778–5

Liquid-Crystalline Polymers
Edited by R. A. Weiss and C. K. Ober
ACS Symposium Series 435; 508 pages; ISBN 0–8412–1849–8

Polymers in Aqueous Media: Performance Through Association
Edited by J. Edward Glass
Advances in Chemistry Series 223; 575 pages; ISBN 0–8412–1548–0

Polymeric Materials: Chemistry for the Future
By Joseph Alper and Gordon L. Nelson
110 pages; clothbound, ISBN 0–8412–1622–3;
paperback, ISBN 0–8412–1613–4

For further information and a free catalog of ACS books, contact
American Chemical Society
Distribution Office, Department 225
1155 16th Street, N.W., Washington, DC 20036
Telephone 800–227–5558